WATER SUPPLY

WATER SUPPLY

Anthony Pingnam

CONSTRUCTION + PLUMBING Skills Series

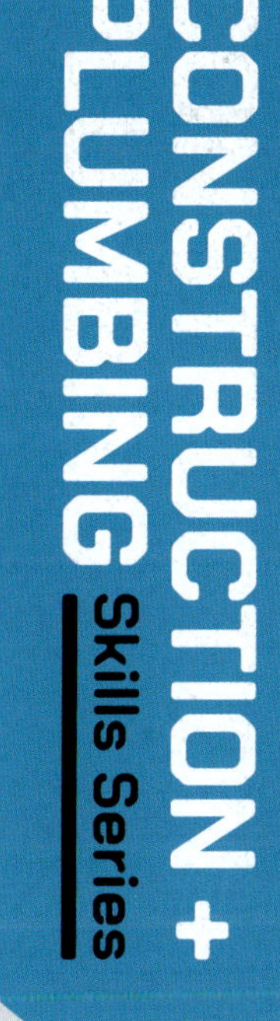

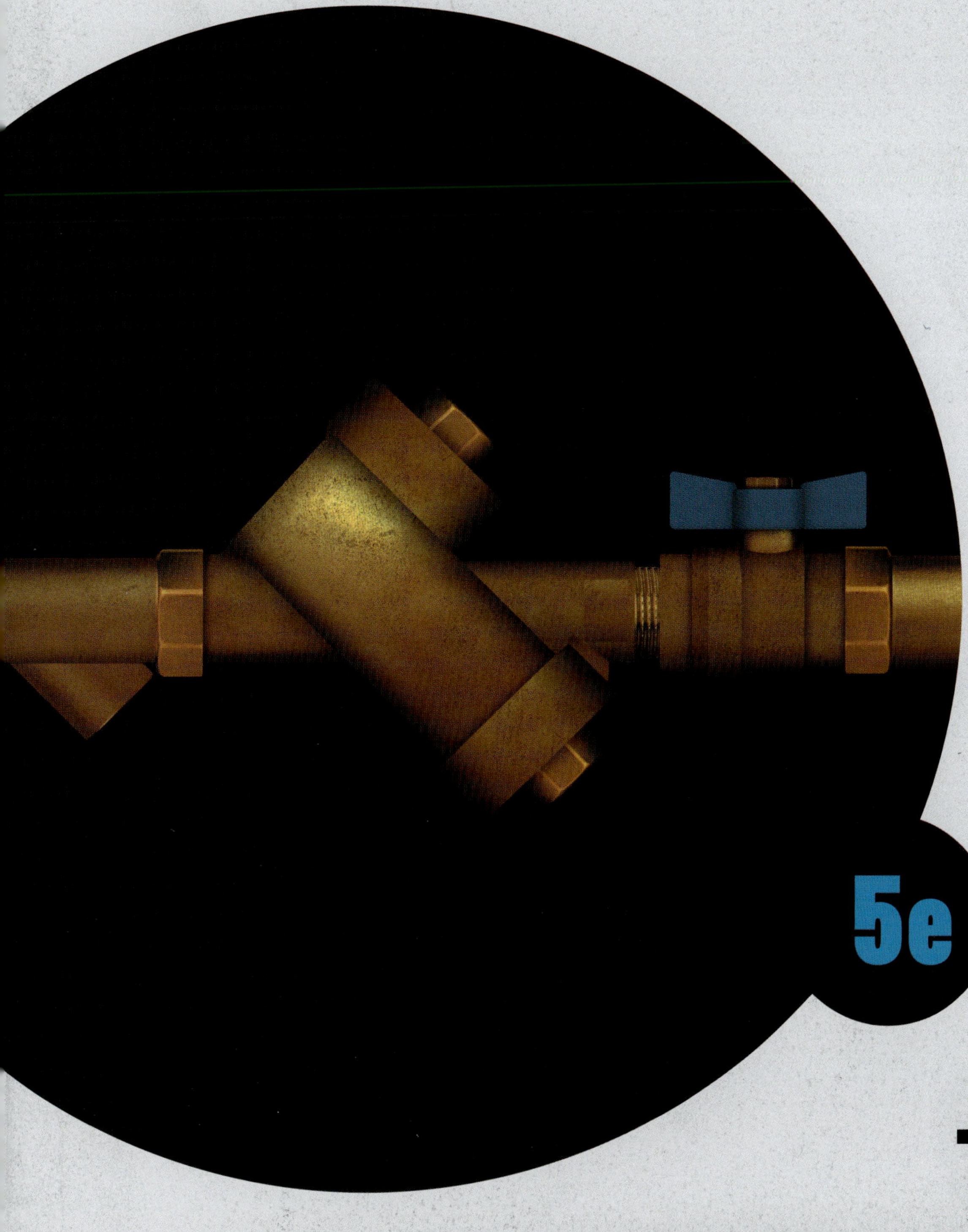

5e

TAFE NSW

Water Supply
5th Edition
Anthony Pingnam

Product manager: Sandy Jayadev
Content Developer: Lucy Wadelton
Content project manager: Sharmilee Govindan
Project designer: Danielle Maccarone
Text designer: Cengage Creative Studio
Cover design: Cengage Creative Studio
Cover Illustration: Antonia Pesenti – Studio Fable
Editor: Julie Wicks
Proofreader: Anne Mulvaney
Permissions/Photo researcher: Catherine Kerstjens/Lumina Datamatics
Typeset by KGL

Fourth edition published in 2021.

For product information and technology assistance,
in Australia call **1300 790 853**;
in New Zealand call **0800 449 725**

For permission to use material from this text or product, please email
aust.permissions@cengage.com

National Library of Australia Cataloguing-in-Publication Data
ISBN: 9780170475150
A catalogue record for this book is available from the National Library of Australia

Cengage Learning Australia
Level 7, 80 Dorcas Street
South Melbourne, Victoria Australia 3205

For learning solutions, visit cengage.com.au

Printed in China by 1010 Printing International Limited.
1 2 3 4 5 6 7 27 26 25

BRIEF CONTENTS

CONTENTS

Guide to the text

As you read this text you will find a number of features in every chapter to enhance your study of Plumbing and help you understand how the theory is applied in the real world.

CHAPTER OPENING FEATURES

The **Chapter overview** introduces the topics that are covered in the chapter.

Identify the key concepts you will engage with through the **Learning objectives** at the start of each chapter.

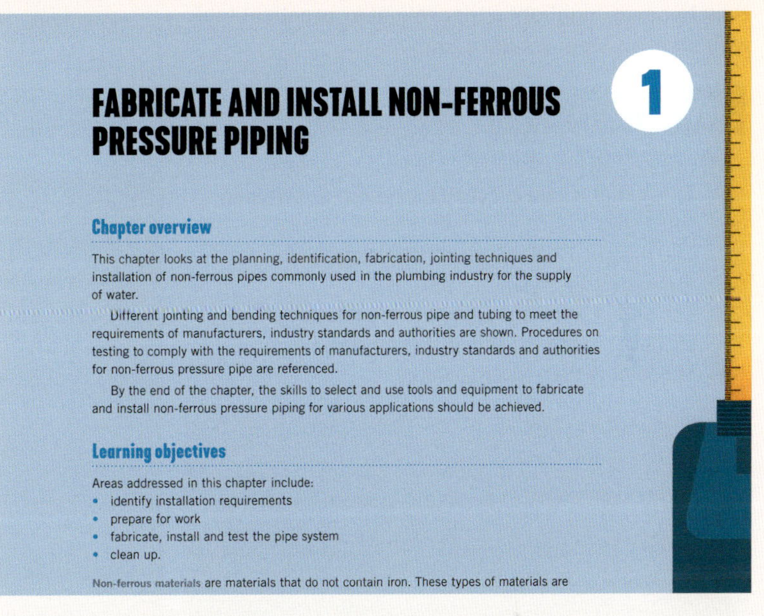
1

FABRICATE AND INSTALL NON-FERROUS PRESSURE PIPING

Chapter overview

This chapter looks at the planning, identification, fabrication, jointing techniques and installation of non-ferrous pipes commonly used in the plumbing industry for the supply of water.

Different jointing and bending techniques for non-ferrous pipe and tubing to meet the requirements of manufacturers, industry standards and authorities are shown. Procedures on testing to comply with the requirements of manufacturers, industry standards and authorities for non-ferrous pressure pipe are referenced.

By the end of the chapter, the skills to select and use tools and equipment to fabricate and install non-ferrous pressure piping for various applications should be achieved.

Learning objectives

Areas addressed in this chapter include:

- identify installation requirements
- prepare for work
- fabricate, install and test the pipe system
- clean up.

Non-ferrous materials are materials that do not contain iron. These types of materials are

FEATURES WITHIN CHAPTERS

Engage actively with the learning by completing the practical activities in the **Learning task** boxes.

LEARNING TASK 

1 Where can information on the specified materials be obtained?
2 What is the purpose of quality assurance procedures?

The **Standards** box highlight where Plumbing Standards are addressed, to strengthen knowledge and hone research skills.

From experience boxes explain the responsibilities of employees, including skills they need to acquire and real-life challenges they may face at work to enhance employability skills on the job site.

FROM EXPERIENCE

Having the knowledge and skills to choose the correct installation method for a selected material earns customer confidence, creating future work opportunities.

FEATURES WITHIN CHAPTERS

Caution boxes highlight important advice on safe work practices for plumbers by identifying safety issues and providing urgent safety reminders

Depending on the size of the construction site, the hot works permit may be provided by the site supervisor or by a specific WHS officer working on the site.

Green tip boxes highlight the applications of sustainable technology, materials or products relevant to plumbers and the plumbing industry.

GREEN TIP

All copper tube offcuts should be scrapped to a metal recycler. It provides extra income and prevents waste in landfill.

How to boxes highlight a theoretical or practical task with step-by-step walkthroughs.

HOW TO

MAKE EXPANDED JOINT FITTINGS

1 The end of a copper tube is annealed or softened (see Figure 1.6).
2 The tube end is then expanded using a **tube expander** (see Figures 1.7 and 1.8).
3 The end of another copper tube is cleaned, then slipped into the expanded section (see Figure 1.9).
4 The joint is then silver soldered.

FIGURE 1.6 Annealed copper tube end, ready to be expanded

END-OF-CHAPTER FEATURES

At the end of each chapter you will find several tools to help you to review, practise and extend your knowledge of the key learning objectives.

Review your understanding of the key chapter topics with the **Summary**.

SUMMARY

- Non-ferrous materials are materials that do not contain iron.
- They are long-lasting because they are less corrosive than ferrous materials.
- Quality assurance helps to ensure a high standard of workmanship and processes, therefore reducing mistakes and problems.
- Always wear the appropriate personal protective equipment (PPE).
- Keep tools and equipment well maintained.
- Take the time to thoroughly read the plans and specifications before starting work.
- A site visit is vital prior to starting work to check for access and pipework location.
- Copper and plastic pressure piping are the most used non-ferrous materials.
- It is important to know the different jointing methods and the limitations of different materials.
- Ordering materials accurately reduces waste and saves time.

END-OF-CHAPTER FEATURES

After you have worked through the chapter, reinforce the practical component of your training with the **Get it right** section.

GET IT RIGHT

1 Which photo shows the correct use of the crimping tool?

2 Why is this method important?

GET IT RIGHT 1

Worksheets give you the opportunity to test your knowledge and consolidate your understanding of the chapter competencies.

WORKSHEET 1

To be completed by teachers
Student competent
Student not yet competent

Student name:

Enrolment year:

Class code:

Unit of competency code/title: CPCPCM Fabricate and install non-ferrous pressure piping

Task: Review 'Identify installation requirements' to the heading 'Compression joints' and answer the following questions.

1 Why is non-ferrous pressure piping less corrosive than ferrous pressure piping?

2 State four common uses for non-ferrous pipes.

3 Describe how copper tubes are classified by type and colour.

WORKSHEETS 1

Worksheet icons indicate in the text when a student should complete an end-of-chapter worksheet.

COMPLETE WORKSHEET 1

Guide to the online resources

FOR THE INSTRUCTOR

Cengage is pleased to provide you with a selection of resources that will help you prepare your lectures and assessments when you choose this textbook for your course. Contact your Cengage learning consultant for more information.

MINDTAP

Premium online teaching and learning tools are available on the MindTap platform – the personalised eLearning solution.

MindTap is a flexible and easy-to-use platform that helps build student confidence and gives you a clear picture of their progress. We partner with you to ease the transition to digital – we're with you every step of the way.

The *Series MindTap for Plumbing* is full of innovative resources to support critical thinking, and help your students move from memorisation to mastery! Includes:

- Plumbing Series eBooks
- Instructional Videos
- Worksheets
- Revision Quizzes
- And more!

MindTap is a premium purchasable eLearning tool. Contact your Cengage learning consultant to find out how *MindTap* can transform your course.

INSTRUCTOR'S RESOURCE PACK

Premium resources that provide additional instructor support are available for this text, including a Word-based Testbank, PowerPoints and all artwork from the text. These resources save you time and are a convenient way to add more depth to your classes, covering additional content and with an exclusive selection of engaging features aligned with the text. The Instructor Resource Pack is included for institutional adoptions of this text when certain conditions are met.

The pack is available to purchase for course-level adoptions of the text or as a standalone resource. Contact your Cengage learning consultant for more information.

SOLUTIONS MANUAL

The solutions manual includes solutions to Learning Tasks, End-of-chapter worksheets and solutions to Get it Right case studies.

POWERPOINT™ PRESENTATIONS

Use the chapter-by-chapter **PowerPoint slides** to enhance your lecture presentations and handouts by reinforcing the key principles of your subject.

ARTWORK FROM THE TEXT

Add the digital files of graphs, tables, pictures and flow charts into your course management system, use them in student handouts, or copy them into your lecture presentations.

WORD-BASED TEST BANK

This bank of questions has been developed in conjunction with the text for creating quizzes, tests and exams for your students. Deliver these through your LMS and in your classroom.

COMPETENCY MAPPING GRID

The downloadable Competency Mapping Grid demonstrates how the text aligns to the Certificate III in Plumbing

WORKSHEETS

Download the end-of-chapter worksheets which encourage your students to revise and further discuss the main points from each topic.

FOR THE STUDENT

MINDTAP

MindTap is the next-level online learning tool that helps you get better grades!

MindTap gives you the resources you need to study – all in one place and available when you need them. In the *MindTap Reader*, you can make notes, highlight text and even find a definition directly from the page.

If your instructor has chosen *MindTap* for your subject this semester, log in to *MindTap* to:

- Get better grades
- Save time and get organised
- Connect with your instructor and peers
- Study when and where you want, online and mobile
- Complete assessment tasks as set by your instructor

When your instructor creates a course using *MindTap*, they will let you know your course key so you can access the content. Please purchase *MindTap* only when directed by your instructor. Course length is set by your instructor.

FOREWORD

The plumbing industry in Australia serves as a fundamental pillar of modern infrastructure, providing essential services in water supply, fire protection, sanitary plumbing, drainage, gas installations, roof plumbing, and mechanical services. As one of the nation's largest trade sectors, plumbing supports residential, commercial, and industrial needs, ensuring the efficient operation and safety of buildings and communities.

In an era of rapid technological advancement and increasing emphasis on sustainability, the industry must continuously adapt to evolving standards and best practices. To facilitate this progress, the vocational education and training (VET) sector plays a critical role in equipping professionals with the necessary expertise to uphold and advance industry standards.

This textbook has been meticulously developed to align with the Construction, Plumbing, and Services Training Package, offering comprehensive knowledge and structured activities that reflect the industry's evolving requirements. It is designed to provide learners with the foundational and advanced competencies essential for conducting safe and effective operations within the water supply services sector. By engaging with this material, both aspiring and experienced professionals will be well-positioned to enjoy a rewarding and enduring career in plumbing.

I would like to express my sincere appreciation to Anthony Pingnam for his invaluable contribution of time and expertise in ensuring the relevance and accuracy of this publication. His dedication has significantly enriched the content, reinforcing its applicability to the current and future needs of the plumbing industry.

Ian McNiven
Director of Learning & Teaching – Plumbing Services
Construction & Energy Faculty
TAFE NSW

PREFACE

Providing clean drinking water to our community is a massive responsibility for all involved. The plumber has a key part to play in installing water supply systems to residential, commercial, industrial and civil sites. Members of the plumbing industry take great pride in achieving the high standards set in Australia by remaining well informed on evolving work practices, new materials and new products.

This book is a collective effort by many plumbing teachers aiming to educate and inspire young apprentices in the industry. Its intention is to underpin their knowledge and develop problem-solving skills, combining safe work practices with professional skills and a thorough understanding of how and why procedures are done a certain way.

The text is aligned to the relevant training package competencies, with many references to AS/NZS 3500, which is also regularly referred to in the Plumbing Code of Australia. Due to some state or territory variations, it may be necessary for teachers to amend the information accordingly.

The passing of knowledge from elders is a time-honoured practice. The master and apprentice relationship is a tradition that goes back thousands of years and is still relevant with the importance of mentoring in the workplace today. This book seeks to uphold that tradition in these current times.

ABOUT THE REVISING AUTHOR

Anthony Pingnam, Grad. Dip. Adult Education, Cert. IV TAE, Cert. IV and Cert. III Plumbing Drainage and Gasfitting, teaches all streams in Plumbing, Drainage and Gasfitting at Certificate IV and Certificate III level. He has been teaching at TAFE NSW for 15 years. Prior to that, Anthony owned and operated a plumbing business for 25 years, undertaking all facets of plumbing in the commercial, industrial and residential sectors – including design, tendering and contract management from concept to completion – and employing and training many apprentices throughout that time. With this experience, becoming a plumbing teacher seemed a natural progression to enable him to impart his knowledge and mentor apprentices.

ACKNOWLEDGEMENTS

This book would not be possible without the efforts from colleagues in the plumbing industry who have contributed their time and resources to write and update chapters in the previous editions. Many thanks to Warren Breese, Gary Cook, Peter Smith, Shaun Kristen, Terry Bradshaw, Peter Towell and Steven Tuckwell for their efforts. Thank you to Chris Mahon for his invaluable contribution to this edition.

The authors and Cengage would like to thank the following reviewers for their advice and valuable feedback:

- Dean Carter – TAFE NSW
- Robert Gilman – TAFE NSW
- Clive Hastie – SR TAFE
- Clive Stander – MPA Skills
- Phil Underwood – FireFox Training

Additionally, we would like to extend our thanks to those who reviewed the previous editions of this text.

Finally, Anthony would like to thank his wife Marissa for her patience, understanding and photographical expertise.

UNIT CONVERSION TABLES

TABLE 1 Length units

Millimetres *mm*	Metres *m*	Inches *in*	Feet *ft*	Yards *yd*
1	0.001	0.03937	0.003281	0.001094
1000	1	39.37008	3.28084	1.093613
25.4	0.0254	1	0.083333	0.027778
304.8	0.3048	12	1	0.333333
914.4	0.9144	36	3	1

TABLE 2 Area units

Millimetre square mm^2	Metre square m^2	Inch square in^2	Yard square yd^2
1	0.000001	0.00155	0.000001
1000000	1	1550.003	1.19599
645.16	0.000645	1	0.000772
836127	0.836127	1296	1

TABLE 3 Volume units

Metre cube m^3	Litre *L*	Inch cube in^3	Foot cube ft^3
1	1000	61024	35
0.001	1	61	0.035
0.000016	0.016387	1	0.000579
0.028317	28.31685	1728	1

TABLE 4 Mass units

Grams *g*	Kilograms *kg*	Pounds *lb*	Ounces *oz*
1	0.001	0.002205	0.035273
1000	1	2.204586	35.27337
453.6	0.4536	1	16
28	0.02835	0.0625	1

TABLE 5 Volumetric liquid flow units

Litre/second *L/sec*	Litre/minute *L/min*	Metre cube/hour m^3/hr	Foot cube/minute ft^3/min	Foot cube/hour ft^3/hr
1	60	3.6	2.119093	127.1197
0.016666	1	0.06	0.035317	2.118577
0.277778	16.6667	1	0.588637	35.31102
0.4719	28.31513	1.69884	1	60
0.007867	0.472015	0.02832	0.01667	1
0.06309	3.785551	0.227124	0.133694	8.019983

TABLE 6 High pressure units

Bar *bar*	Pound/square inch *psi*	Kilopascal *kPa*	Megapascal *mPa*	Kilogram force/ centimetre square *kgf/cm²*	Millimetre of mercury *mm Hg*	Atmospheres *atm*
1	14.50326	100	0.1	1.01968	750.0188	0.987167
0.06895	1	6.895	0.006895	0.070307	51.71379	0.068065
0.01	0.1450	1	0.001	0.01020	7.5002	0.00987
10	145.03	1000	1	10.197	7500.2	9.8717
0.9807	14.22335	98.07	0.09807	1	735.5434	0.968115
0.001333	0.019337	0.13333	0.000133	0.00136	1	0.001316
1.013	14.69181	101.3	0.1013	1.032936	759.769	1

TABLE 7 Temperature conversion formulas

Degree Celsius (°C)	(°F – 32) × 0.56
Degree Fahrenheit (°F)	(°C × 1.8) + 32

TABLE 8 Low pressure units

Metre of water *mH_2O*	Foot of water *ftH_2O*	Centimetre of mercury *cmHg*	Inches of mercury *inHg*	Inches of water *inH_2O*	Pascal *Pa*
1	3.280696	7.356339	2.896043	39.36572	9806
0.304813	1	2.242311	0.882753	11.9992	2989
0.135937	0.445969	1	0.39368	5.351265	1333
0.345299	1.13282	2.540135	1	13.59293	3386
0.025403	0.083339	0.186872	0.073568	1	249.1
0.000102	0.000335	0.00075	0.000295	0.004014	1

INTRODUCTION: PROPERTIES OF WATER

Chapter overview

This chapter looks at the general properties of water.

Learning objectives

Areas addressed in this chapter include:

- the origins and chemistry of water
- properties of water and its characteristics
- filtration and purification methods and the properties of 'hard' and 'soft' water, including sources of contamination and impurities
- drinking water (potable) supplies and protection measures
- hydrostatics and hydraulics.

Background

A plumbing apprentice was asked by his TAFE instructor where he thought water originated from. The apprentice replied that it came from the tap. While that is partly true, a much greater depth of knowledge about water and its properties is required as a professional working in the water industry.

As plumbers, we are entrusted with helping to maintain the general health of the people in our community. We strive to achieve this by installing and maintaining a suitable water supply system, as well as installing and maintaining an appropriate sanitary plumbing and drainage system to dispose of our waste products. In order to achieve this aim, an understanding of water systems and the plumber's role in installing and maintaining a safe and healthy water supply is necessary. Water is one of our most basic needs for survival. It is necessary for sustaining human, animal and plant life on Earth. Like many of nature's commodities, it is to be treated with respect and used economically.

In Australia, the provision of safe drinking water (potable) is regulated by guidelines produced by the National Health and Medical Research Council (NHMRC) in collaboration with the Natural Resource Management Ministerial Council (NRMMC).

GREEN TIP

Water is precious and necessary for all living things. It certainly must not be wasted.

Where does water come from?

Water, from sources such as rivers, lakes and the ocean, is heated by the sun and converted into water vapour, which then rises back into the atmosphere as a gas. This process is called **evaporation**.

Plants take in water through their root systems and pass moisture back into the atmosphere via their leaves in a process called **transpiration**. Animals and humans pass moisture back into the atmosphere when they perspire (sweat).

Evaporation and transpiration form part of the **natural water cycle** (see **Figure I.1**). The cycle is completed when the water vapour is cooled and condenses to form clouds; when the **condensation** reaches dewpoint, the water vapour will form rain droplets and return to the surface as **precipitation**. The water may also fall in other forms; for example, as snow when the temperature is low enough.

See the References at the end of the chapter and investigate history and education links to find out more about where our drinking water comes from.

The chemistry of water

Water is a combination of two elements, hydrogen and oxygen, in the proportion of two to one. Its chemical symbol is H_2O, which in chemistry means that two atoms of hydrogen have combined with one atom of oxygen to form one molecule of water (see **Figure I.2**).

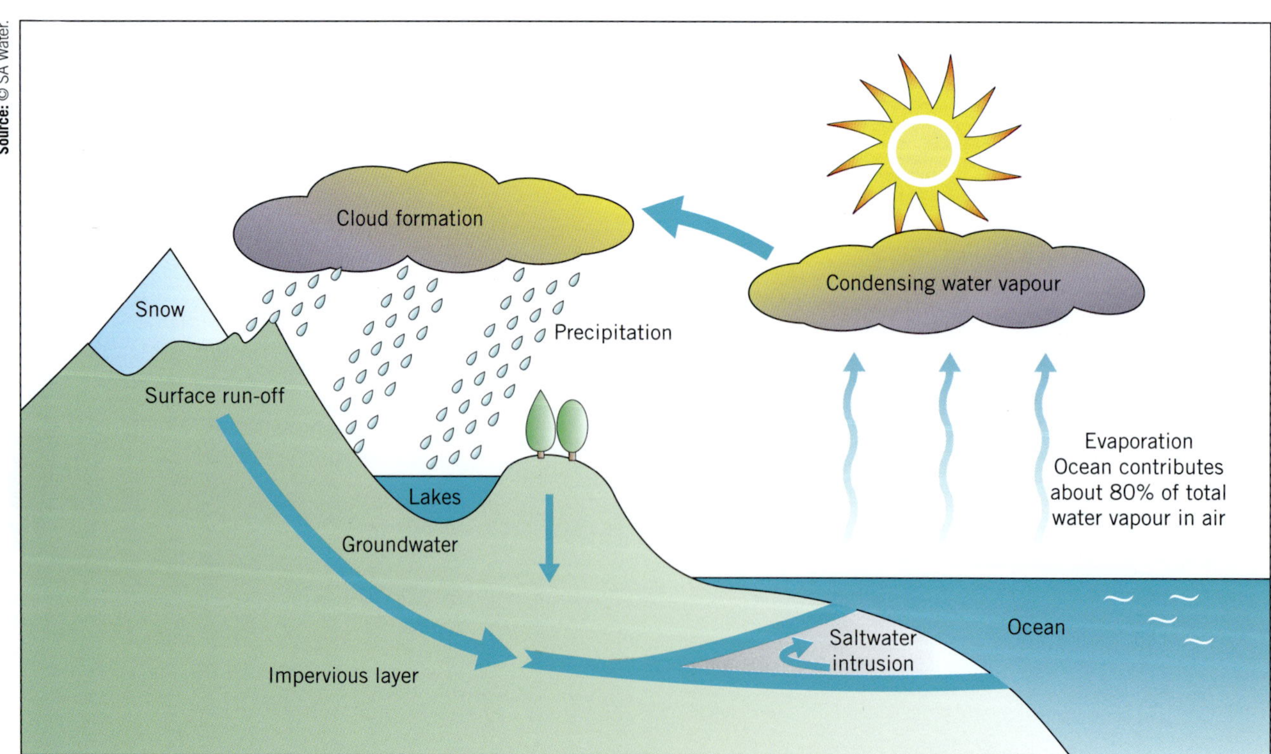

FIGURE I.1 The natural water cycle

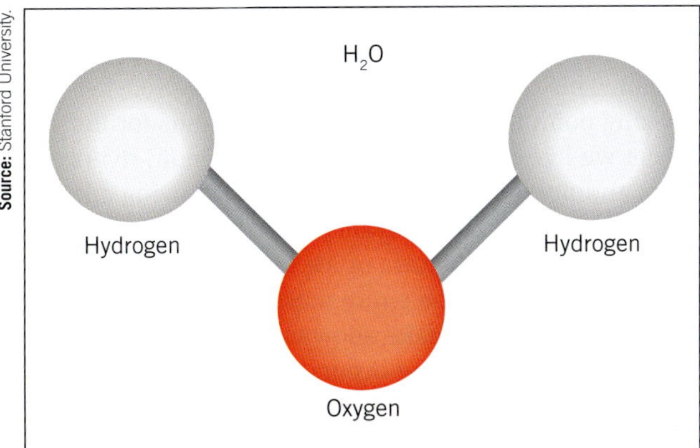

FIGURE I.2 The molecular structure of water as a liquid

The symbol represents only the composition of pure water such as is obtained through distillation; in this state it is a tasteless, odourless, transparent and colourless liquid, although when seen through considerable depth it can take on a bluish-green colouring.

States of water

Water may occur in three physical conditions or states:

1 In the solid state as ice. This condition starts at 0°C and continues at lower temperatures.
2 In the liquid state between the temperature range of 0°C and 100°C, where it is generally referred to as water (see **Figure I.2**).
3 In the gaseous state as steam at temperatures of 100°C and higher.

When it changes from one form to another, water is described as being in a changing state (see **Figure I.3**). Water changing from solid to liquid is said to be **melting**. When it changes from liquid to gas it is *evaporating*. Water changing from gas to liquid is called *condensation* (an example of this is the 'dew' that forms on the outside of a glass of cold drink). **Deposition** is when water changes from gas directly to solid form. When water changes directly from solid to gas the process is called **sublimation**.

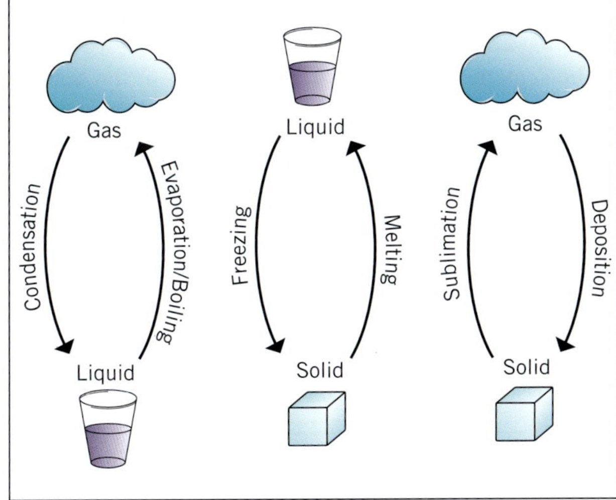

FIGURE I.3 The changing states of water

Depending on certain conditions, water can exist in both its solid and liquid states at 0°C.

Water hardness

As water evaporates, impurities are left behind. The purest water occurring in nature is probably rainwater, which leaves the rain cloud over areas where there is no atmospheric pollution. So rainwater in densely populated or highly industrialised areas will have some impurities.

The rain falling to Earth from the cloud collects different types of impurities located in the atmosphere. Water has the capacity to absorb gases and to dissolve many solids. In its downward travel to the ground, as a result of collecting impurities, and depending on the type of ground over which it falls and travels across, the chemical composition of water is altered.

The terms 'soft water' and 'hard water' are used to indicate the capacity of water to dissolve soap and form suds. Hardness is caused by salts of magnesium and/or calcium. The 'softer' the water, the easier it is for suds to form from soap; the 'harder' the water, the more difficult it becomes to do so.

Hard water is not a serious health risk, but it can cause stomach upsets. It causes problems in the textile industry from the scum creating discolouring of fabrics. It also causes problems in boilers and heat exchangers where the scale caused by the salts coming out of solution may damage the boiler. The city of Adelaide has experienced these problems for many years due to the water supply having above-normal levels of calcium and magnesium salts. Rainwater that has fallen on granite or other volcanic rock types will usually be soft, clear and sparkling because such rocks have little effect on water quality. If it falls on boggy or peat-type soil, the water may remain relatively soft, but may become slightly acidic and take on a dirty-brown appearance. Although not pleasant in appearance, such water may still be potable.

On the other hand, water may fall on, or penetrate through, deposits of lime or magnesia. As these deposits dissolve in water, they make the water 'hard'. Such water tends to be a little more sparkling than in the other cases mentioned, and although its capacity to dissolve soap is much reduced, it may nevertheless be quite suitable for normal domestic use and for drinking.

There are two types of water hardness:

1 **Temporary water hardness** is caused by bicarbonates of calcium or magnesium.
2 **Permanent water hardness** is caused by calcium sulphates or magnesium sulphates.

Temporary hardness can be removed by simply boiling the water. When boiled, the carbonates are left at the bottom in the form of minute particles.

In certain areas plumbers may have to install water-softening devices to help provide water that is satisfactory for domestic usage. The typical water softener is a mechanical appliance that is plumbed into the home's water supply system. All water softeners use the same operating principle: they trade the minerals for something else, which is usually sodium. Permanent hardness can also be removed by a **zeolite process**, which works by replacing the magnesium and calcium salts, which cause hardness of water, with sodium salts.

FROM EXPERIENCE

Plumbers need to choose the correct water filter for the right application.

The pH scale

The term 'pH' stands for 'power of hydrogen' and is a measurement of the level of acidity or alkalinity of the water. The **pH scale** starts at 0 and ends at 14. A reading of 0 indicates a strong acidic content in the water, whereas a reading of 14 indicates a strong alkaline content. Swimming pools use this scale to balance the water levels for safe use. The midway point of 7 is known as neutral. Water should have a pH level of slightly higher than 7 to avoid acidity damage to piping systems (see **Figure I.4**).

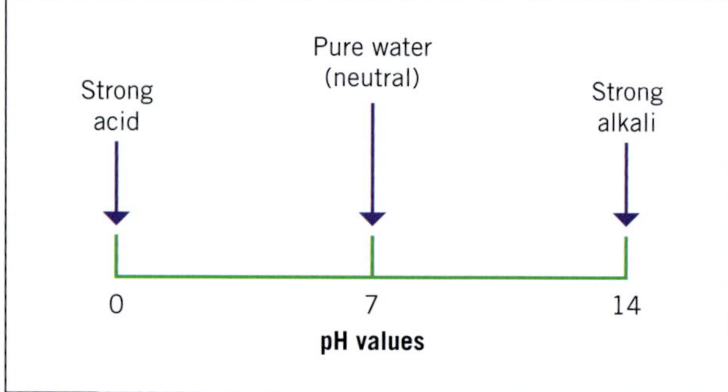

FIGURE I.4 pH values

LEARNING TASK I.1

1 Explain the evaporation process.
2 What should the pH level of potable water be?

Characteristics of water

Water density varies depending on temperature. Water is most dense at 3.984°C (say 4°C) at sea level. At this temperature, it weighs approximately 1000 kg per cubic metre. Therefore, it is common to say that 1 L of water has a mass of 1 kg. This is an important consideration when installing **water services** and storage tanks.

Any decrease or increase in temperature will be relative to the increase in volume and decrease in weight. This change in density as water approaches freezing point has wide-ranging implications for the planet we live on. Without this phenomenon, rivers, lakes and oceans would freeze solid, destroying all life within them. The buoyancy effect of ice on water creates an insulating mass of ice on the surface of the water that prevents the entire mass from freezing.

A plumber must make allowances for this increase in volume when freezing occurs. The volume can increase by as much as 9% at atmospheric pressure and even more in a pressurised pipeline. If the increase in volume on freezing is not prevented, an increased pressure of up to 25 megapascals (MPa) may be generated in water pipes, which will in turn burst pipes in winter conditions.

It is important that plumbers allow for expansion and contraction in pipework to avoid pipe bursts, especially in freezing conditions.

Water pressure

The branch of science that deals with fluids at rest is called **hydrostatics**, and that which deals with fluids in motion is called **hydraulics**. It is necessary for plumbers to know something about these two terms and how they affect the trade. For instance, a storage tank filled with water is subject to hydrostatic pressure and water supply reticulation system is subject to the laws of hydraulics.

Hydrostatics

The intensity of pressure exerted by water is directly proportional to the depth of the water and is measured in **kilopascals (kPa)**. When water is static, the depth at any particular point is called the **static head** (head is another way of saying 'pressure'). Static head is vertical height measured in metres. If a number of containers of water are connected together by tubes below the surface of the water and the upper ends are not sealed, the water will stand at the same level in each, regardless of their size or shape. Since pressure at the same head is uniform, the pressure at the bottom of all the containers is the same, although they may not hold the same amount of water (see **Figure I.5**).

FIGURE I.5 Containers of varying shapes on the same level have the same outlet pressure and head level

A head of water of 1 metre has a pressure value of 9.80638 kPa. Mathematically, however, this is an awkward number to use for calculations. It is much more useful to round up and regard 1 m head of water as having a value of approximately 9.81 kPa; or, if the head is in metres, we can use the formula:

P (in kPa) = H (in metres) × 9.81

Every 1 m of static water head corresponds to approximately 10 kPa (actual pressure = 9.81 kPa). Approximate conversions are:

10 mm = 0.1 kPa
100 mm = 1 kPa
1000 mm = 1 m = 10 kPa
10 m = 100 kPa

For any given head, water transmits pressure equally in all directions. This is illustrated by a firefighter's hose, which first lies flat and then becomes cylindrical when filled with water and under pressure.

If a closed water container is supplied with water from a tank, the pressure on the top, sides and bottom of the container will differ only in proportion to the difference in head at each particular location. In Figure I.6, the pressure at take-off point 3 of the container is caused by head A. The pressure at take-off point 1 of the container is caused by heads A + B + C. The average pressure anywhere on the sides of the container can be calculated from the vertical height or head.

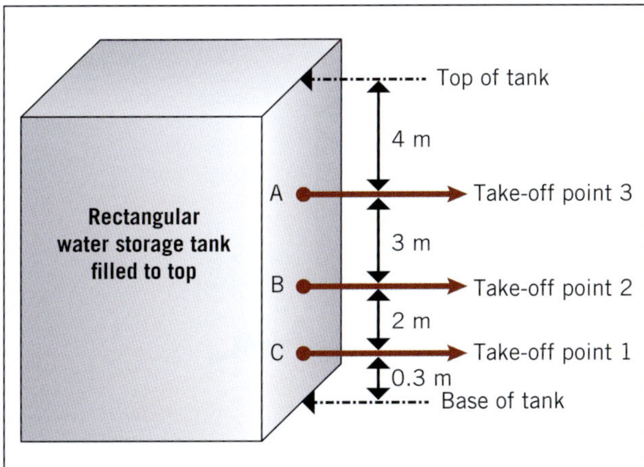

FIGURE I.6 Water pressure in a water storage tank

If a large-diameter pipe and a small-diameter pipe are fed with water at the same head, the pressure (kPa) in each will be the same, but the larger pipe will deliver more water because of its greater volume.

Hydraulics

A high static pressure does not necessarily mean that a large volume of water will be discharged from the end of a pipe connected to a source of supply. Friction between the moving water and the internal surface of the pipe, amount of bends and length of run is a major factor in the restriction of the volume of water delivered by a pipe. In general:

- The greater the velocity of water, the greater the friction and noise transmitted from the pipe. Greater velocity is generated using smaller pipe.
- Tight radius bends will create friction and noise transmission.
- The rougher the internal surface of the pipe, the greater the friction and noise transmitted from the pipe.
- The longer the pipe run, the greater the pressure loss due to more friction.

FROM EXPERIENCE

It is important to consider pressure, pipe size, length of pipe, change of direction and internal pipe bore smoothness when selecting piping materials.

LEARNING TASK I.2

1 How much pressure is generated by 2 m head of water?
2 What would the water pressure be at an outlet if the height of the water supply above is 16 m?

COMPLETE WORKSHEET 2

Filtration and purification

Water supply authorities are required to provide a water supply that meets guidelines set down in the *2011 Australian Drinking Water Guidelines* (ADWG). Typical procedures and processes used to satisfy these guidelines are outlined below.

Water from reservoirs and storage basins may contain organic matter and a variety of dissolved minerals. To make it fit for consumption, these may have to be removed by special treatment. Water purification can be achieved using some or all of the following processes:

- *Screening*. The water passes through a screen to remove large objects that may have entered the pipe from the dam.
- *Aeration*. In this process, air is passed through the water to eliminate soluble iron compounds and other impurities.
- **Flocculation**. A coagulant (such as ferrous chloride) is added to the water, causing small suspended particles to aggregate or collect as a 'floc', which then becomes heavy and settles to the bottom. Mixing the coagulant with the water via an impeller speeds up this process.

- *Filtration*. The water is passed through one or more filters that retain solids and allow only clear water to pass through.
- *pH correction*. Lime is added to the water to adjust the pH level to between 7 and 7.5.
- *Softening*. This is the removal of magnesium and calcium salts from the water. These salts are the cause of water hardness.
- *Chlorination*. Chlorine is added to eliminate bacteria from the water that may be dangerous to health. Through this process, the water is sterilised.
- *Sterilisation*. Sterilisation of the water supply by the addition of chlorine kills and inhibits the growth of harmful organic bacteria.
- *Fluoridation*. Fluoride is added at the direction of the Australian Government Department of Health to aid in the prevention of tooth decay.

A typical flowchart of the water supply and filtration process is shown in **Figure I.7**.

Disease and contaminants

A number of organisms or substances that cause human disease can contaminate water supplies. The most serious of these are micro-organisms, which can have immediate and devastating effects on our health. Some chemical contaminants also cause human disease, the effects of which depend on the type and level of exposure. These dangers are why we have rules and regulations that place limits on how, where and why we install water supply systems.

Disease-causing organisms

Pathogenic (disease-causing) micro-organisms in drinking water pose the greatest potential threat to human health. Worldwide, more than three million people per year, many of them children under five years of age, die from waterborne and sanitation-related diseases. Most of these deaths occur in the developing world, where many communities have no access to clean or treated water or to adequate

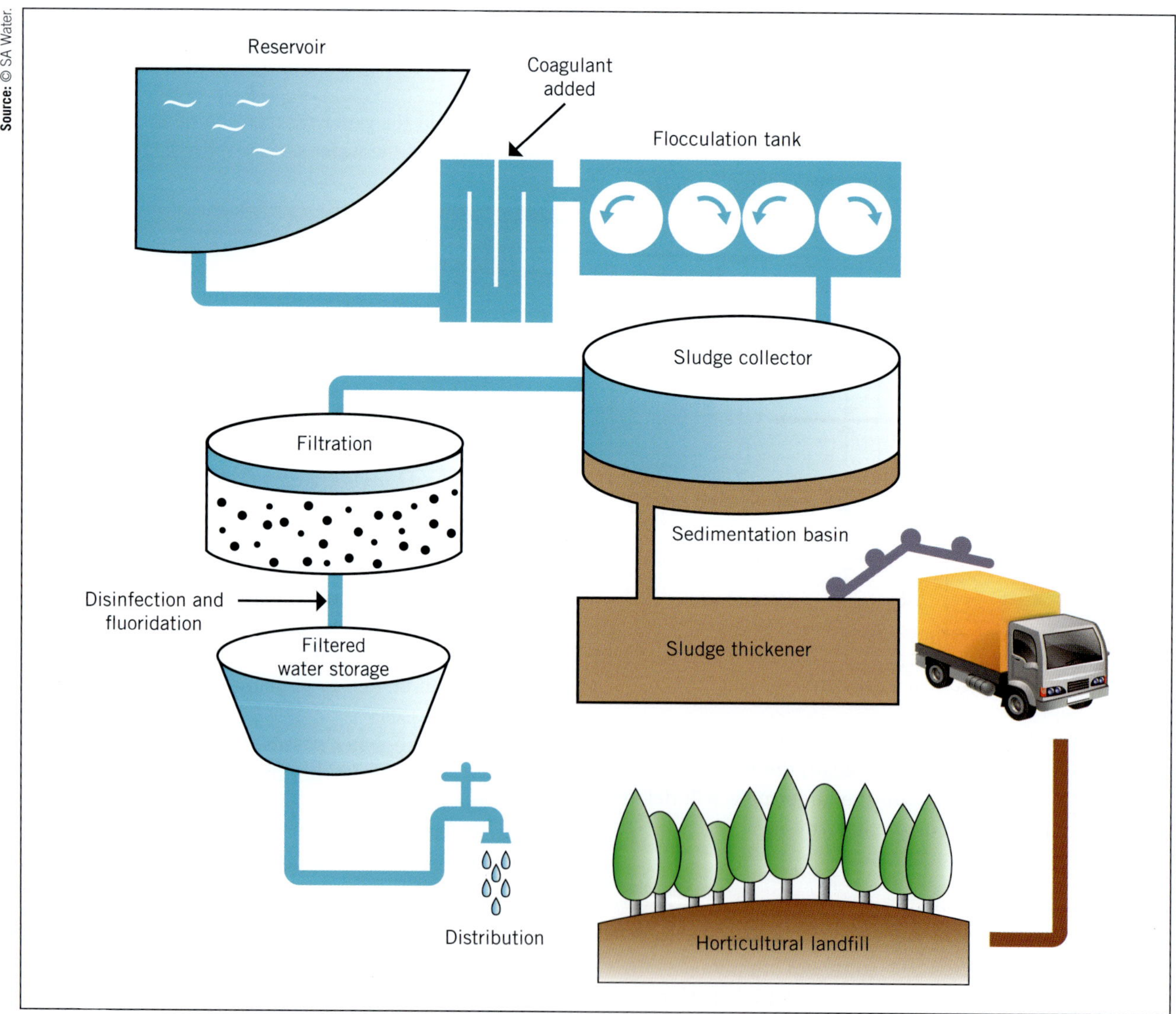

Source: © SA Water.

FIGURE I.7 Typical flowchart of the water supply and filtration process

sanitation. While the potential threat remains, in most parts of Australia waterborne disease is controlled by good water management. However, in some parts of Australia water quality remains a problem, especially in remote communities.

Micro-organisms include bacteria, viruses and protozoa, only a few of which cause disease. However, micro-organisms in human and animal faeces are responsible for most waterborne diseases. In some parts of the world, waterborne diseases such as dysentery, hepatitis, cholera and typhoid cause severe, and at times fatal, diarrhoea. *Cryptosporidium* and *Giardia* were brought to attention in Australia by the 1998 Sydney 'water crisis', where Sydney residents were put on a 'boil water' alert. *Cryptosporidium* and *Giardia* are protozoans – parasites that consist of a single cell and are a problem for the water supply industry because they are widespread in surface water, can survive for long periods and are difficult to treat.

Toxic substances

It is important to be aware of certain toxic substances that can occur in our waterways with harmful effects to our health.

Blue-green algae

Cyanobacteria, better known as blue-green algae, exists in rivers and dams if the water is stagnant, in warmer weather and if the ratio of nitrogen to phosphorous concentration is low.

It is a health hazard because of the toxins it releases. Some toxins result only in a skin rash, but others are more serious, causing liver and nerve damage. The toxins are released into the water and can remain even when the bacteria themselves have been removed. If contamination occurs, it may not be sufficient to boil the water. Boiling destroys the cells, but not all of the toxins. Therefore, special treatment is required to remove the toxins from water contaminated by blue-green algae.

Chemicals

Pathogenic micro-organisms have fairly immediate effects, but health effects from potentially harmful chemical and radioactive contaminants in drinking water become evident only after long exposure (typically many years). For example, low levels of arsenic in drinking water might increase the incidence of skin, lung or bladder cancer in a population that had been drinking the water for many years.

Chemicals of concern for drinking water include some naturally occurring chemicals such as nitrates, selenium and uranium; agricultural chemicals such as pesticides and fertilisers; and the chemical by-products formed when water is treated with a disinfectant (these disinfection by-products are discussed in detail in the section on water filtration and purification). However, the amount of these chemicals in our drinking water is generally very small – much lower than the levels that would be considered harmful to health. Indeed, we are exposed to higher levels of these chemicals in our environment and our food (although they are well below what is considered a harmful level).

Radioactive contaminants

The health effect most strongly associated with radioactive contaminants is cancer. Extremely low levels of radiation are a naturally occurring characteristic of water in our environment. Drinking water is likely to contribute only a very small proportion of a person's overall natural exposure to radiation.

GREEN TIP

Water authorities are responsible for maintaining an efficient water treatment process to prevent disease and contaminants entering the system.

LEARNING TASK 1.3

1 Name three waterborne diseases.
2 Does boiling water remove the toxins in water contaminated with blue-green algae?

COMPLETE WORKSHEET 3

SUMMARY

- The natural water cycle is made up of the combination of evaporation, transpiration, condensation and precipitation taking place over the natural water systems and land.
- The chemical symbol for water is H_2O – two atoms of hydrogen and one atom of oxygen.
- There are three states of water – liquid, solid and gas.
- Hard water is high in mineral content – calcium and magnesium.
- The pH scale measures the acidity and alkalinity of water from 0 (acid) to 14 (alkaline).
- One metre head of water has a pressure value of 9.81 kPa ($P = H \times 9.81$).
- Water purification (treatment) is a process of screening, aeration, flocculation, filtration, pH correction, chlorination, sterilisation and, in most cases, fluoridation.

REFERENCES

To find out more about your local authority, visit:

Icon Water (ACT): **https://www.iconwater.com.au/**

Melbourne Water: **https://www.melbournewater.com.au**

NHMRC: **https://www.nhmrc.gov.au/about-us/publications/australian-drinking-water-guidelines**

PowerWater (NT): **https://www.powerwater.com.au**

QldWater: **https://www.qldwater.com.au**

SA Water: **https://www.sawater.com.au**

Sydney Water: **https://www.sydneywater.com.au**

Water Corporation (WA): **https://www.watercorporation.com.au**

WORKSHEET 1

To be completed by teachers	
Student competent	☐
Student not yet competent	☐

Student name: ______________________________

Enrolment year: ______________________________

Class code: ______________________________

Task: Review 'Where does water come from?' and answer the following questions.

1 Explain the following terms in your own words:

a Evaporation

____

____

____

____

b Condensation

____

____

____

____

c Precipitation

____

____

____

____

2 What is the chemical symbol for water?

____

3 Name the three different states (forms) of water. State the temperature range of these three forms.

a ______________________________

b ______________________________

c ______________________________

4 The pH scale is a measurement of the amount of acid or alkaline in a substance. Neatly draw a diagram that indicates pH values.

5 What number on the pH scale represents 'neutral'?

____

6 What are the two main types of mineral salts found in hard water?

____

____

7 Temporary hardness can be removed by which method?

____

8 Permanent hardness can be removed by which method?

____

9 Why should water have a pH level of slightly higher than 7?

____

____

WORKSHEET 2

To be completed by teachers

Student competent ☐

Student not yet competent ☐

Student name: ______________________

Enrolment year: ______________________

Class code: ______________________

WORKSHEETS

INTRO

Task: Review the section 'Characteristics of water' and answer the following questions.

1 At what temperature is water most dense?

____

2 What is the mass (in kilograms) of 1000 L of water?

____

3 Explain in your own words why ice floats in water.

____

____

4 Describe in your own words the problems that can occur when water freezes within pipework systems.

____

____

5 What problems occur from high velocity within a pipe?

____

____

6 How can you reduce velocity?

____

____

7 What can cause friction within a pipe?

____

____

8 What effect do tight radius bends have in a piping system?

____

WORKSHEET 3

Student name: ______________________

Enrolment year: ______________________

Class code: ______________________

To be completed by teachers	
Student competent	☐
Student not yet competent	☐

Task: Review 'Filtration and purification' and answer the following questions.

1 Briefly describe in your own words each of the following water purification processes:

a Screening

b Aeration

c Flocculation

d Filtration

e pH correction

f Softening

g Chlorination

h Sterilisation

i Fluoridation

2 Name the two parasites that contaminated Sydney's water supply in 1998.

FABRICATE AND INSTALL NON-FERROUS PRESSURE PIPING

1

Chapter overview

This chapter looks at the planning, identification, fabrication, jointing techniques and installation of non-ferrous pipes commonly used in the plumbing industry for the supply of water.

Different jointing and bending techniques for non-ferrous pipe and tubing to meet the requirements of manufacturers, industry standards and authorities are shown. Procedures on testing to comply with the requirements of manufacturers, industry standards and authorities for non-ferrous pressure pipe are referenced.

By the end of the chapter, the skills to select and use tools and equipment to fabricate and install non-ferrous pressure piping for various applications should be achieved.

Learning objectives

Areas addressed in this chapter include:

- identify installation requirements
- prepare for work
- fabricate, install and test the pipe system
- clean up.

Non-ferrous materials are materials that do not contain iron. These types of materials are most commonly used in the plumbing industry to carry or store water as they are more resistant to corrosion (rust) than pipe materials that do contain iron, and therefore have a longer lifespan.

Tradespeople working in the plumbing services sector are required to identify and install a wide selection of pipe materials. Plans and specifications must be interpreted to allow the selection of the appropriate material for a given job. Tradespeople also must have the knowledge and skill to select and use the correct method/s for fabricating, joining, clipping and testing the particular pipe material using a variety of tools and techniques as per the manufacturer's specifications, industry standards and authority requirements.

Identify installation requirements

The plans and specifications for the job must be obtained to ensure the work carried out is compliant. Information from several sources is necessary to identify the requirements for the correct installation of non-ferrous pressure piping. This involves authorities' requirements, manufacturer's specifications, Plumbing Code of Australia (PCA), Australian Standards, plans and specifications, and site inspection.

Plumbing and gasfitting standards, regulations and codes have sections that indicate where a material may or may not be used and state the restrictions and/or limitations on the use of a given material. The Australian Standards and authorities' requirements work in with the plans and specifications for the job. The specifications will help to eliminate having to choose the appropriate materials and pipework fabrication for a specific task.

AS/NZS 3500.1 PLUMBING AND DRAINAGE: WATER SERVICES

Common uses of non-ferrous pressure piping

Non-ferrous piping materials commonly used in plumbing include:

- copper
- copper alloys (e.g. brass)
- aluminium
- plastics (e.g. polyethylene, polybutylene and cross-linked polyethylene).

Non-ferrous materials are used in plumbing for a wide variety of other purposes, including:

- heated water
- cold water
- gas
- waste
- mechanical service pipes for heating and cooling buildings
- compressed air
- medical gases
- steam and condensate
- fuel oils
- irrigation.

When fabricating and installing pipework it is essential that the plumber knows the correct method/s for:

- joining pipes
- bending pipes
- supporting/bracketing pipes
- testing the installation for soundness.

FROM EXPERIENCE

Having the knowledge and skills to choose the correct installation method for a selected material earns customer confidence, creating future work opportunities.

Access and determine requirements from codes, standards and specifications

The Plumbing Code of Australia (PCA), AS/NZS 3500 and AS/NZS 5601 specify the types of non-ferrous piping systems, the jointing systems and what plumbing applications they are used for, as well as the grade they must comply with when used with different types of systems.

It is also important to check job specifications, manufacturer's instructions and the local council regulations for specific requirements of piping material and jointing methods. Planning and checking these ensures the correct installation of piping materials and jointing methods. Always check that the material, fittings and jointing systems are approved as per Australian Standards.

AS/NZS 3500.1 PLUMBING AND DRAINAGE: WATER SERVICES

Identify and apply workplace policies and procedures, work health and safety (WHS) and environmental requirements

It is important to know the work health and safety (WHS) requirements for each job. A risk assessment must be carried out identifying the hazards, accessing the risks and implementing control measures to ensure the safety of everyone.

The WHS requirements and quality assurance policy should also consider the potential impact of works on the environment.

Be sure to check that personal protective equipment (PPE) is fit for purpose and is maintained in accordance with the manufacturer's instructions. Remember to wear the appropriate PPE. This can include:

- safety boots
- safety glasses
- ear plugs or ear muffs
- dust masks or respirators
- gloves
- high-vis workwear.

The work area must be risk assessed before commencing the job. Keeping the site clean and tidy will ensure a safer and hassle-free job. It will also save time in the long run due to fewer obstacles in the way.

Consideration must be given to the potential impact of works on the environment. Steps should be taken to ensure hazardous chemicals are not released into the waterways and harmful fumes from welding non-ferrous metals, depending on the situation, are captured and treated in accordance with organisation policy and procedure.

A hot works permit may be necessary when silver soldering. This describes the location and detail of the task with the control measures in place for fire protection.

Quality assurance

Most companies have quality assurance procedures in place to promote a good standard of workmanship and efficient operations. This helps to prevent mistakes from happening.

The requirements of this policy can include:

- controls and procedures in the workplace
- use and maintenance of equipment
- interpreting specifications
- quality of materials
- handling procedures.

LEARNING TASK 1.1

1. Where can information on the specified materials be obtained?
2. What is the purpose of quality assurance procedures?

Prepare for work

Understanding plans and specifications is important to meet the job requirements for fabricating non-ferrous pressure piping. Plans and specifications will explain where the pipes and equipment are to be installed.

A site inspection prior to starting work is important, as it may identify a clash of services not shown as per plan. Checking the site will help to ensure the installation will not damage or interfere with surrounding structures, therefore saving time and money. Also, access for deliveries, equipment and machinery can be determined.

When ordering materials, it is important to make an accurate 'take-off' of the materials from the plans and specifications to avoid excess waste of materials and time. Always check your order on delivery to make sure it is complete and not damaged.

Organising and planning tasks

Fabricating and installing non-ferrous pressure piping for the supply of water can involve trench excavation, working at heights and working in confined spaces, so it is vital the correct procedures are followed. This involves planning and sequencing the order of tasks. Some preliminary tasks are:

- visiting the site to assess access for deliveries, plant and equipment
- ensuring all fees and permits have been dealt with; for example, a road opening permit, footpath opening permit and water main drilling fee
- carrying out a Before You Dig Australia enquiry
- being certain a thorough risk assessment has been done with a completed safe work method statement (SWMS) and a hot works permit, if required.

It is important to notify other trades or the site supervisor of the tasks planned so there is minimal disruption and everyone can work together towards a common goal.

Setting up silver brazing equipment

Set up oxygen-acetylene equipment in accordance with NSW codes of practice for welding processes.

Take steps to prepare the work area for welding work in accordance with the NSW codes of practice for welding processes. *Note:* refer to unit CPCPCM2052 Weld mild steel using oxy-acetylene equipment for further information on setup requirements.

Select and check serviceability of appropriate tools and equipment including personal protective equipment (PPE)

Tools, equipment and PPE must be maintained and kept in good working order. They should be cleaned after use and inspected for any damage. Electrical tools and leads must be regularly tagged by a qualified person to ensure that they are safe to use. Ladders and steps should be inspected for damage before using. All tools, equipment and PPE should be stored out of the weather. All equipment used for silver soldering, such as oxygen and acetylene equipment, should be inspected for defects and checked for leaks on set up. Refer to unit CPCPCM2052 Weld mild steel using oxy-acetylene equipment for further information on setup requirements.

LEARNING TASK 1.2

1. Why is a site visit important before starting work?
2. The job is to run a 20 mm copper water line from a house to the water meter to replace the existing corroded galvanised line. The trench has been dug and the bedding material has been supplied. The water meter is 11 m from the house and the existing fittings on the water meter can be reused. All joints will need to be brazed using silver solder.

   Create a materials list noting the materials, tools, equipment and PPE required to complete this task.
3. Name four services non-ferrous piping is used for.

Fabricate, install and test the pipe system

The non-ferrous pressure pipes and fittings mainly used in the supply of water are copper, brass and a range of polymers (plastics). Information on these materials, the jointing techniques and testing requirements are listed as follows.

Set out and install pipework according to plans and specifications

It is important to refer to drawings, specifications and job instructions when setting out pipework and connection points to fixtures. This ensures the fixture installation will be straightforward and trouble free.

Pipework must be installed as per AS/NZS 3500 for Plumbing and Drainage and AS/NZS 5601 for Gas Installations to comply. The manufacturer's specifications must be adhered to for product warranty. Local council requirements need to be followed as there are varied requirements in different regions.

Copper and copper alloys

The copper tube widely used in the plumbing industry is made to AS/NZS 1432 Copper Tubes for Plumbing. It is available in 6 m hard drawn lengths for diameters up to 150 mm and annealed 18 m coils for diameters from 8 mm to 32 mm. Copper tube is long-lasting, corrosion-resistant and has a smooth bore (it reduces friction), and it does not adversely affect the quality of the fluid flowing through it.

Blue-water phenomenon

Blue water is a phenomenon of copper tube that randomly affects large parts of the east coast of Australia, as well as New Zealand, the United States and Europe. It is characterised by the production of voluminous blue-green corrosion products in the tube bores and is primarily associated with cold, soft, unbuffered waters of high pH (alkalinity) and negligible levels of disinfectant residual.

It has been discovered that for corrosion to occur in Australia, the prevailing water chemistry must have certain characteristics. Tubes, of whatever surface condition, do not cause blue water or pitting to perforation in Adelaide, Brisbane or anywhere else where the source water is moderately 'hard', which means the water is high in calcium and magnesium. Minor surface pitting can be found in these waters, but rarely is a serious corrosion problem.

The extent of copper corrosion on the east coast of Australia, Perth, Tasmania, New Zealand and elsewhere encompasses a range of 'soft' water chemistries. All areas of contamination, however, seem to share significant characteristics:

- soft water, with total dissolved solids less than 300 mg per litre
- low alkalinity (typically less than 30 mg per litre)
- the ability to leach lime from cement-lined cast-iron pipes, leading to a relatively high pH at the end of the distribution system
- almost invariably cold water, or temperatures lower than 50°C if warm water is involved
- varying levels of chloride and sulphate, though corrosion is known to occur with both these analytes at less than 7 mg per litre
- low residual disinfectant.

Although extensive research has been conducted on this subject, a precise and definitive diagnosis is yet to be determined. However, the evidence has shown that blue water can at least be controlled by a corrosion management program that provides buffering of the supply on either a localised or regional basis, together with maintenance of an adequate disinfectant residual.

Copper tube

Copper tube manufactured to AS 1432 is classified into four types: A, B, C or D. Each type refers to the wall thickness category of the tube, with Type A having the thickest wall and Type D the thinnest. Each type has colour-coded writing: Type A – green, Type B – blue, Type C – red and Type D – black. Copper tube is sized from the outside diameter of the tube (OD), unlike steel pipe, which is measured from the inside diameter (ID). Copper tube can be easily annealed (softened) by heating, which makes bending and expanding tube ends easy.

Copper tube may also be purchased pre-insulated with a polyvinyl chloride (PVC) polylag sleeve. This will reduce some heat loss, provide clearance for expansion and protect copper tube when laid in corrosive soils. It is also available with a purple PVC sleeve for use on recycled water services.

GREEN TIP

All copper tube offcuts should be scrapped to a metal recycler. It provides extra income and prevents waste in landfill.

Jointing copper

There are numerous methods used to join copper tubes. These include silver soldering (brazing), soft soldering, olive compression, croxed compression, flare compression, press-fit and push lock fittings. Refer to AS/NZS 3500.1 for jointing limitations, particularly soft solder, as there are limitations in the latest standard stating that soft-soldered joints are not to be used on new work but only on repairs to existing work.

AS/NZS 3500.1 PLUMBING AND DRAINAGE: WATER SERVICES

AS/NZS 5601.1 GAS INSTALLATIONS

Silver-soldered/brazed joints

Silver soldering (also known as **silver brazing**, brazing or hard soldering) is a jointing process that relies on capillary attraction to draw solder into joints while intergranular penetration of the materials also takes place. This is where the heat allows the molten brazing rod to form a very strong bond with the material's surface. Silver soldering requires more heat than soft soldering and so, generally, oxygen/acetylene or oxygen/LPG equipment is used to heat the joints. For the capillary action to take place, the surfaces must be clean and closely aligned.

There are three types of joints that can be silver soldered (see Figure 1.1):

- a tube end into a capillary fitting: (a) and (b)
- a tube end into a formed branch in another tube: (b)
- a tube end into an expanded tube: (c).

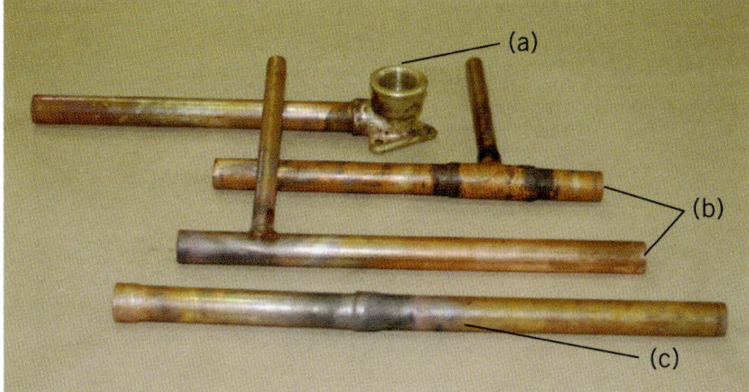

FIGURE 1.1 Types of joints

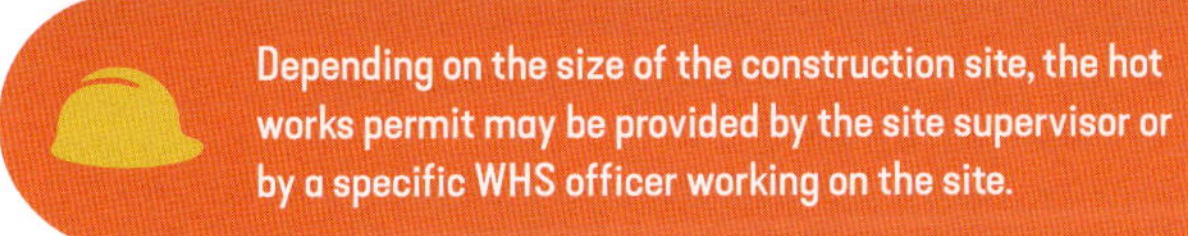

To prepare for jointing, the outside surface of the tube and the inside surface of the fitting must be clean and free of oxides to ensure proper adhesion of the solder. New tubes and fittings are generally clean enough to make sound silver-soldered joints without further cleaning. If the fitting or tube-end surfaces are dirty, the surfaces should be cleaned with abrasive cloth or steel wool. All internal and external burrs must be removed to prevent friction loss.

No **flux** is required when silver soldering copper to copper joints. However, a general-purpose silver brazing flux must be used when silver soldering copper tube to a brass fitting.

The filler rod used contains copper, phosphorus and silver. Other additives may include tin, zinc and cadmium. The silver content of the rod varies with the type of work you are performing, in accordance with Australian Standards and the specifications of a given project. The higher the silver content, the more expensive the rod will be. High-silver-content rods flow better for jointing purposes with additional strength. The silver content of a rod is indicated by the colour of the end of the rod (see Table 1.1).

TABLE 1.1 Silver-soldering filler rod specifications

Rod tip colour	Silver content	Temperature range (°C)
Yellow	2%	645–704
Silver	5%	645–740
Brown	15%	645–700
Gold	40%	660–780
Pink	50%	688–744

Source: © 2009 MM Kembla.

Capillary fittings

Capillary fittings may be made from copper or brass (see Figure 1.2). Capillary fittings are made for silver soldering and soft soldering. There are a large number of shapes and sizes of capillary fittings available. Therefore, plumbers must know how to describe each fitting when ordering them from plumbing suppliers or when communicating with co-workers. Table 1.2 lists some of the terminology used when describing fittings.

FIGURE 1.2 Samples of some of the plain capillary fittings available

Many plumbers still use an old numbering system to describe capillary fittings (see Figure 1.3). For example, a number 12 is an elbow and a number 24 is a tee. So a plumber who knows the number for a given fitting may ask for a '15 mm no. 24' if they want a 15 mm tee. If a plumber does not know the number allocated to a fitting, they must know how to describe

TABLE 1.2 Terminology used to describe fittings

Terminology	Description
External thread male iron (MI)	The externally threaded end of a fitting
Internal thread female iron (FI)	The internally threaded end of a fitting
Elbow bend	A fitting with a sharp 90° change of direction
Tee	A fitting allowing the connection of three pipe ends
OD	Outside diameter – used when describing the capillary end
BP/lugged	Used to describe fittings that have a backplate used to screw or fix the fitting securely against a wall, post, timber or other solid object
Dezincification-resistant (DR)	Indicates that the fitting is made from brass of a type (DR) that resists a form of corrosion called dezincification. Brass plumbing fittings must be made from DR brass. A longer radius fitting will allow better flow.

each fitting. Imperial measurements are still used occasionally when ordering fittings (20 mm = ¾ inch; 15 mm = ½ inch).

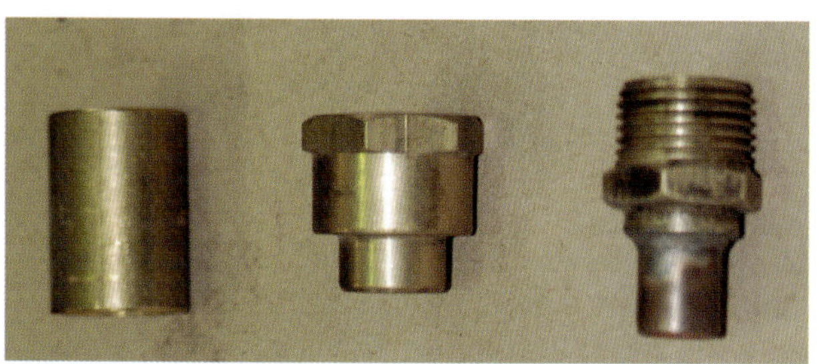

FIGURE 1.3 Capillary fittings
From the top:
20 mm OD × 20 mm OD brass socket or 20 mm no. 1 DR
15 mm OD × 15 mm internally threaded brass connector or 15 mm no. 2
15 mm OD × 15 mm externally threaded brass connector or 15 mm no. 3

HOW TO

DESCRIBE CAPILLARY FITTINGS

The method used to describe capillary fittings is as follows:

1. When describing the end of a fitting, you should state the size and the type of end (OD, externally or internally threaded).
2. For a fitting with two ends, where one end is capillary (OD) and the other end threaded (externally or internally), ask for the capillary end first and then the threaded end. If both ends are capillary, ask for the larger end first, if there is a difference in size (see Figure 1.4).
3. For fittings with three ends, the ends along the straight line are asked for first, stating the capillary (OD); or if the end sizes are different, state the larger capillary (OD) first, then the other end, followed by the branch (see Figure 1.5).

FIGURE 1.4 Brass no. 1 and copper no. 1R (socket and reducing socket)

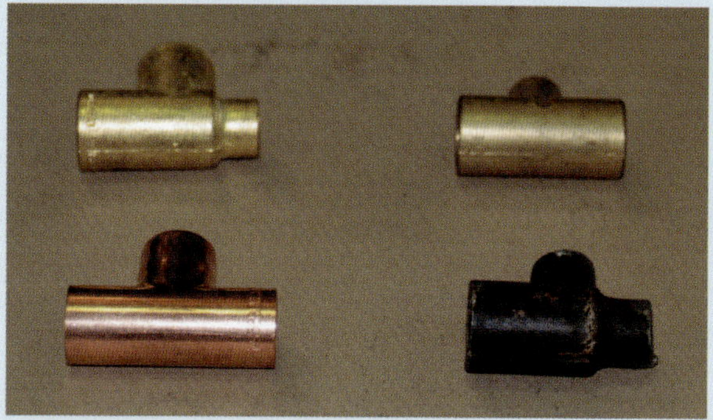

FIGURE 1.5 Capillary fittings with three ends
Clockwise from top left:
20 mm × 15 mm × 20 mm brass tee or 20 × 15 brass no. 26
20 mm × 20 mm × 15 mm brass tee or 20 × 15 brass no. 25
20 mm × 15 mm × 15 mm brass tee or 20 × 15 brass no. 27
20 mm copper tee or 20 no. 24

Expanded joint fittings

The second type of joint for silver soldering is an expanded joint. This is a joint made from the pipe without the need to supply a socket fitting. **Expanded joint fittings** must be silver soldered and have the advantage of only soldering one joint instead of two, as needed for a socket joint.

HOW TO

MAKE EXPANDED JOINT FITTINGS

1 The end of a copper tube is annealed or softened (see **Figure 1.6**).
2 The tube end is then expanded using a **tube expander** (see **Figures 1.7** and **1.8**).
3 The end of another copper tube is cleaned, then slipped into the expanded section (see **Figure 1.9**).
4 The joint is then silver soldered.

FIGURE 1.6 Annealed copper tube end, ready to be expanded

FIGURE 1.7 Tube expander fitted into annealed tube end, ready to be expanded

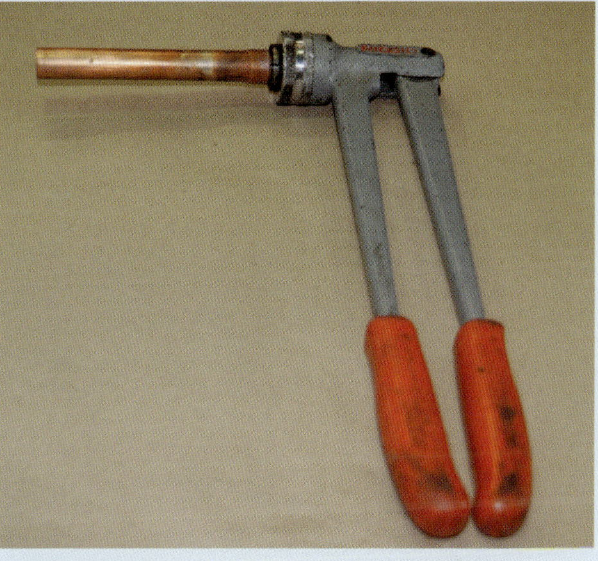

FIGURE 1.8 Tube expander handles closed together, expanding tube end

FIGURE 1.9 Copper tube slipped into expanded tube end, ready for silver soldering

Branch-formed joints

The final type of joint for silver soldering is a **branch-formed joint**. This is a joint formed with a special tool that creates a branch connection. This replaces the need for a tee fitting and is very handy when retro fitting as only one soldered joint is needed instead of three for a tee fitting. It is also very cost effective.

Heat source

Oxygen/acetylene or oxygen/LPG equipment is generally used for silver soldering. LPG or MAPP gas may be used, but these heat sources take much more time to heat the joint to a temperature that will allow the silver-soldering filler rod to flow easily.

HOW TO

MAKE BRANCH-FORMED JOINTS

To make branch-formed joints, a tee joint is made in a copper tube as follows:

1 The piece of copper tube is marked, lightly centre-punched and drilled where the branch is to be formed. Care should be taken to drill the correct size hole. Do not anneal the copper as the tube will deform when using the branch-forming tool.

>>

2 The hooked end of the branch former is inserted into the hole in the copper tube.
3 The bonnet of the branch former is held firmly against the copper tube while the branch is formed by using a ratchet to turn the hooked end, which pulls through the drilled hole, forming the sides of the branch.

Branch forming may not be used to form tees in the same size pipe (e.g. DN 20 pipe with a DN 20 branch). The branch must be one pipe smaller (e.g. DN 20 pipe with a DN 15 branch).

Figure 1.10 shows the tools for making a branch, and **Figure 1.11** shows a branch being formed.

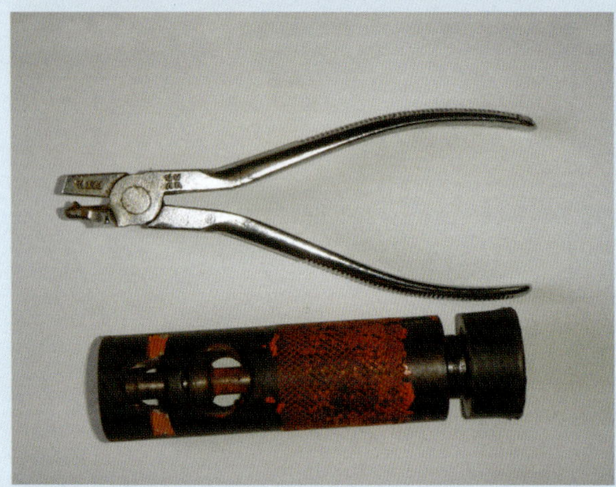

FIGURE 1.10 Notching tool and dimple pliers used in branch forming

FIGURE 1.11 Branch being formed

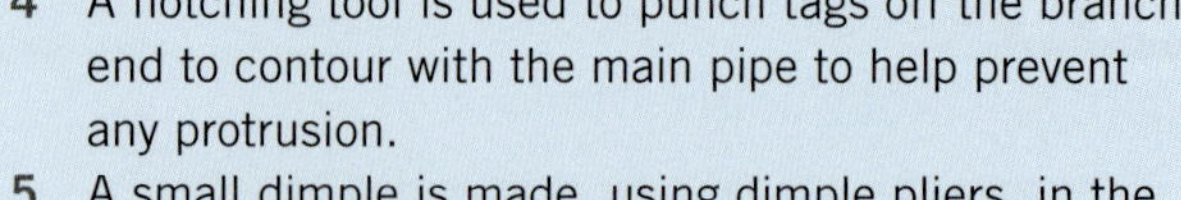

4 A notching tool is used to punch tags off the branch end to contour with the main pipe to help prevent any protrusion.
5 A small dimple is made, using dimple pliers, in the pipe end being used as the branch to prevent the branch from being inserted too far into the joint (see **Figure 1.12**).
6 The branch pipe is now inserted into the formed branch ready to be silver soldered (see **Figure 1.13**). Note the position of dimples (on the high side) to prevent the pipe branch from penetrating into the main pipe.

FIGURE 1.12 Notching tool in use

FIGURE 1.13 Branch pipe inserted into formed branch, ready for silver soldering

Procedure for silver-soldered joints

The procedure to make a quality silver-soldered joint must be carried out in a step-by-step process as listed below. *Note:* refer to unit CPCPCM2052 Weld mild steel using oxy-acetylene equipment for further information on setup requirements.

HOW TO

MAKE A SILVER-SOLDERED JOINT

The procedure for jointing silver soldering is as follows:

1 Cut the tube square (tube cutters are a purpose-built tool) and remove the internal burr.
2 Clean the pipe and fitting surfaces if oxidised or dirty.
3 Apply flux to the outside of the pipe (only necessary when using silver-soldering brass fittings).

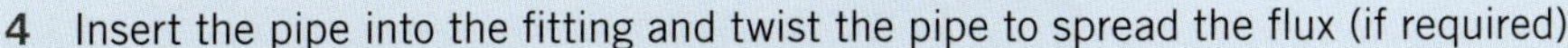

>>

4 Insert the pipe into the fitting and twist the pipe to spread the flux (if required).
5 Support the joint so that it does not move during soldering.
6 If there are flammable/painted surfaces near the joint, protect them with a heat-resistant mat.
7 Use a neutral flame.
8 Heat the joint until it is a dull maroon colour.
9 Touch the end of the silver soldering stick against the joint. When the joint is hot enough, the filler rod will melt and begin to flow. Keep the flame moving at all times; if the flame is held in the one spot for too long, you may melt the copper tube or damage the fitting.
10 Move the heat around the joint, melting the brazing rod into the joint.
11 Allow the joint to cool until the solder solidifies before moving it.
12 Wash off the flux with water and a damp rag (if necessary).

Once the joint has been soldered it is important to shut down oxygen/acetylene equipment in accordance with the workplace policy and procedures and the NSW codes of practice for welding processes. *Note:* refer to unit CPCPCM2052 Weld mild steel using oxy-acetylene equipment for further information on shutdown requirements.

The advantages of silver soldering are as follows:

- There is visible proof of a sound joint.
- It is simple and quick.
- It is suitable for gas services.
- It is strong.

The disadvantage is that it softens the copper tube near the joint so the pipe could easily bend, possibly kink and break if not correctly supported.

When silver soldering, tinted safety glasses should be worn to protect the eyes from the flame brightness, sparks and any possible debris.

Soft-soldered capillary joints

This jointing method uses capillary action in which a liquid (in this case, molten solder) is drawn into the space between two closely aligned surfaces (the outside of the tube and the inside of the fitting). Soft solder has a low melting point, so much less heat is required compared with silver soldering. This method of jointing is usually used for jointing copper to brass fittings and copper to copper fittings (see **Figure 1.14**).

Soft solder for use on water supply pipework is 'lead free'. Traditional 50/50 lead/tin solders are no longer used due to health problems associated with the use of lead. Lead-free soft solders are made from combinations of tin, silver, antimony and copper, with tin making up at least 95% of the solder.

To prepare for jointing, the outside surface of the tube and the inside surface of the fitting must be clean and free of oxides to ensure proper adhesion of the solder. The surfaces should be cleaned with abrasive cloth or steel wool before flux is applied.

Flux is used to:

- clean and remove oxides
- prevent new oxide from forming during soldering
- help the solder to flow.

A paste flux containing zinc chloride is applied to the surfaces being soldered.

Fittings may be made from copper or brass. There are two types of fittings: solder ring and plain (without solder ring) fittings.

Solder ring fittings

For a **solder ring fitting**, a ring of solder is placed into a groove around the fitting. This ring of solder has enough solder in it to make a sound joint. When liquefied by the heat, the solder ring flows into the capillary space around the fitting, making a sound joint (see **Figure 1.15**).

FIGURE 1.14 Soft solder, flux and a brass capillary fitting soldered onto a copper tube

FIGURE 1.15 Solder ring fitting

Plain fittings

Plain fittings (without solder ring fittings) have no solder ring (see Figure 1.16). Solder is added to the joint by the plumber.

FIGURE 1.16 20 mm × 15 mm copper elbow (no. 12R) and 20 mm × 15 mm × 20 mm copper tee (no. 26)

Heat source

LPG, MAPP gas or air acetylene heating equipment are recommended for soft soldering. Oxygen and acetylene equipment must be used with care as it is easy to overheat the joint and burn the solder or melt other joints nearby and possibly cause leaks.

FROM EXPERIENCE

When soldering near existing soft-soldered joints, wrap them with a wet rag or spray cool gel over them so they do not melt and leak.

Jointing procedure for soft-soldered capillary fittings

The procedure for jointing soft-soldered capillary fittings (see Figure 1.17) is similar to silver soldering, with a few exceptions.

Justin Kase z12z/Alamy Stock Photo

FIGURE 1.17 Soft soldering a capillary joint

HOW TO

MAKE A SOFT-SOLDERED JOINT

The tube and fittings must be thoroughly cleaned with steel wool or an abrasive cloth to a shiny finish for all joints.

1 Flux is required on all joints.
2 Apply heat to the joint. Concentrate heat on the lap so the solder is drawn into the joint. Be careful not to overheat the joint.
3 Strike the solder on the joint without the flame. When the solder melts, be sure it flows around the whole joint. If there are several joints, solder them at the same time to avoid further disturbance.
4 Remove the heat and allow the joint to cool. The joint must not be disturbed until it has cooled and all parts of a fitting must be soldered at the same time.
5 Wipe off excess flux with a wet rag.

The advantage of a soft-soldered joint is it does not **anneal** (soften) the copper tube. Disadvantages include:

- It cannot be used on gas services.
- It cannot be used on new joints for water services (only on existing soft-soldered joints with no more than 0.1% lead content).
- It cannot be used with annealed (coiled) tube.

COMPLETE WORKSHEET 1

Compression joints

In the plumbing industry, **compression joints** are widely used and offer the advantages of:

- easy installation
- economy
- the ability to make sound joints in water-charged pipework
- a variety of sizes and styles and the ability to join a wide range of pipe materials
- the ability to disassemble pipework without damage.

There are different kinds of compression joints, and they can be divided into two main groups:

- **manipulative joints**, where the tube is manipulated (has its shape altered) as part of the joint-making process (e.g. by croxing or flaring the tube)
- **non-manipulative joints**, where nothing is done to the tube or pipe except to cut it to the correct length (e.g. using a copper, brass or nylon olive).

Kinco nuts and olives (non-manipulative)

A brass nut compresses a nylon, copper or brass olive onto the pipe and into the joint (see Figure 1.18). This joint may also be croxed; otherwise, it could fail under high pressure or high temperature if it is not installed correctly. **Kinco nuts** and **olives** (non-manipulative) are used to join copper tube to brass fittings and for the connection of water supply pipework to fixtures such as toilets, basins and kitchen sinks. These joints must meet the requirements as outlined in AS 3688–2005 Water supply – Metallic fittings and end connectors.

AS 3688–2005 WATER SUPPLY – METALLIC FITTINGS AND END CONNECTORS

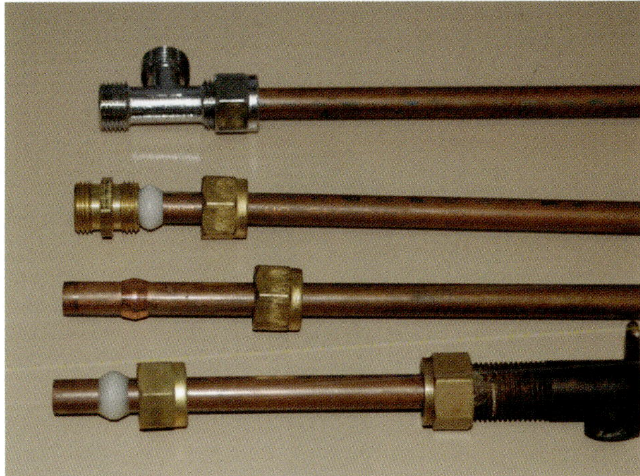

FIGURE 1.18 Kinco nuts and olives (non-manipulative)

Flare-type fittings (manipulative)

For **flare-type fittings** (manipulative), the joint is made by slipping a flare nut onto the tube first, then annealing the square end of the tube and flaring it with a flaring tool. The fitting can then be tightened onto the 60° angled end of the fitting.

Flare-type fittings are used to join copper tube to brass fittings or valves (see **Figure 1.19**). They are widely used on gas and water services. These joints must meet the requirements as outlined in AS 3688–2005 Water Supply – Metallic Fittings and End Connectors.

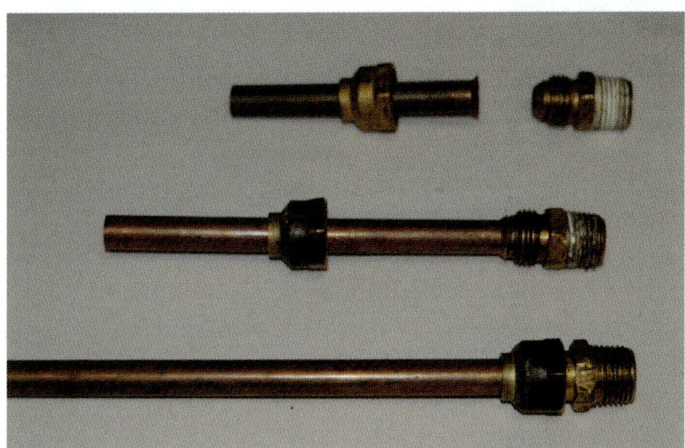

FIGURE 1.19 Flare-type fittings (manipulative)

A flaring tool is used to flare ends of copper tube (see **Figure 1.20**). Jointing compound may be used on water joints to enhance the seal, but not on gasfittings.

FIGURE 1.20 Flaring tool

Kinco nut and croxed joint (manipulative)

A **croxed joint** (manipulative) is made by slipping a nut onto the tube first and then placing the croxing tool inside the end of the tube (see **Figure 1.21**). The croxing tool (see **Figure 1.22**) rotates and tightens a ball bearing to form an olive in the tube itself. Hemp or a teflon tape grommet is wrapped around the crox to seal the joint between the crox and the male pipe thread being connected. This form of connection is generally used to join copper supply pipework to fixtures.

It is good practice to use the croxing tool with nylon olives to lock them in position.

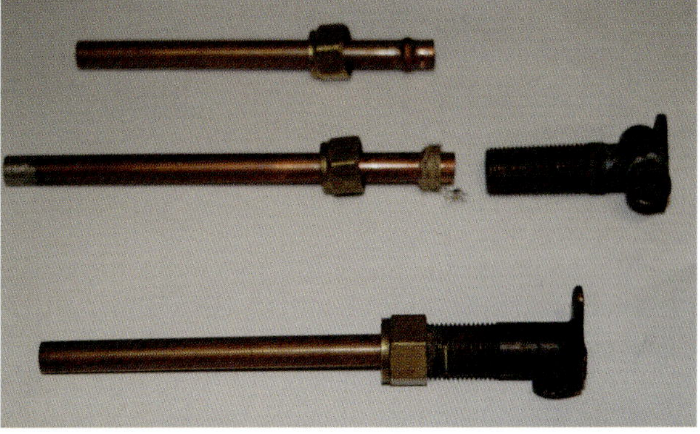

FIGURE 1.21 Making the joint
Top: Copper tube and nut showing formed crox
Centre: Hemp grommet wrapped around crox ready for assembly
Bottom: Assembled crox joint

FIGURE 1.22 Croxing tool

Push lock fittings (non-manipulative)

This method relies on a fitting being pushed over the tube end and the joint being made watertight by an **O-ring**. The tube and fitting are held together by a barbed clip on the inside of the fitting that bites into the outside wall of the tube. **Push lock fittings** (non-manipulative) may be used to join copper, crossed-linked polyethylene, polyethylene and polybutylene tubes (see **Figure 1.23**).

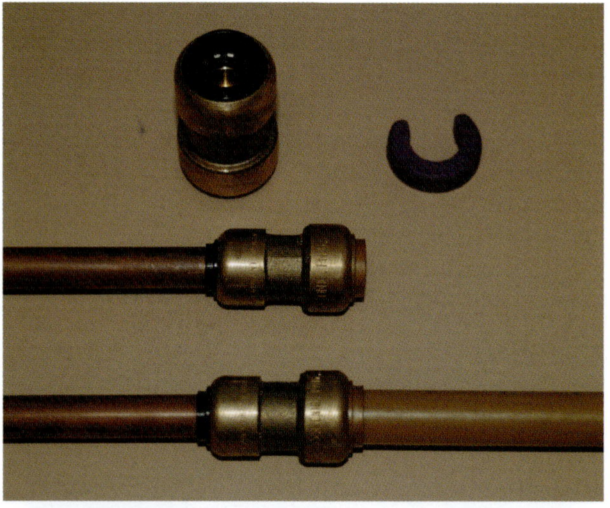

FIGURE 1.23 Push lock fittings (non-manipulative)

Both the inside and the outside of these tube ends must be cut straight, then de-burred and a **witness mark** applied to the tube end to indicate that the joint has been enlarged for its full depth. A witness mark is placed on the tube or pipe to check the depth that the pipe should go into the fitting. This helps to ensure that the joint is strong.

Prohibited locations of compression fittings

Kinco nuts and flare fittings must not be located:

- under concrete slabs
- where strain may be placed on the joint
- in wall cavities.

Kinco nuts and olives must not be used on gas installations.

Copper press-fit

Copper press-fit is a very popular way of jointing copper tube and fittings for gas and water services because no heat is required and it is a fast process. Fittings of up to 50 mm have a rubber ring inserted into a groove. Larger fittings have a rubber ring and a stainless steel grip ring. The grip ring gives extra strength to the joint by stopping the pipe from pulling out of the fitting. To make the joint, the pipe is inserted into the fitting and the jaws of a battery-operated pressing tool are positioned over the fitting. The tool is then used to form the joint by crimping the fitting onto the pipe (see **Figure 1.24**).

FIGURE 1.24 Copper press-fit tool

Large services

Where copper is used on large-diameter services such as hydrants or to supply water to a multistorey property, other methods of mechanical jointing, such as flange joints and roll-grooved joints, are commonly used. **Flanged jointing** (see **Figure 1.25**) is commonly used on large copper tube applications, particularly when connecting a copper branch line to a cement-lined cast iron or cement-lined ductile iron main and/or for the connection of valves and ancillary items. Copper flanged adapters comprise a steel or brass flange and a copper socket that has a roll groove or expanded joint moulded into it. A rubber gasket is positioned between the two flanges, which when bolted together provides a watertight joint.

FIGURE 1.25 Copper adapter flanged fitting

There is also a mechanical jointing method (see Figures 1.26 and 1.27) that incorporates the use of a grooved pipe, clamp and rubber/neoprene seal. This method is mostly used in above-ground installations and is named a roll-groove joint.

Making a roll-grooved mechanical joint

When making the joint, the end of the pipe is either grooved using a specialised tool or has been pre-grooved by the manufacturer. When the pipe is cut into shorter lengths, it is usually roll-grooved onsite by the installer.

FIGURE 1.26 Mechanical roll-grooved joint

FIGURE 1.27 Roll-groove fittings

HOW TO

HOW TO MAKE A ROLL-GROOVED MECHANICAL JOINT

To make the joint, the following procedure should be observed:

1. Roll-groove the ends of the pipes to be joined to the manufacturer's specifications.
2. Lubricate the inside edges of the rubber gasket with the lubricant supplied by the manufacturer.
3. Pull the rubber ring completely over the spigot end of one of the pipes to be joined until it is all the way over the end of the pipe.
4. Bring together the pipe ends to be joined so that they are just touching and in line with each other.
5. Feed the rubber ring back over the joint so that it is over both pipe ends and the centre of the joint is in the centre of the ring.
6. Apply a lubricant to the outside of the rubber gasket.
7. Place both pieces of the clamp over the joint and rubber gasket and make sure they are lined up with each other in the groove.
8. Place the nuts and bolts into the clamp and then tighten each alternately to the specified torque while ensuring the gasket is not pinched.

When making a roll-grooved joint in galvanised mild steel pipe, a similar procedure is followed, although a different roll-grooving tool is used. It should be noted that when making roll grooves for copper tubing, grooving tools intended for steel pipe must *not* be used or joint failure may occur.

A variety of roll-grooving machines are available from plumbing and mechanical services suppliers (see Figures 1.28 to 1.31). There are two distinctively different types: a manual tool with a handle and a power-driven tool. There are several different brands and models within these categories. When purchasing a roll-grooving tool, or when using an existing machine, it is important that the correct roll-depth setting is used. Only roll sets specially made to form grooves to the Australian Standard for copper tube are suitable.

COMPLETE WORKSHEET 2

FIGURE 1.28 Victaulic roll-grooving tool (hand use)

FIGURE 1.29 Manual roll-grooving tool

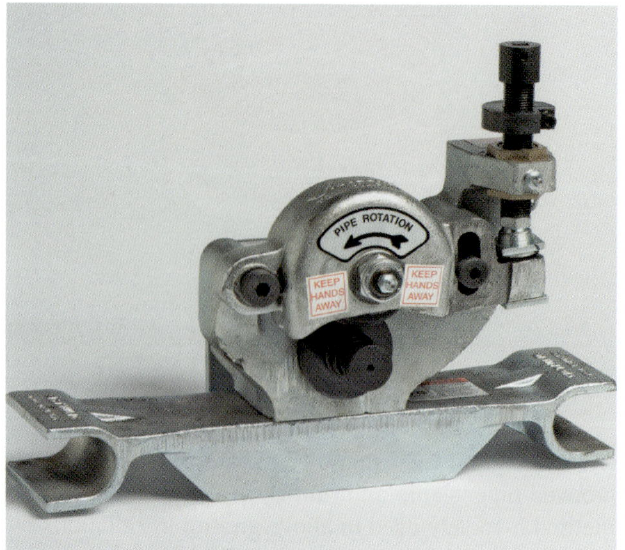

FIGURE 1.30 Roll-grooving attachment for a power-threading machine

FIGURE 1.31 Power-threading/grooving machine

Plastic pressure pipes (polymers)

Plastic pressure pipes and fittings are the most common material being used by plumbers in Australia today due to reduced cost and installation time. When installing plastic pipes for hot and cold water services, a number of restrictions apply. Plumbers should consult their codes and standards to ensure that they meet these requirements. Some general restrictions apply to plastic water supply pipes and fittings; they cannot be used:

- for a meter assembly or a standpipe
- within 1 m of a water heater
- in direct sunlight without further protection.

Other limitations are discussed in later chapters.

The types of plastic pipes available to plumbers include:

- polyethylene (PE)
- polybutylene (PB)
- cross-linked polyethylene (PE-X)
- unplasticised polyvinyl chloride (PVC-U)
- composite cross-linked polyethylene/aluminium/cross-linked polyethylene (PE-X/AL/PE-X)
- composite cross-linked polyethylene/aluminium/polyethylene (PE-X/AL/PE)
- polypropylene (PP).

Similar to copper tube, plastic pipes are also colour-coded to identify the intended use. Purple is used for recycled water and green for rainwater. Plastic pipe is generally available in 5 m straight lengths, as well as 25 m, 50 m and 100 m coils.

AS/NZS 3500.1 PLUMBING AND DRAINAGE: WATER SERVICES

AS/NZS 5601.1 GAS INSTALLATIONS

Jointing plastic pressure pipes

The following types of joints are useful for various applications and different types of polymer products. Some require special tools and some are more cost effective than others. These are explained in more depth.

Types of joints for plastic pipes include:

- push to connect fittings
- crimp ring fittings
- heat fusion
- compression sleeve.

Push to connect assembly

Push to connect fittings (see Figure 1.32) generally consist of the fitting body, an O-ring (which forms a seal between the pipe and the fitting) and a grab ring that secures the pipe in the fitting. This type of fitting is used on PE, PB and PE-X piping systems.

FIGURE 1.32 Assembled push to connect elbow

The pipe is cut to length and pushed into the fitting. An internal sleeve that prevents the pipe from distortion is built into most fittings. If it is not, a support sleeve is positioned inside the pipe end prior to pushing the pipe into the fitting. Most manufacturers place marks on the outside of the pipe to help indicate the depth that the pipe has to be pushed into the fitting. If there are no marks, a witness mark should be placed on the pipe prior to assembling the joint.

Care must be taken when handling pipe to be used with push to connect fittings to ensure that the outer surface of the pipe is free from damage such as scratches. The O-ring may not seal properly on the outside surface of the pipe if the pipe's surface is not smooth.

Crimp ring assembly

Crimp ring fittings generally involve pushing the pipe over the end of the fitting and inside a copper, stainless steel or soft metal ring. It is important to measure and make a witness mark on the pipe to check the pipe is fully engaged into the fitting before crimping (see Figure 1.33). A crimping tool is then used to compress the crimp ring and the pipe end onto the fitting (see Figures 1.34 and 1.35). Manual and battery-charged crimping tools are available. This type of fitting is used on PE, PB, PE-X and composite pipes such as PE-X/AL piping systems.

Source: Reliance Worldwide Corporation (RWC).

FIGURE 1.33 Pipe, fitting and copper crimp ring

FIGURE 1.34 Battery-operated crimping tool

Heat fusion

The **heat fusion** method of pipe jointing involves heating and melting the inside of a fitting and the outside of the pipe and pushing the pipe into the fitting to form a joint (see Figure 1.35). This method of pipe jointing is suitable for hot and cold water services, heating services, cooling services and industrial uses. The pipe material used is PP.

FIGURE 1.35 Manual-operated crimping tool

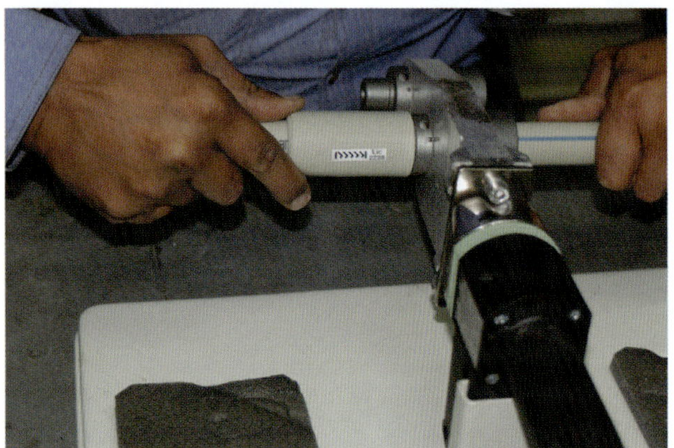

FIGURE 1.36 Heat fusion

The pipe end is pushed into the socket side and the fitting socket is placed over the spigot side of the heat fusion welding machine until both have melted (see Figures 1.36 and 1.37). Manufacturers' instructions give tables indicating for how long the pipe and fittings of a given size must be heated. In most cases, manufacturers also offer training for the use of their products.

Compression sleeve

The **compression sleeve** jointing method (see Figure 1.38) is suitable for use with PE-X and PE-X/AL/PE-X. This piping system is suitable for heating systems, gas, heated and cold water services. A brass compression ring is fitted and the tube is expanded using a similar tool to the one used for copper. The expanded tube end is then placed over the fitting. A compression tool is used to slide the compression ring over the expanded tube end and fitting (see Figures 1.39 and 1.40).

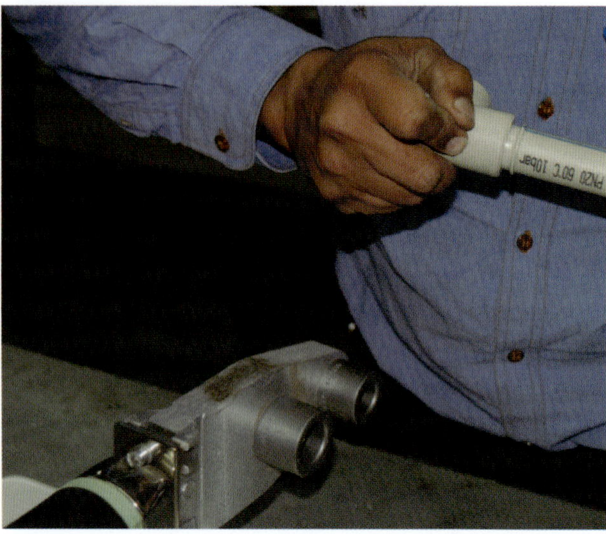

FIGURE 1.37 The pipe end and the fitting are pushed together and held together until the joint sets

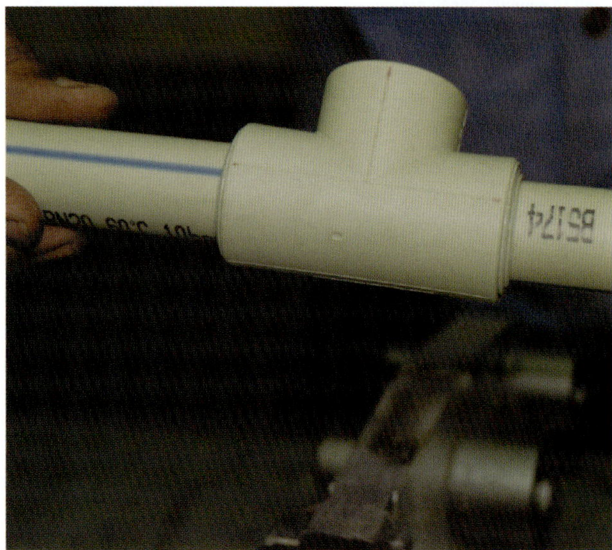

FIGURE 1.38 Completed heat fusion joint

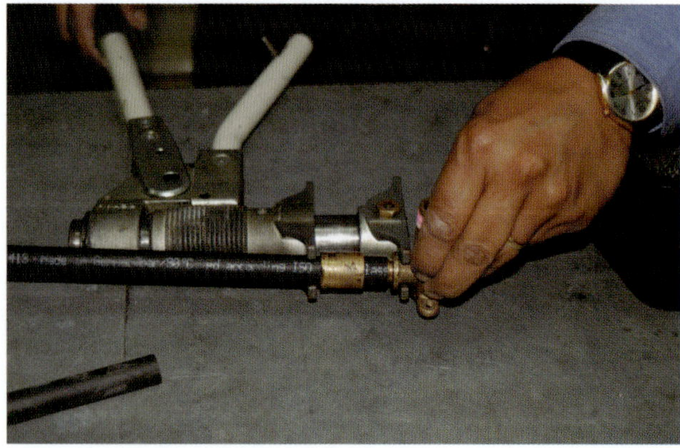

FIGURE 1.39 Compression sleeve

This photograph shows the pipe being placed on the fitting, ready for completion of the joint.

FIGURE 1.40 Finished joint with the tool removed

PVC-U pipe and fittings (pressure pipe)

This pipe material is generally used for water services, irrigation systems and water mains. Generally, PVC-U pipe is joined by using a solvent cement; however, larger pipes used for water mains may be joined using rubber rings and gibault joints. When making a joint with solvent cement, the pipe is cut square, burrs are removed and the outside of the pipe end and the inside of the fitting socket are cleaned with priming fluid. They are then coated with PVC solvent cement of a type suitable for pressure fittings, assembled and held in place until the solvent sets.

Large services

Table 1.3 lists a number of plastics/polymers commonly used on large diameter services.

Jointing methods for PVC-U, PVC-M and PVC-O pipes

Polyvinyl chloride (PVC) pipes can be joined using the following methods:

- rubber ring jointing
- cold solvent cement welding (except PVC-O pipes, which should not be joined with solvent cement)
- flanged joints
- mechanical couplings.

Rubber ring jointing in PVC pipes

Elastomeric rubber ring jointing methods (see Figure 1.41) are similar to those employed in cement-lined cast iron/cement-lined ductile iron pipes.

TABLE 1.3 Materials suitable for use in the construction of water mains and their characteristics

Material	Characteristics
Unplasticised polyvinyl chloride (PVC-U)	■ Lightweight ■ Does not corrode ■ Can be damaged easily (although with recent developments in plastics this disadvantage has been reduced) ■ May be installed in direct sunlight only under certain circumstances ■ Susceptible to becoming oval shaped (squashing) when high external soil/traffic loads are present
Polyvinyl chloride modified (PVC-M)	■ Similar characteristics to PVC-U, although PVC-M is stronger, more flexible, lighter and possesses better flow characteristics than PVC-U pipe
Orientated polyvinyl chloride (PVC-O)	■ Similar characteristics to PVC-U, although PVC-O is stronger, more flexible, lighter and possesses better flow characteristics than other PVC pipes ■ Solvent cement jointing is unsuitable
Polyethylene (PE)	■ Similar characteristics to PVC-U, although PE possesses high-impact strength, damage resistance, abrasion resistance, flexibility and high-flow capacity; is available in longer lengths (which could be suitable for trench-less construction); and is weather resistant ■ Three grades of polyethylene pipe are available: – MDPE (medium density polyethylene) – HDPE (high density polyethylene) – HPPE (high performance polyethylene)
Acrylonitrile-butadiene-styrene (ABS)	■ Similar characteristics to PVC-U, although ABS has high-impact strength and abrasion resistance, and is lightweight and weather resistant ■ Suitable for installations where the pipes are exposed to direct sunlight (although this may be subject to regulations, standards and local water utility requirements)
Glass-reinforced plastic (GRP)	■ Lightweight ■ Available in long standard lengths of 6 m, 12 m and 18 m ■ High strength and flexibility ■ High flow characteristics ■ Abrasion and corrosion resistant

FIGURE 1.41 Ring formed into a heart shape to easily fit inside a collar

HOW TO

HOW TO MAKE AN ELASTOMERIC RUBBER RING JOINT

When making the joint, the following steps should be followed:

1. Ensure all jointing surfaces are completely dry before jointing.
2. Ensure the ring and the collars are free from dirt, dust and grit by wiping them out with a clean rag. Care must be taken during dusty or windy conditions to ensure the joint is kept clean and free of dust.
3. Prepare the rubber ring for insertion by flexing it into a heart shape.
4. Insert the ring into the groove in the socket/collar. Ensure the ring is fitting snugly in the groove and not protruding into the pipe or socket.
5. Lower the pipe into the trench in accordance with proper handling procedures.
6. Ensure the pipe is aligned correctly and able to enter the socket straight and freely.
7. Lubricate the inside surface of the rubber ring with a thin film of the manufacturer's recommended lubricant. Only lubricant supplied by the manufacturer should be used. The lubricant can be applied using a glove, rag or brush.
8. With the same lubricant, lubricate the spigot of the pipe as far as the witness mark, paying particular attention to the chamfer.
9. Realign the pipe and enter the spigot carefully into the socket until it makes contact with the rubber ring.
10. Using the appropriate approved mechanical assistance while keeping the pipe straight along the centreline (axis) of both pipes being joined, push the spigot into the socket until the witness mark reaches the face of the socket while still remaining just visible.

Assembly of elastomeric joints

Assembly of elastomeric joints in PVC piping is usually a quick and easy process. However, as the joint is a compression-type joint and a fair amount of force is usually required, it is recommended that it should not be made without the use of some form of assisting mechanism. The joint will be more easily made and will help satisfy WHS requirements for addressing strain hazards by employing one of the following methods (depending on local conditions and the size of the pipe). Water services joined with rubber rings must have thrust blocks installed at bends more than 5°, tees, end caps, reducers, valves and inclines in excess of 1:5.

Remember to keep your back straight when manually lifting to avoid back strain.

The crowbar/fulcrum technique

As illustrated in **Figure 1.43**, the joint can be pushed home using a crowbar or other suitable lever against the socket of the pipe. Note the timber block placed between the lever (as in **Figure 1.44**) and the pipe socket to protect it from damage. This leverage method should be used only on pipes no larger than 150 mm.

FIGURE 1.42 PVC-U pipe collar with rubber ring installed

FIGURE 1.43 The crowbar/fulcrum technique

The excavator method

This method is mainly used on large-diameter pipe. When an excavator or backhoe is available, it can be used to complete the joint by pushing the spigot home (see Figure 1.44). To prevent damaging the pipe, a timber block is placed between the bucket and the collar. The joint is brought together in the usual way, ensuring it is properly aligned, and then the pipe is gently nudged forward with the machine's bucket until the joint is completed. Again, excessive force should not be used and care should be taken to ensure joints in the line that have already been made are not closed up past their witness marks, as this may affect allowances for movement and expansion.

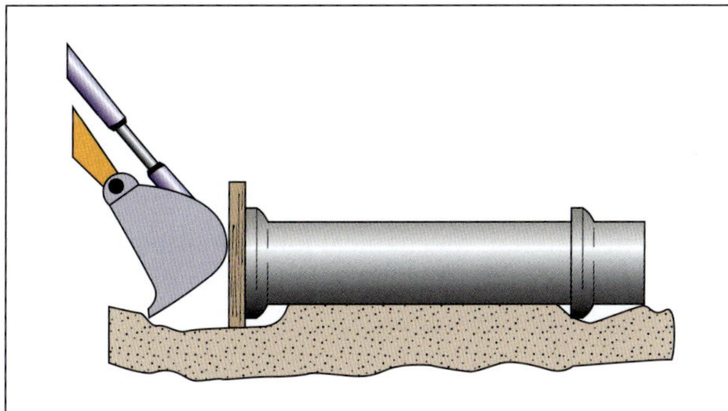

FIGURE 1.44 The trench excavator method

The excavator/backhoe slewing method

Using the excavator/backhoe slewing method (see Figure 1.45), the pipe spigot is first positioned in the usual way, ensuring it is properly aligned and ready for jointing. A sheathed webbing lifting sling is then placed around the collar end of the pipe and pulled tight. The other end of the sling is attached to the excavator/backhoe's lifting hook, as shown in Figure 1.45. The bucket is then slowly slewed in the direction of the arrow, along the same axis as the pipe, drawing the pipe home into the collar and so completing the joint. Care should be exercised not to use excessive force or swing of the bucket as this may cause the pipe to come off its centreline, tilt or lift and damage the joint.

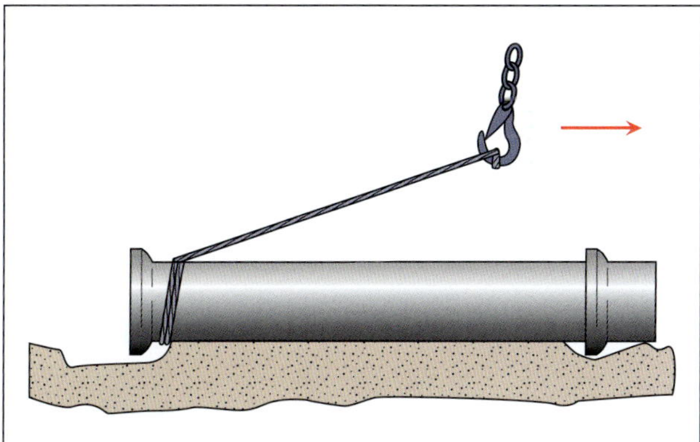

FIGURE 1.45 The excavator/backhoe slewing method

FROM EXPERIENCE

To avoid skin contact with solvents or cleaning fluid, gloves and glasses should be worn during this process. It is also essential that the solvent cement jointing process is carried out in a well-ventilated area (see Figure 1.46).

HOW TO

MAKE A SOLVENT CEMENT JOINT IN PVC PIPE AND FITTINGS

The process for solvent cement jointing in PVC pipes is as follows:

1 Ensure that the end of the pipe is cut square and chamfered to the recommended tolerance, and that all burrs are removed. Chamfering of the spigot is of utmost importance as it promotes the even distribution of the solvent cement within the joint. A square end will push the glue off the collar and out of the joint.
2 Wipe the surface of the pipe end (spigot) and inside the collar with a clean dry cloth.
3 Inspect the pipe and collar/socket for any imperfections such as scratches and gouges. If the scratches or gouges exceed 10% of the pipe wall thickness, then the pipe end should be discarded or cut off until a clean undamaged spigot is achieved.
4 Dry fit the joint. Do not force the pipe into the socket. If the pipe will not go into the socket, or bind up before reaching the full depth of the socket, the tolerances of the pipe or collar may not be correct. If this is the case, the joint must be abandoned as the finished product may not result in a successful full-strength, watertight joint.

>>

>>

5 Place a witness mark on the pipe spigot at a distance equal to the socket depth less 10 mm. Do not scratch the witness mark into the surface of the pipe. Felt-tipped pens only should be used.
6 Wipe over both the spigot and the socket surfaces with a clean dry cloth for a second time to remove any dust immediately before making the joint.
7 Using a swab (usually supplied) or a dedicated clean rag moistened with a cleaning fluid recommended by the manufacturer (usually methyl ethyl ketone, or MEK), thoroughly wipe the spigot up to, but not beyond, the witness mark.
8 Ensuring the solvent cement is properly stirred or shaken before use, evenly apply solvent cement to the spigot and the socket with a clean dedicated brush. Do not use the same brush or swab as the one used for the cleaning fluid. Do not apply solvent cement onto the pipe beyond the first witness mark.
9 Once both socket and spigot are freshly and evenly coated with solvent, and while ensuring the axis of both the pipe and the socket are in line, immediately push the pipe, in a smooth even motion, into the socket to a depth where the witness mark on the spigot reaches the face of the socket. It is essential that the joint is completed as quickly as possible, while the solvent cement is still wet on the spigot and inside the collar. Do not attempt to make the joint if the solvent cement has dried out.
10 Wipe any excess solvent cement from around the socket and, where possible, from inside the joint. Ensure joints are supported and do not use cleaning fluid to clean up excess solvent cement.
11 Do not disturb joints for at least five minutes after jointing. Pipe testing should not be carried out above working pressures until 24 hours after jointing.
12 To ensure the integrity of the solvent joint, the joint and the pipe must be kept as close to normal room temperature as possible. In hot conditions, shading of the jointing work area for a minimum of one hour can reduce the temperature to a more desirable level and make for an easier jointing process. In hot or wet conditions, a cover over the jointing work area will prevent direct sunlight or precipitation from compromising the jointing process.

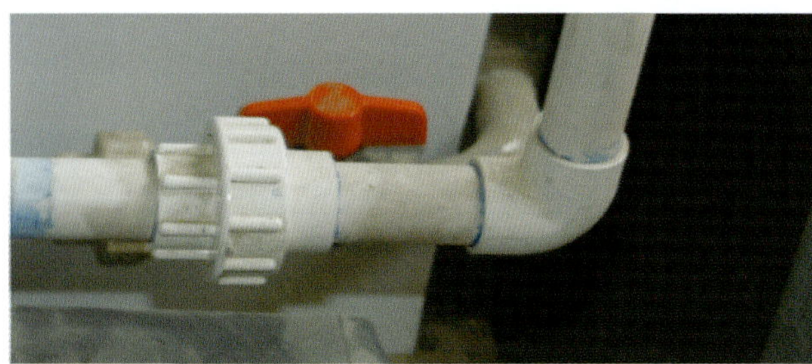

FIGURE 1.46 PVC union and elbow

FROM EXPERIENCE

Applying solvent cement to the socket first reduces the risk of the spigot coated with solvent cement getting any dirt stuck to it. Also, the solvent cement dries faster on the spigot.

Different drying and curing times apply depending on the pipe size, ambient temperature and type of glue, so it is important to read the manufacturer's instructions.

Tip: Always screw the lid on the primer and solvent cement immediately after each joint to prevent spillage.

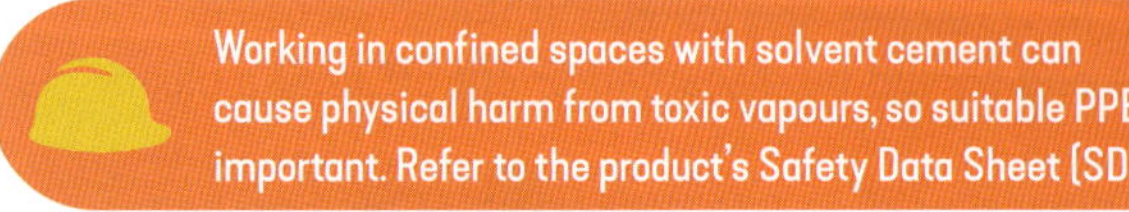

Working in confined spaces with solvent cement can cause physical harm from toxic vapours, so suitable PPE is important. Refer to the product's Safety Data Sheet (SDS).

Jointing methods for PE pipe

Joining lengths of PE can be achieved by the following techniques.

Mechanical compression couplings

Mechanical compression couplings (see **Figure 1.47**) are used for pipe sizes up to DN 110. Mechanical joints are generally specified for temporary services and where welding is impractical. Mechanical assembly requires the use of fittings, generally working on the compression principle. Mechanical compression fittings can also be used for joining PE to pipes or fittings of a different material. Mechanical jointing fittings are generally self-restraining, thus eliminating the need for thrust blocks and restraints.

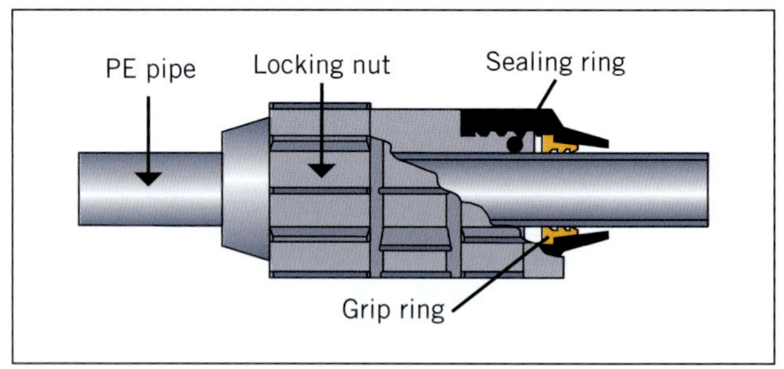

Source: Vinidex Pty Ltd, Polyethylene Design Manual ©.

FIGURE 1.47 Typical mechanical compression coupling used for joints in PE pipe up to DN 110

Electro fusion

Electro fusion (or **socket fusion**) **joints** (see **Figure 1.48**) are normally used for pipe sizes DN 90 and above. Fittings are available in the size range DN 16 to DN 710. Manufacturers offer training in the use of their products, and larger sizes may be available.

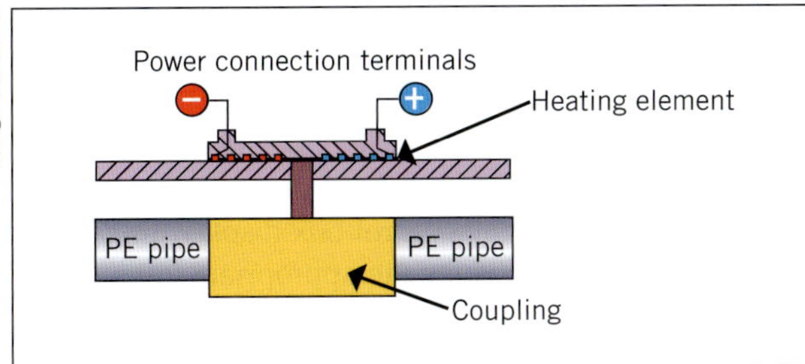

Source: Vinidex Pty Ltd, Polyethylene Design Manual ©.

FIGURE 1.48 Cross-section of a typical electro fusion joint

Electro fusion welding is when the cleaned/shaved ends of pipe are inserted into a socket that is heated to create a bond between the socket and the pipe. Resistive heating wires are embedded in the socket. These resistive heating wires are connected to a control box and an electrical current is then passed through the wires. This applied heat melts the interior of the socket and the exterior of the pipe to form the welded joint.

Butt fusion

Butt fusion joints (see Figure 1.49) are normally used for pipe sizes DN 90 and above, and are often chosen as the preferred jointing method for larger water mains. However, butt fusion welding is the primary technique used for large pipe sizes, such as 250 mm diameter and above, and is usually the only jointing method used for pipe diameters greater than 500 mm.

FIGURE 1.49 Butt welder machine

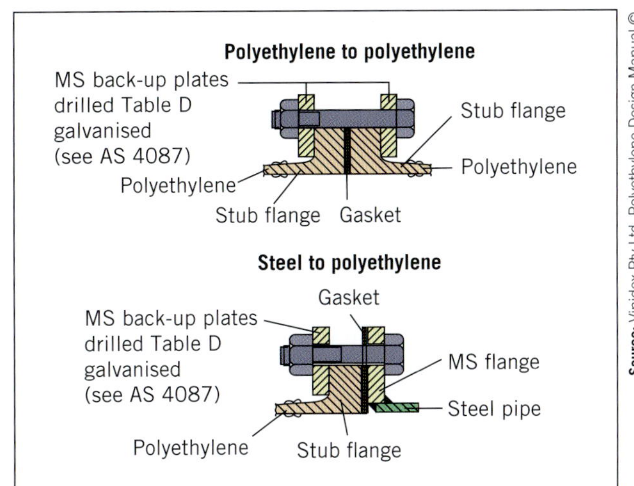

Source: Vinidex Pty Ltd, Polyethylene Design Manual ©.

FIGURE 1.50 Cross-section of a typical PE flanged joint

In **butt fusion welding**, also known as **hotplate welding**, a heated plate is placed between the squared and cleaned/shaved ends of the pipes being joined. The plate heats the pipe ends and is quickly withdrawn, and the molten pipe ends are then pushed together. The manufacturer's recommended pressure is used when the pipes are brought together so that molecular segments can diffuse across the contact surfaces, producing what is known as an interment contact.

Flanged joints (PE flanged stubs)

Flanged PE stubs to join to PE pipe by electro fusion or butt fusion are available in sizes DN 90 to DN 450 (see Figure 1.50). Flanged PE stubs to join PE pipe by mechanical compression are available in sizes up to DN 160. The flange requires a corrosion-resistant backing plate of Class 316 stainless steel for below-ground applications or hot-dipped galvanised steel for above-ground installations. The nuts and bolts used to make the flanged joint should be the same material as the backing plate. Class 316 stainless steel nuts and bolts should be identified by an A4 marking.

Flange gaskets are to comply with WSA 109. Hydrant installations require full-face flanges, with bolting details complying to AS/NZS 4087.

AS/NZS 4087 METALLIC FLANGES FOR WATER WORK PURPOSES

Making an electro fusion joint in PE pipe

For the electro fusion jointing method to be effective, it is essential to pay particular attention to preparation of the jointing surfaces. It is equally important to pay attention to the removal of the oxidised surface along the pipe spigot for a distance equal to, if not slightly more than, the depth of the socket, as well as ensuring both the spigot and the insides of the socket jointing surfaces are clean. The pipe is prepared for jointing by removing a layer of material to a specified depth, thus removing the oxide and allowing fusion of a clean, oxide-free pipe spigot.

An electro fusion joint is made as follows:

1 To allow for efficient jointing, ensure there is enough space to permit safe access to the work area. If the joint is to be completed in situ, such as in a trench, there should be a free space with a minimum clearance of 150 mm around the joining area. More clearance may be necessary when joining larger pipes.
2 Because the temperature of the jointing process is so critical, cover the open end of the pipe at the opposite end from the fitting joint to make sure airflow through the pipeline does not happen during the heating and cooling cycles of the jointing process.
3 Ensure the pipe ends being joined are cut square to the axis of the pipe and remove any burrs and **swarf**.
4 So as to remove any traces of dirt, mud and/or other contamination, wipe the pipe surface being prepared for the jointing process with an **isopropanol** wipe (an alcohol-based cleaning solution). Methylated spirits, acetone, MEK and other solvents are *not* suitable for preparing the joining surface. Each wipe should be used only once.
5 Ensure the joint area is completely dry and the alcohol left by the wipe has completely evaporated before proceeding with the jointing process.
6 Check the pipe for roundness (**ovality**). If necessary, use an appropriate re-rounding tool to remove any ovality and ensure the pipe is round and within specified tolerances.
7 While still in its bag, place the socket of the fitting alongside the pipe end; place a witness mark on the pipe at half the fitting depth plus about 20 mm. This will ensure that the scraped area on the pipe can be visually checked once the joint is completed.
8 Ensure the pipe clamps are of the correct size for the pipes being joined.
9 Using an appropriate scraper tool, remove the entire surface along the pipe from the spigot up to the witness mark. Do not use tools such as metal files, rasps or emery paper as they are not suitable end preparation tools.
10 Insert the pipe ends into the fitting so that they are in contact with the centre stop. If there is no centre stop, set up an accurate witness mark on the pipe spigot, as previously explained. Ensure the pipe ends are correctly aligned along their axis to avoid any stress during the jointing process. This is of particular importance when using coiled pipes.
11 Place the pipes into the correct-size pipe clamps or other suitable support mechanism to ensure the pipe joint cannot move. Also ensure the fitting is satisfactorily supported to prevent it sagging during the electro fusion procedure (see **Figure 1.51**).
12 The required jointing time is usually indicated on the fitting or specified on the data sheet generally supplied with the fitting. Check that the correct time is shown on the control box display (see **Figure 1.52**); if not, enter the appropriate jointing time into the control box timer. (Automatic control boxes are available that alleviate the need to enter fusion times. They use a barcode system to set the appropriate times for each size and type of fitting.)

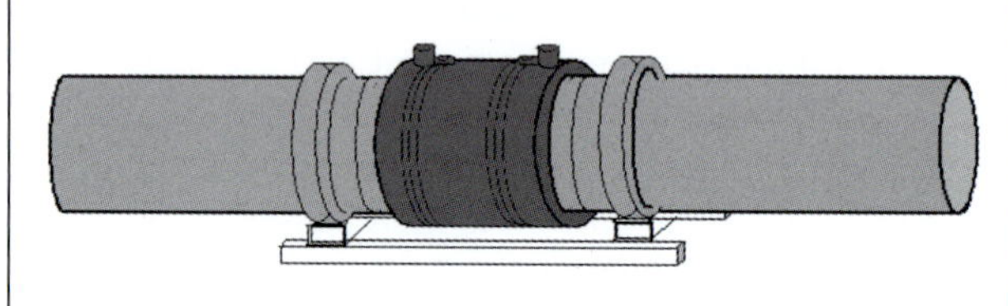

FIGURE 1.51 Pipes inserted into the socket fitting and clamped ready for welding

Source: Plastics Industry Pipe Association of Australia, *Industry Guidelines, POP 001, Electrofusion jointing of PE pipe and fittings for pressure applications*, Issue 7.3, 2015, http://www.pipa.com.au

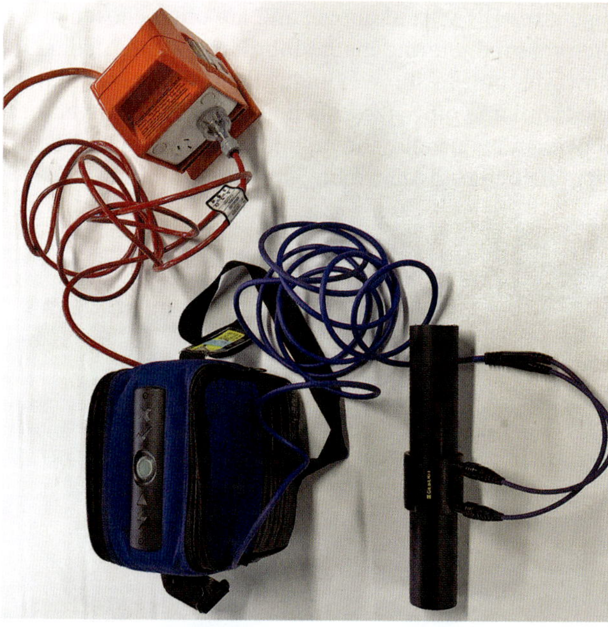

FIGURE 1.52 Electro fusion welder

13 Activate the electro fusion machine by pressing the start button. Observe the machine display to ensure the heat cycle is progressing.
14 If the electro fusion cycle terminates before the specified time has elapsed, check for faults as indicated by the control box warning lights or display. Check for a possible cause of the breakdown, such as inadequate fuel in the generator or power supply failure. If the cycle is terminated early, do not attempt a second fusion cycle until the entire joint has cooled down to less than 45°C. Some manufacturers may recommend replacement of the fitting rather than attempting a second electro fusion weld.

15 The completed joint should be left in the clamps for cooling. The time needed will be specified on the fitting or its data sheet, or shown in the display of the automatic control box.

Source: Plastics Industry Pipe Association of Australia, *Industry Guidelines POP 001, Electrofusion jointing of PE pipes and fittings for pressure applications,* Issue 6, 2005, http://www.pipa.com.au.

COMPLETE WORKSHEET 3

Bending tube

There are several methods available to plumbers to create changes of direction in tubes. These methods include:

- elbows
- bends
- lever-type benders
- bending springs
- sand bending
- supported bends for plastic tube
- long-radius curves.

Time and money can often be saved by bending the pipe rather than buying and installing elbows and bends. Bending the pipe will also result in fewer joints and so less chance of leaks. Sometimes it may just look better to bend the tube, such as when using gold or chrome tube. Perhaps the building design may require you to make bends of a radius that fits the building better. And at other times you may not have fittings readily available and so have to do without.

To ensure good flow, it is important when bending tube that the pipe remains round throughout the length of the bend. Bending machines, springs and sand are all used to ensure that pipes do not lose their round shape when bent. They all support the side walls of the pipe being bent, preventing them from closing in or kinking. An example of the sides of a pipe closing in is when a garden hose is bent sharply. Two opposite sides of the pipe squeeze in, and the two sides at 90° to these move out. It is possible to reduce the flow out of the end of the hose or, if you bend the hose 180° back onto itself, you can stop the flow completely.

To ensure good flow, it is essential that a plumber uses bends that are free from defects.

Lever-type benders

The most common method used by plumbers to make bends in copper tube up to and including DN 20 is with a tube bender (see Figure 1.53). The bend is made by fitting the copper tube into the bender and, by pulling the levers, bending the tube to the angle required (up to 180°). The sides of the tube are held round throughout the bend by the tube bender. Benders may also be used on composite pipe. The radius of the bend when using tube benders is not adjustable.

FIGURE 1.53 Bend in copper tubing made with lever-type benders

Bending springs

Support can be given to tubes during bending by using internal or external **bending springs** (see Figure 1.54). They can be used on copper tube and composite pipes.

External springs

For copper tubes, the tube where the bend is required must be annealed (softened) and cooled down. Slide the spring over the annealed tube and bend the tube until it is bent a little past the required angle. Then bend the tube back to the required angle. (The overbending and bending back of the tube makes it easier to slide the spring off the tube after the tube has been bent.) Slide the spring off the tube.

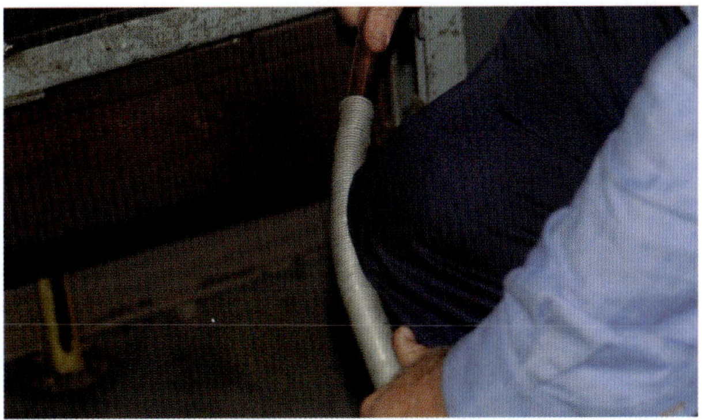

FIGURE 1.54 Bending copper tube using an external spring

Internal springs

These are suitable for making bends near the ends of the tube only. For copper tube, the area to be bent must be annealed. This is not required for composite tube. Slide the spring into the tube. Again, overbend the tube and bend it back to the required angle, then slide the spring out.

FROM EXPERIENCE

Internal springs are handy to use when pulling a bend on copper tube tails protruding from a wall instead of using a fitting.

Sand bending

Sand bending uses sand packed tightly inside the pipe (usually copper tube or steel pipe) to keep the pipe round during the bending process (see **Figure 1.55**). This method would be used to custom-make bends to a given radius for reasons such as aesthetics (looks).

Dry sand is used as any moisture would turn to steam while heating. The steam would expand and blow the plugged end, which could possibly burn a person.

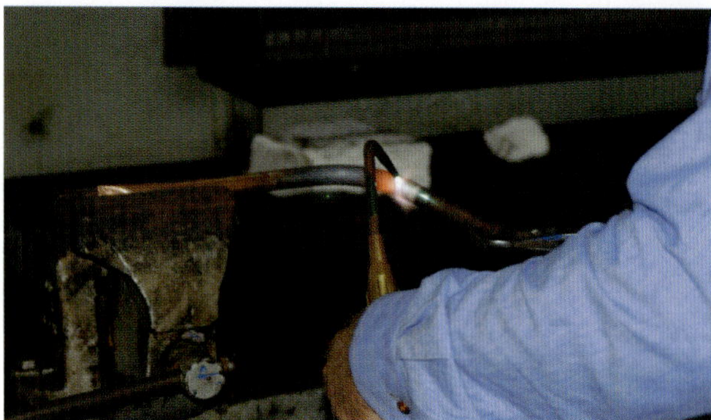

FIGURE 1.55 Sand bending copper

The first thing to be done when making a sand bend is to select a bend radius. This may be as small as three to five times the pipe diameter for copper tubes up to DN 20. Larger tubes will require a larger bend radius.

When sand bending copper tube, the section of the tube to be bent is the only section that is heated. The length of the section of tube to be bent must be calculated. It is called the **heat length**. This is done by selecting the bend radius and using the formula:

$$HL = \frac{2 \times R \times 3.1416 \times A}{360}$$

where

HL = heat length of the bend in mm

R = radius of the bend in mm

3.1416 = the constant pi

A = the angle the pipe is to be bent through

360 = the number of degrees in a circle

Supported bends for plastic tubes

Most plastic tube manufacturers make a pipe-bending guide for use with their pipes. These guides are used to make relatively short-radius bends in plastic pipes (see **Figure 1.56**).

Long-radius curves

All tubes can be bent without support using long-radius curves (see **Figures 1.57** and **1.58**). The problem with these is that they are not aesthetically pleasing, so they are commonly used in unseen areas, such as within walls. This is particularly so with plastic tubes where the cost of the extra tube used is small in comparison to the cost of a fitting and the time taken to install the fitting (including the added possibility of a leak).

EXAMPLE 1.1

CALCULATE THE HEAT LENGTH FOR A 90° BEND IN A DN 15 COPPER TUBE WITH A BEND RADIUS OF 90 MM

Formula: $HL = \frac{2 \times R \times 3.1416 \times A}{360}$

where

HL = heat length of the bend in millimetres = ?

R = radius of the bend in millimetres = 90

3.1416 = the constant pi

A = the angle the pipe is to be bent through = 90

360 = the number of degrees in a circle

Workings: $HL = \frac{2 \times 90 \times 3.1416 \times 90}{360}$

$= \frac{50893.9}{360}$

$= 141.372$

Answer: HL = 141mm

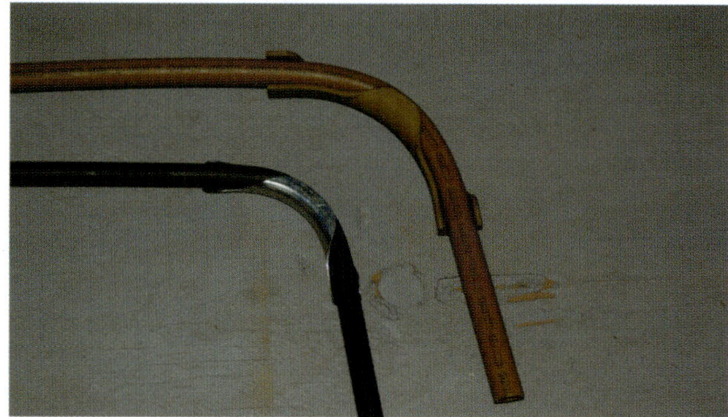

FIGURE 1.56 Metal and plastic support guides used with plastic pipes

FIGURE 1.57 Long curve within a wall frame using plastic tube

FIGURE 1.58 Alternative using an elbow

Clipping and fixing

Appropriate clips and fixings that are recommended by the manufacturer should be used. This will ensure the clips are compatible with the pipe material, preventing any corrosive effect and therefore being longer lasting.

Refer to AS/NZS 3500.1 for the recommended vertical and horizontal spacing of clips for different pipe materials.

AS/NZS 3500.1 PLUMBING AND DRAINAGE: WATER SERVICES

Testing

Once the pipe system has been installed, it must be tested to make sure it doesn't leak. The testing must be carried out to regulatory authority requirements, standards and codes of practice. The test results must be recorded in a format required by the regulatory authority.

A test bucket is required for cold-water and hot-water installations. A test bucket is a piston pump pressurising the water service as the pump handle is manually moved up and down. Alternatively, an electric pump can be used for larger services.

More information on testing requirements is covered in Chapters 3 and 4.

AS/NZS 3500.1 PLUMBING AND DRAINAGE: WATER SERVICES

LEARNING TASK 1.3

1 Name three jointing methods for copper pipe.
2 Name three types of polymer pipe.
3 Name three jointing methods for polyethylene pipe.
4 What is the advantage of a long-radius bend?
5 What is the most common method used by plumbers to make bends in copper tube?
6 What is the least used method by plumbers to make bends in copper tube?

Clean up

Cleaning up the work area and disposing of the materials at the end of the job is just as important as setting up the job. This ensures a good reputation with clients.

Clear the work area

- Clean the work area by disposing of rubbish in the appropriate bins and sweep the floor. This reduces trip hazards and keeps the work area orderly.
- Any excess material should be recycled or reused wherever possible to prevent waste and to save costs.
- Any state or territory legislation as well as workplace policies and procedures must be adhered to regarding the reuse and recycling of materials.

Clean tools and equipment

- Cleaning tools and equipment as they are packed away makes the job easier for when they are used next time. Store tools and equipment securely in a clean dry place.
- Check for tools and equipment for serviceability and report any damage to the supervisor.
- Be sure to sign off the required documents upon completion of the work. The old saying 'The job's not finished till the paperwork is done' is as relevant today as ever!

EXAMPLE 1.2

REDUCE WASTE

Any lengths of copper tube and brass fittings should be kept for recycling as it is worth quite a lot in scrap metal. Polymer pipe should also be recycled in accordance with workplace policy and procedures.

LEARNING TASK 1.4

1 What type of documents need to be submitted upon completion of work?
2 Why is recycling excess material important?

COMPLETE WORKSHEET 4

SUMMARY

- Non-ferrous materials are materials that do not contain iron.
- They are long-lasting because they are less corrosive than ferrous materials.
- Quality assurance helps to ensure a high standard of workmanship and processes, therefore reducing mistakes and problems.
- Always wear the appropriate personal protective equipment (PPE).
- Keep tools and equipment well maintained.
- Take the time to thoroughly read the plans and specifications before starting work.
- A site visit is vital prior to starting work to check for access and pipework location.
- Copper and plastic pressure piping are the most used non-ferrous materials.
- It is important to know the different jointing methods and the limitations of different materials.
- Ordering materials accurately reduces waste and saves time.

REFERENCES

ABB: **https://www.baldor.com/brands/baldor-reliance**
Crane Enfield Metals: **http://www.cranecopper.com.au**
Iplex Pipelines: **http://www.iplex.com.au**
Reliance Valves: **https://www.rmc.com.au**

GET IT RIGHT

1 Which photo shows the correct use of the crimping tool?

2 Why is this method important?

3 What is the advantage of a battery-operated crimping tool compared with a manual crimping tool?

WORKSHEET 1

To be completed by teachers	
Student competent	☐
Student not yet competent	☐

Student name: ______________________

Enrolment year: ______________________

Class code: ______________________

Unit of competency code/title: CPCPCM Fabricate and install non-ferrous pressure piping

Task: Review 'Identify installation requirements' to the heading 'Compression joints' and answer the following questions.

1 Why is non-ferrous pressure piping less corrosive than ferrous pressure piping?

____

____

____

____

2 State four common uses for non-ferrous pipes.

____

____

____

____

3 Describe how copper tubes are classified by type and colour.

____

____

4 State three reasons for insulating copper tube.

____

____

____

5 Complete the following statement in relation to capillary action: 'Capillary action is where a ______________ is drawn into the ______________ between two closely ______________ surfaces.'

WORKSHEETS 1

6 List three methods that may be used to join copper tube.

______

______

______

7 List the three joint types commonly used when silver soldering copper tube.

______

______

______

8 Is the following statement true or false? 'No flux is required when silver soldering a joint between a copper tube and a copper capillary fitting.' Circle the correct answer.
T F

9 How is the silver content of a silver-soldering filler rod indicated?

______

10 What heat source is commonly used by plumbers when silver soldering?

______

______

11 Complete the table below.

Terminology	Description
MI	
	A fitting with a sharp 90° change of direction
DR	
Tee	
	Outside diameter
FI	

12 Is the following statement true or false? 'When ordering a capillary tee, you should ask for the branch first and then the ends along the straight line.' Circle the correct answer.
T F

13 Identify each of the fittings in the following picture showing capillary fittings with three ends.

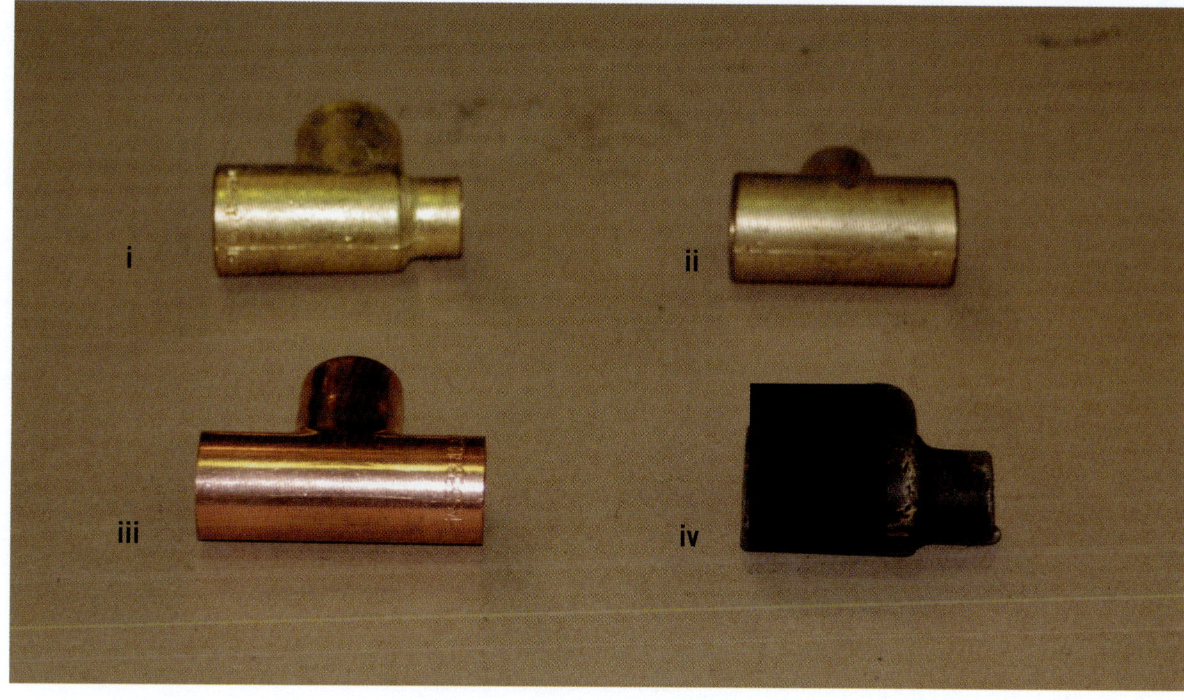

____

____

____

____

14 According to AS/NZS 3500.1, what are the limitations for soft soldered joints on water services?

____

15 State the maximum amount of lead content permitted in soft solder used on water services.

____

16 Is soft soldering permitted on gas services?

____

WORKSHEET 2

To be completed by teachers	
Student competent	☐
Student not yet competent	☐

Student name: ______________________

Enrolment year: ______________________

Class code: ______________________

Unit competency code/title: CPCPCM Fabricate and install non-ferrous pressure piping

Task: Review 'Compression joints' and 'Large services' and answer the following questions.

1 List three advantages of compression joints used in plumbing.

____

____

____

2 What are the two main types of compression joints?

____

____

3 What are Kinco nuts and olives commonly used for?

____

____

____

4 What tool is used to secure an olive in position?

____

____

5 State two uses for flare-type fittings.

____

____

6 How is a flanged joint connected?

____

____

7 List three locations where compression fittings must not be used.

____

____

____

8 Explain in your own words the differences between manipulative and non-manipulative joints.

____

____

____

____

9 Is the following statement true or false? 'Copper tube must be annealed prior to joining using the press-fit system.' Circle the correct answer.
T F

10 Is the following statement true or false? 'Roll-grooved joints are commonly used on large diameter services such as fire services.' Circle the correct answer.
T F

11 Is the following statement true or false? 'When making roll grooves for copper tubing, grooving tools intended for steel pipe must *not* be used or joint failure may occur.' Circle the correct answer.
T F

12 What type of joint is shown in the picture below?

Source: *Copper Tube Handbook*, Copper Development Association Inc., McLean, Virginia.

____

WORKSHEET 3

To be completed by teachers	
Student competent	☐
Student not yet competent	☐

Student name: ______________________________

Enrolment year: ______________________________

Class code: ______________________________

Unit competency code/title: CPCPCM Fabricate and install non-ferrous pressure piping

Task: Review the sections from 'Plastic pressure pipes (polymers)' to 'Bending tube' and answer the following questions.

1 Is the following statement true or false? 'When using push to connect type fittings, a special tool is needed.' Circle the correct answer.
T F

2 Explain the term 'witness mark' and describe its purpose.

____

____

____

3 Describe the heat fusion method of jointing pipes.

____

____

____

4 Is the following statement true or false? 'A special tool is required when using the compression sleeve method for jointing plastic pipes.' Circle the correct answer.
T F

5 Is the following statement true or false? 'A battery-operated crimping tool is more suitable to operate in a confined space than a manual crimping tool.' Circle the correct answer.
T F

6 State three locations where polymer pipes are not permitted for use.

____

____

____

7 List five types of polymer pipes used by plumbers.

____

____

____

____

____

8 Is the following statement true or false? 'Water services joined with rubber rings must have thrust blocks installed at bends more than 5°, tees, end caps, reducers, valves and inclines in excess of 1:5.' Circle the correct answer.
T F

9 Which type of PVC pipe cannot be solvent cement-welded?

____

10 Name three jointing methods of polyethylene pipe (PE).

____

____

____

11 Is lubricant required on an elastomeric (rubber ring) joint when joining large diameter PVC-U pressure piping?

____

12 What is generally the minimum size that polyethylene pipe is butt welded?

____

WORKSHEET 4

To be completed by teachers	
Student competent	☐
Student not yet competent	☐

Student name: ______________________

Enrolment year: ______________________

Class code: ______________________

Unit competency code/title: CPCPCM Fabricate and install non-ferrous pressure piping

Task: Review the section from 'Bending tube' to the end of the section '1.4 Clean up' and answer the following questions.

1 List seven methods used to create changes of direction in pipework.

____

____

____

____

____

____

____

2 What is the test pressure and time for water services as per AS/NZS 3500.1?

____

____

3 What is the horizontal spacing for clips on 20 mm copper tube as per AS/NZS 3500.1?

____

4 Is the following statement true or false? 'Short-radius bends create more friction loss than long-radius bends.' Circle the correct answer.
T F

5 What is the minimum bend radius suitable for bending 20 mm copper tube?

____

6 Why must dry sand be used when sand bending?

____

____

7 Internal bending springs are only suitable to make bends in which section of pipe?

____

____

WORKSHEETS 1

8 What safety provision must be adhered to prior to removing the sand from a heat bend?

____

____

9 What role does a template have in a heat bend?

____

____

10 What is the formula used to calculate the heat length for a sand bend?

____

____

____

11 Why is it important to leave the workplace clean and tidy at the end of the day?

____

____

____

12 What is the purpose of recycling material?

____

____

____

INSTALL WATER PIPE SYSTEMS

2

Chapter overview

The aim of this chapter is to address the skills and knowledge required to install different types of large water services/mains using various approved materials and techniques with reference to the relevant codes and standards. It also deals with the fabrication, corrosion protection, jointing methods and support requirements of large water main services used for the purpose of distributing drinking (potable) and recycled water above and below ground. This also applies to the installation of large fire hydrant services within a property.

Learning objectives

Areas addressed in this chapter include:

- identify installation requirements
- prepare for work
- install large water service
- clean up.

A pipeline that conveys water throughout the community and is owned by a water authority or local council is known as a **water main**. A further definition of a water main can also be obtained from AS/NZS 3500.0.

Water mains form an integral part of community infrastructure, so it is imperative that they are designed and installed correctly to minimise maintenance issues.

The Goldfields Water Supply Scheme in Western Australia is one of the longest water mains in the world (see Figure 2.1). It connects Mundaring Weir (near Perth) with the Mount Charlotte Reservoir at Kalgoorlie.

Adwo/Shutterstock.com

FIGURE 2.1 Section of Goldfields Water Supply Scheme pipeline

Construction of the 530 km-long pipeline commenced in 1898 and was completed in 1903. This was built to supply water to the communities in the fast-growing goldfields in Western Australia known as ‘Eastern Goldfields’. It supplies water to Coolgardie and Kalgoorlie, as well remote communities. The amount of steel used in the pipeline’s construction was greater than that of any other steel structure elsewhere in the world at the time. The water main has been operating for 120 years, supplying water to more than 100 000 people in over 33 000 households, as well as mines, farms and other various industries.

Early water mains were made from timber (see **Figure 2.2**) and then moved to more modern materials, such as cast-iron, ductile-iron, steel and various polymers.

FIGURE 2.2 Timber water main

Identify installation requirements

This learning outcome aims to help understanding of how to access, read and determine large water service installation requirements from job specifications, relevant Australian Standards, codes, manufacturers' instructions and jurisdictional requirements.

Additionally, it aims to help identify and apply workplace, work health and safety (WHS) and environmental requirements. Review Chapter 4 Carry out interactive workplace communication in *Basic Plumbing Skills* in relation to accessing relevant information.

Access and read plans and specifications

Plans and specifications usually contain information such as the address of the job, a general layout of the project, geotechnical reports and the engineer's detailed drawings. They also contain other important details for the job, such as pipe sizing, the location of the proposed works and the materials to be used. Plans and specifications are usually prepared and supplied in collaboration between the client, architect, hydraulic consultant and engineer. In some cases, such as work on public utilities, a standard set of specifications are used from the Water Services Association of Australia (WSAA). The standard AS/NZS 3500.1 Plumbing and Drainage: Water Services is used for large water services on private properties.

AS/NZS 3500.1 PLUMBING AND DRAINAGE: WATER SERVICES

WSA 03-2011 WATER SUPPLY CODE OF AUSTRALIA VERSION 3.1

As the job progresses, a set of 'as constructed/workshop' drawings may be required and should be marked up and kept on file as the installation progresses. Upon completion of the contracted works, the subcontractor is usually required to submit a set of 'as built' drawings. These drawings contain information relating to the actual 'as installed' position of the completed pipework, valve and ancillary locations, termination points, hydrant points, air valves, scouring points, flushing points and other information.

Work health and safety (WHS) requirements

Installation of water mains is a complex task and usually requires the use of heavy machinery and lifting equipment. Also, excavation works, concrete placement, support structures, access chambers and thrust blocks are part of the installation process.

There must be a proper risk management system in place. The hazards must be identified and the risk factors assessed, with the appropriate control measures put in place to prevent the likelihood of accidents. A job safety analysis (JSA) statement and a safe work method statement (SWMS) must be completed and understood by all parties concerned before work commences. These risk assessments must be updated regularly as the job progresses and different situations arise.

Site inspection

Inspecting the site before starting work is essential to determine the specific installation requirements. It is necessary to understand the topography of the land and determine the available access for plant and machinery, such as excavators, tipper-trucks and cranes.

A site inspection will also establish available areas for the storage of materials, plant and equipment, and temporary amenities, as well access for deliveries.

Working in and around excavations

Working in and around excavations can be hazardous due to the risk of collapse or falling into trenches. To reduce or alleviate the risks of collapsing, trenches should be properly supported when they are in unstable ground and/or deeper than 1.5 metres.

Confined space management

Because it may happen frequently when working on water mains piping systems, it is important that consideration is given to the possibility of being in or entering a confined space. A confined space is determined by the hazards associated with a particular set of circumstances; that is, restricted entry or exit, hazardous atmospheres or risk of engulfment.

Therefore, a confined space exists when several factors are present simultaneously, and is not simply defined as a confined space because work is performed in a physically restrictive location.

Note: For further information and requirements for working in confined spaces, contact your WHS state authority.

Working at heights

Because water mains can be installed above or below ground, the need to observe safe practices in line with working at heights may arise. Deep excavations may create hazards similar to those of working at heights. When entering a deep excavation, there should be appropriate fall protection and a proper access system, such as a ladder or stairs. When working on support structures or pipework at height above ground, appropriate scaffolding, lifting equipment and fall arrestor measures should be employed.

Excavation work, confined spaces and working at heights are all high-risk activities, so a safe work method statement (SWMS) must be completed before starting work.

Handling and storage

Manufacturers can supply information regarding the handling and storage of water mains pipes and ancillaries. Generally, the mechanisms used to move pipes should be suitable for the weight and the type of pipe material being used.

The method and apparatus used should not damage the pipe when it is transported or lifted for loading or installation. Mostly, slings, chains and cables are used to lift pipes and ancillaries (see **Figure 2.3**). However, some manufacturers recommend the use of and can supply specialised lifting equipment, such as hairpin mandrels.

Some manufacturers specify minimum requirements and/or place restrictions on pipe-lifting methods – for example, not using cables to lift glass-reinforced plastic pipes. This is because cable slings may damage the pipe due to the point stresses they create.

When lifting fibrous reinforced plastic pipes, pliable straps, slings or rope should be used. Do not use steel cables or chains to lift/secure pipes.

When handling single pipes, they can be lifted with only one support point, although for safety reasons the use of two support points is the preferred method as it makes the pipe easier to control.

Manufacturers generally state the maximum number of pipes that are supplied in packs. This is because of the potential damage (in the form of pipe shape distortion) that may occur when too many pipes are placed in a stack. Manufacturers also state the preferred lifting methods that suit their product.

When pipes are being transported, they should be supported as per the manufacturers' recommendations and securely fixed to the transport vehicle. Most pipe manufacturers prepare the pipes for transport with the use of timber supports, blocks and cradles.

Flat vectors/Shutterstock.com

FIGURE 2.3 The sling method of lifting pipes

Various pipe materials will have limitations placed on them as to the stacking method, height and/or number of pipes that can be placed in a pack or stack. The manufacturer recommendations must be consulted when stacking pipes onsite as incorrect storage can damage the pipe, which will limit the service life of the product and/or void the warranty.

Pipe storage and stacking

A suitable storage area must be selected, which should have a firm foundation that is capable of supporting the pipe stack. It should have adequate access via a suitable temporary road. To allow for the slinging of the uppermost pipe, safe access to the top of the stack is essential. During inclement weather conditions, consideration should be given to the provision of a suitable lightweight platform, which will further enhance safe access to the top of the stack for the operator and reduce the risk of a slipping hazard.

Depending on the material, pipes should be stacked on a suitable timber base, such as raised timber battens, and chocked appropriately to avoid the pipes rolling. Sharp edges that may damage the pipe should be removed from the ground as well as the timber base. The most common way of stacking pipes is through the use of the square, parallel and pyramid methods (see **Figures 2.4A**, **B** and **C**).

It should be noted that pipe storage and stacking methods are largely dependent on the material, its durability, resilience to sunlight and the elements. The stack height should never exceed that which is recommended, and storage methods should never compromise the pipe's ability to support itself while maintaining its roundness.

Therefore, the manufacturers' recommendations must be sought to ensure the most appropriate stacking and storing method is adopted for the material being used.

GREEN TIP

Prevent excess waste by handling and storing materials correctly.

UlgenDes/Shutterstock.com

FIGURE 2.4A Horizontal stacking method

Victor Yarmolyuk/Shutterstock.com

FIGURE 2.4B Parallel stacking method

Elshad Aliyev/Shutterstock.com

FIGURE 2.4C Pyramid stacking method

Environmental management

When carrying out works, it should be remembered each person is responsibe for taking all practical measures to ensure:

- the workplace under their control is free from polluting the environment
- any refuse or waste product is to be removed, controlled or treated to prevent pollution of the environment
- all legislative requirements are being met.

LEARNING TASK 2.1

1 Name the code used for water supply systems for public utilities.
2 What document must be completed before starting a high-risk activity such as working at heights?

Prepare for work

This learning outcome aims to help you understand how to plan and prepare the work site safely, create a materials list and collect materials, and select and check serviceability of appropriate tools and equipment, including personal protective equipment (PPE).

Identifying, locating and marking existing services

Existing services, such as water pipes and mains, sewer pipes and mains, stormwater pipes and mains, gas pipes and mains, communications cables and infrastructure, electricity mains conduits and so on, must be located before work begins. It should be noted that due to the danger to life and amenities, no work should proceed until this process is thoroughly completed, with the services identified and the area marked. It is also handy to carry out the service investigations process at the tendering stage of the contract. This will establish any special requirements that might affect the proposed works, which should be considered so the submitted price can be established accordingly.

If a long time has elapsed between the tender period and the original service investigation and the commencement of the works, say three months, this process should be repeated before starting the job, in case there are any changes to the area affected. The Before You Dig Australia service can assist with this process. However, this service may not identify items such as water and sewer mains; these will need to be obtained from your local water and sewerage authority, usually at a cost.

Traffic control and road opening permits

When working in an area where there is traffic movement either within, entering or leaving a site or in a public or private roadway, properly trained personnel must be employed to arrange traffic management plans and control the flow of traffic in and around the area. Not only should traffic be controlled in this manner, but it may also be a WHS requirement in your state. Failure to implement proper traffic control could be an infringement of the relevant WHS Act applicable to your workplace and result in prosecution or a fine.

When excavation works are to be carried out in a public area, all applications and permits should be obtained from the relevant authority, such as the local council or state roads or transport authority. This usually requires the payment of a restoration fee.

This may also affect the tender price, as each authority has a different scale it uses when calculating this fee. Some fees differ immensely and are dependent on the type and load capacity of the road; that is, if the road has a concrete substrate, it could be very costly to restore.

Local emergency services such as the police, fire brigade and ambulance services may also need to be notified, especially if roads are to be closed, as this may affect the emergency routes.

Many issues can occur when excavating in a public area, so it is critical that effective planning is done to minimise problems.

Ordering, delivery sequencing and storage of material

Consideration should be given to the coordination of deliveries during the works, so materials arrive in good time. Some of the materials used for water mains, such as large gate valves, tees, bends, prefabricated pits and chambers, have long lead times and lack of sequencing may delay the works.

In the case of water mains that form part of the assets of a water supply utility, ordering should be determined by WSA 03-2011 Water Supply Code of Australia Version 3.1 and Water Services Association of Australia (WSAA) purchase specifications or the local water utility agency's requirements. For other water pipe system installations, AS/NZS 3500.1 Plumbing and Drainage: Water Services shall determine the approved materials and jointing methods.

Plant, equipment, tools and PPE

Laying water mains is generally a large undertaking and will require a variety of tools and equipment. Contractors that are regularly involved in this type of work will be well equipped and may either own or hire some of the larger items of equipment and tools required for each project.

All tools and equipment must be checked to ensure they are operating correctly and currently tagged for use. Appropriate PPE must be worn to suit the site condition requirements and to suit the materials being handled.

For details on the more usual items of tools and equipment, please refer to the unit of competency CPCPCM2046 Use plumbing hand and power tools.

Selection of water main materials

The selection of materials depends on several factors (see Table 2.1). In most instances, the preferred materials are listed in the job specifications, although based on cost and serviceability; and, most importantly, after consultation with the client/architect, variations on specified materials may be approved.

It is important that the selected materials are compatible with each other and the environment in which they are installed, particularly when connecting to or extending an existing main. They should also be well suited to, and/or properly protected from, corrosion caused by aggressive soils. The possibility of materials of dissimilar metals reacting with each other should also be considered.

Other factors that can determine the selection of materials are:

- the location of the mains (e.g. trafficable areas, above or below ground)
- the size of the main
- the type of main (e.g. pumping, trunk or distribution)
- whether the new main is to be connected to an existing main
- the cost of the pipes, fittings and installation.

Water mains materials should comply with the WSAA's Product and Material Information Guidance for WSA 03-2011 Water Supply Code of Australia Version 3.1 and the AS/NZS 3500.1 Plumbing and Drainage: Water Services, the requirements of local water authorities and other relevant standards.

AS/NZS 3500.1 PLUMBING AND DRAINAGE: WATER SERVICES

WSA 03-2011 WATER SUPPLY CODE OF AUSTRALIA VERSION 3.1

LEARNING TASK 2.2

1 Name an advantage of using PVC-U pipe for water mains.
2 Why is it important to determine the type of soil in which the pipe is installed?

COMPLETE WORKSHEET 1

Install large water service

Setting out and installing a large water service/main requires proper interpretation of plans and specifications, site preparation for access and storage, and compliance with the current codes and regulations.

Excavating well-graded trenches and using correct jointing techniques with the appropriate valves and ancillaries, as well as adequate bedding and backfill requirements, is necessary for a compliant installation.

Testing pipework must be carried out according to standards and manufacturers' instructions.

All documentation must be completed according to relevant regulatory requirements and submitted upon completion of work.

TABLE 2.1 Materials suitable for use in the construction of water mains and their characteristics

Material	Characteristics
Copper (Cu)	■ Durable ■ Generally delivers years of trouble-free service ■ Extremely resistant to corrosion except in certain circumstances, such as when exposed to aggressive soils ■ Quite expensive ■ Easily damaged, particularly in the annealed state
Cement-lined cast-iron and cement-lined ductile-iron (CICL DICL)	■ Strong and durable ■ Can withstand some physical damage, high internal hydrostatic pressure or large external soil/traffic loads ■ Heavy to handle and can be susceptible to aggressive soils
Polyethylene-coated cement-lined steel (SCL)	■ Steel pipe has characteristics similar to those of CICL and DICL in terms of strength and ability to withstand some physical damage, high internal hydrostatic pressure or large external soil/traffic loads ■ Lighter and easier to handle than CICL and DICL ■ With advances in technology, some steel products with a polyethylene coating do not require wrapping when buried in aggressive soils ■ Available in lengths up to 13.4 metres
Unplasticised polyvinyl chloride (PVC-U)	■ Lightweight ■ Does not corrode ■ Can be easily damaged (although with recent developments in plastics, this disadvantage has been reduced) ■ May only be installed in direct sunlight under certain circumstances ■ Susceptible to becoming oval-shaped (squashing) when high external soil/traffic loads are present
Polyvinyl chloride modified (PVC-M)	■ Similar characteristics to PVC-U, although PVC-M is stronger, more flexible, lighter and possesses better flow characteristics than PVC-U pipe
Orieted polyvinyl chloride (PVC-O)	■ Similar characteristics to PVC-U, although PVC-O is stronger, more flexible, lighter and possesses better flow characteristics than other PVC pipes ■ Solvent cement jointing is unsuitable
Polyethylene (PE)	■ Similar characteristics to PVC-U, although PE possesses high impact strength, damage resistance, abrasion resistance, flexibility, high flow capacity, is available in longer lengths (which could be suitable for trench-less construction) and is weather resistant
Acrylonitrile-butadiene-styrene (ABS)	■ Similar characteristics to PVC-U, although ABS has high impact strength and abrasion resistance, and is lightweight and weather resistant ■ Suitable for installations where the pipes are exposed to direct sunlight (although this may be subject to regulations, standards and local water utility requirements)
Glass-reinforced plastic (GRP)	■ Lightweight ■ Available in long standard lengths of 6, 12 and 18 metres ■ High strength and flexibility ■ High flow characteristics ■ Abrasion and corrosion resistant

Source: SWSU TAFE (2009). Basic Plumbing Services Skills: Water Supply, Pearson, 2009.

Set out and install a large water service

Preparing to install large-diameter pipe involves good planning, site establishment and accurate material ordering, as well as having paid all the relevant fees.

There are many steps involved and much to consider when setting out and installing a large water service. All of these steps – taking delivery, excavation, pipe laying, jointing methods, bedding and backfilling – must be planned well for a successful job.

Taking delivery

When taking delivery of the pipes and fittings, it is essential the loading docket is checked against original order forms, and any shortages and back orders noted. All pipes and fittings should be inspected upon receipt to ensure there is no damage from transit.

Depending on length of storage, the amount of onsite handling and other factors that may influence the condition of the pipes and fittings, it is recommended that the pipe be re-inspected immediately prior to installation.

Bedding and backfill requirements

To provide for an extended service life, water mains require proper support from an appropriate bedding material (see Figure 2.5). When water mains are installed in private property and where they do not form part of a water supply utility's asset, the requirements for bedding and backfill must be followed as stated in AS/NZS 3500.1 Plumbing and Drainage: Water Services.

When water mains are installed as part of the asset of a water supply utility, it should comply with the standards set by the WSAA's WSA 03-2011 Water Supply Code of Australia Version 3.1.

Backfill material should be clean and free of any builder's refuse, such as bricks and rocks. It should also not contain any large earth lumps or excavation material, such as hard clay, which should be broken up into pieces no bigger than 75 mm in diameter.

Minimum cover requirements

The cover requirements for water mains are usually specified. However, at the very least, they should comply with the requirements of AS/NZS 3500.1 Plumbing and Drainage: Water Services.

When water mains are installed as part of the asset of a water supply utility, depth of cover should comply with the standards set by the WSAA's WSA 03-2011 Water Supply Code of Australia Version 3.1.

The minimum cover may also be determined by specifications in accordance with the local water utility agency and road owner's requirements.

Maksim Safaniuk/Shutterstock.com

FIGURE 2.5 Preparing bedding for water mains pipe installation

Removal and disposal of spoil, backfill and bedding materials

When planning the work, the removal of excess spoil needs to be allowed for, especially when the original material cannot be reused as backfill, such as in a roadway where stabilised backfill material is generally specified. The environmental management plan should also allow for and specify the methods that will be used for the removal and storage of the spoil. It is good practice to place the spoil directly into a skip bin or truck for immediate removal from the site. This will help reduce any effect stockpiling may have on the site, roadway and stormwater system through minimising the risk of sediment release.

To determine the cost of excavation or backfilling of a trench, a volume calculation has to be made to obtain the size of the excavation in cubic metres (m^3).

The equation for this is:

$$V = L \times W \times D$$

Where:

V = volume
L = length
W = width
D = depth

Generally, depending on the diameter of the pipe, most trenches are from 300 mm (0.3 metres) to 600 mm (0.6 metres) in width and up to 1.5 metres in depth, unless otherwise specified. After that depth, the trench should be timbered or open cut, and so the width may not necessarily be determined by the diameter of the pipe.

Note: Bedding/backfill material is usually ordered as a cubic metre amount. Because each material has a different weight per cubic metre, if tonnage is required, the supplier usually applies a conversion factor to the cubic metre value to calculate the required tonnage.

Corrosive areas

Corrosive areas are those that contain compounds consisting of magnesium oxychloride (magnesite) or its equivalent, coal wash, ash, sodium chloride (salt), ammonia or materials that may produce ammonia. What may be considered a benign soil is typically a sandy, free-draining soil of low salt content; anything else should be treated as a potentially aggressive/corrosive area.

Protection against corrosion

All metallic water mains need some form of protection against corrosion, both internally and externally.

When copper piping is specified, it usually requires wrapping with a polyethylene sleeve to protect it from aggressive soil environments. In extreme circumstances, pre-lagged copper tube may be specified, and where the pipe is exposed for joining, it is sealed with a petroleum tape. This can be cumbersome and alternative materials are usually used to save costs on labour-intensive installation.

Pre-lagged copper tube is available in sizes up to DN 100. Plain copper tube is available in various sizes and wall thickness up to DN 200 mm.

Cast-iron and ductile-iron mains piping are afforded some external protection from a bituminous pitch coating. However, there is also a need for them to be fitted with a polyethylene sleeve to protect them against aggressive soils (see **Figure 2.6**).

Steel water mains are coated with a fusion-bonded polyethylene coating. Some steel water mains products have polyethylene internal linings or cement lining.

Cathodic protection is a common method that uses a low electrical current to prevent corrosion of metal structures, such as pipelines, tanks, steel-pier piles and offshore oil platforms. For protection of steel pipelines, cathodic protection has been used since the 1930s.

FIGURE 2.6 Water mains repair wrapped with polyethylene sleeving for extra corrosion protection

Cast-iron, ductile-iron and steel mains are coated internally with a cement mortar lining to protect them from corrosion (see **Figure 2.7**).

FIGURE 2.7 Cast-iron, ductile-iron and steel mains are coated internally with a cement mortar lining to protect from corrosion

To aid with corrosion protection by limiting contact with the soil, water mains should also be backfilled with clean sand.

Note: If the plastic sleeve is damaged during main installation or tapping, then it must be repaired to the satisfaction of the water utility. See manufacturers' details for sleeving procedures.

Galvanic corrosion

Where ferrous and non-ferrous pipes or fittings are joined together in a water service, protection against galvanic corrosion shall be provided by:

a fitting a plastic connector or a short length of plastic pipe between the dissimilar metals, for threaded type joints; or

b fitting an insulating gasket between flanges, insulating sleeves along the bolts and insulating washers under the bolthead and nut for flanged type joints.

This is of particular importance when non-ferrous services and mains tappings are connected to CICL/DICL mains.

Jointing methods

Most modern water mains piping systems are joined using elastomeric, neoprene or rubber rings moulded to fit into a machined recess in the socket of the pipe. The seal is then compressed between the spigot and socket ends of each pipe length (see **Figure 2.8**).

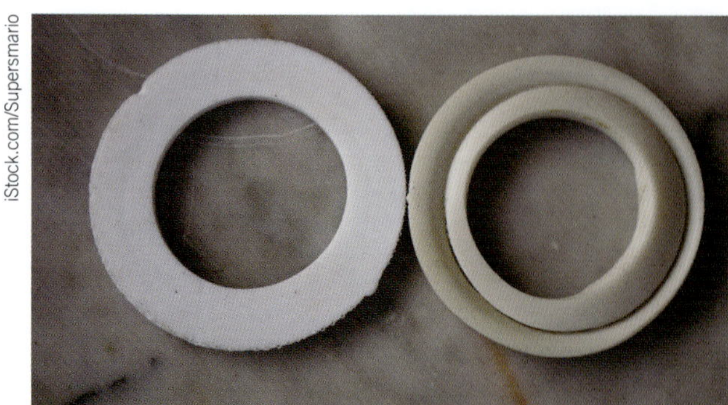

FIGURE 2.8 Typical elastomeric rubber ring used on ductile pipe

When joints are made using the elastomeric rings, lubricant must be used. The lubricant is generally specific to each type of pipe material and it also may contain some form of antibacterial ingredient. The selected lubricant should be suitable for use in potable water supply systems and be compatible with the compound used to make the sealing ring.

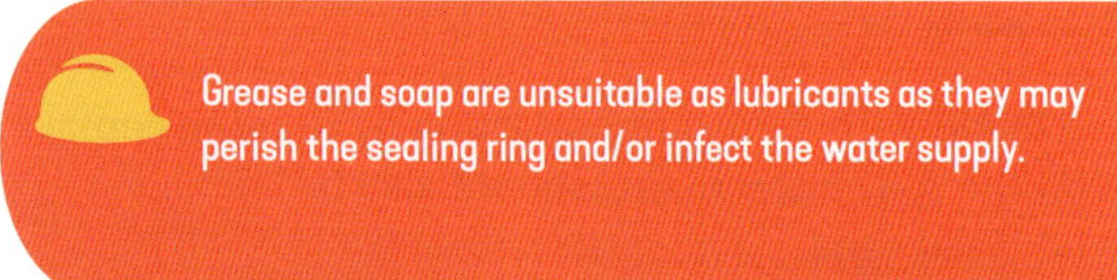

Grease and soap are unsuitable as lubricants as they may perish the sealing ring and/or infect the water supply.

Jointing methods for cast-iron/ductile-iron pipe

Cast-iron and ductile-iron pipes can be joined using the following methods:

- screwed flange joints
- cast-in flange joints
- push-in compression elastomeric joining systems
- mechanical couplings.

Jointing method for steel pipe

Except for welded joints, the jointing methods for steel piping are similar to those used in CI/DICL pipe and include:

- rubber ring joints
- roll groove joints
- welded joints
- flanged joints
- mechanical couplings.

The type of joint selected depends on the pipe diameter, pressures and application.

Note: a duly qualified and certified person must perform all welding when installing steel pressure piping systems.

Jointing methods for PVC-U, PVC-M and PVC-O pipe

PVC pipes can be joined using the following methods:

- cold solvent cement welding (PVC-O pipes should not be joined with solvent cement)
- rubber ring jointing
- flanged joints
- mechanical couplings.

Rubber ring jointing in PVC pipes

Assembly of elastomeric joints in PVC piping is usually a quick and easy process (see Figure 2.9). However, as the joint is a compression-type joint and a fair amount of force is usually required, it is recommended that it should not be made without the use of some form of assisting mechanism. The joint will be more easily made, and will help satisfy WHS requirements for addressing strain hazards, by employing one of the following methods (depending on local conditions and the size of the pipe). Water services joined with rubber rings must have thrust blocks installed at bends more than 5°, tees, end caps, reducers, valves and inclines in excess of 1:5.

There are several different methods that can be used: the crowbar/fulcrum technique; the fork tool method; the excavator method; the excavator/backhoe slewing method; the come-along method.

PVC pipes connected to ductile-iron fittings

PVC-U pipes may be joined to ductile-iron push fit and compression elastomeric gasket fittings. The

FIGURE 2.9 PVC-U pipe collar with rubber ring installed

standard factory witness marks on the spigots of a new length of PVC-U pipe are not applicable when joining to ductile-iron fittings. So there is proof of full insertion, a new witness mark should be placed on the spigot of the PVC-U pipe, which is marked at an appropriate depth that will suit the ductile-iron fitting. When making the joint, the spigot of the pipe should be inserted fully into the fitting until it reaches the stop. A short piece of pipe can be added, which will give an extra collar to provide for any movement due to expansion and soil loads.

Solvent cement jointing for PVC pipes

When making a joint with solvent cement, the pipe is cut square, burrs are removed, and the outside of the pipe end and the inside of the fitting socket are cleaned with priming fluid. They are then coated with PVC solvent cement of a type suitable for pressure fittings, assembled and held in place until the solvent sets.

Jointing methods for polyethylene pipe

Joining lengths of polyethylene (PE) can be achieved by the use of the following techniques:

- mechanical compression couplings
- electro-fusion welded
- butt welded.

Jointing methods for acrylonitrile-butadiene-styrene (ABS) pipe

Pipes and plain-ended fittings can be joined by the following methods:

- sockets – cold solvent cement welded
- elastomeric sealed (rubber ring joint)
- backing plated flanges
- shoulder-style coupling (e.g. Victaulic)
- threaded adaptors
- tapping saddles
- unions
- mechanical couplings (e.g. Gibaults, Straub, Wang).

Elastomeric sealed sockets

There are two types of elastomeric sealed sockets fitting available:

- One side of the fitting is the cold solvent cement weld type and the other side an elastomeric seal type joint.
- Both sides of the fitting are an elastomeric seal-type joint.

Cold solvent cement welded sockets

Solvent cement welded sockets are the quickest, most economical joining method for ABS pipes and plain-ended fittings, and are available for all pipe sizes.

Solvent cement welding eliminates the need for thrust blocks as the longitudinal stress is taken in the pipe wall.

Joining ABS to PVC

The solvent cement joining of these two dissimilar materials is not recommended.

A rubber ring socket or a mechanical connection such as a threaded connection or a flange is recommended where it is necessary to join ABS to other materials.

Jointing methods for glass-reinforced plastic (GRP) pipe

Glass-reinforced plastic pipes (GRP) are joined in similar ways to other products, such as cast or ductile iron, including:

- double bell couplings (elastomeric socket joiners)
- GRP flanges
- flexible couplings
- lay-up joints.

Set out the works to specification

Once the information is gathered from the plans and specifications, it must be physically set out. Several methods can be used. In the case of directional borers, the coordinates need to be entered into the machinery to ensure the specified route is followed. When excavating an open trench, measurements need to be taken from the set-out pegs and benchmarks that have been set by the surveyor.

The excavation and trench work then requires marking out and levels calculated. One method of marking out the direction of the trenching is with the use of a simple tried-and-proven method of a string line set along the route of the trench and then marked with lime powder or line marking paint.

Determine the location, alignment direction, level and grade of mains pipe systems

Generally, the location, alignment direction, level and grade of the mains piping systems are determined by a number of factors:

- the purpose or use of the main: gravity main, trunk main, distribution main, pumping main
- the topography of the land and whether the main is to be laid through mountainous country, desert, in a busy city street or out in the bush
- the locality of the main: for example, the state, territory, city or towns' infrastructure it is to be installed in and for; this is dictated by where the work must be performed and is not a variable factor other than for the finer location of the main
- its proximity to other services, proximity to adjacent buildings and rivers
- whether the main is to be installed in the footpath or road, above or below ground, and the depth of cover.

Figure 2.10 shows a pipe being installed in a remote location. Locality is not usually a variable when installing water mains pipes and the design must suit local conditions, including the topography of the land.

iStock.com/Photon-Photos

FIGURE 2.10 Pipe installed in a remote location. Note the trace wire on the plastic main so the pipe can be detected with a pipe locator.

Positioning and finer location and alignment of the main

Positioning of the mains can be as easy as measuring off an existing building or as complex as having to establish a waypoint in the wilderness (a position in relation to global coordinates known as longitude and latitude, usually expressed in degrees, minutes and seconds). Techniques to do this vary from engaging the services of a qualified surveyor who can set up set-out pegs and temporary levelling benchmarks to, with advancements in accuracy, the use of global positioning systems (GPS) units.

Advise the plant operator of excavation requirements

Different plant operators have different preferences as to where the line is placed for them to dig. Some operators want the line in the centre and others like it to be placed on either the right or left side of the trench, depending on the side of the machine that has the most visibility of the bucket when being operated.

The selection of the appropriate bucket size is also very important and possibly the first question a plant operator will ask. Bucket sizes come in a number of standard widths and cubic metre capacity; depending on the size of the excavator, they are generally 300 mm, 450 mm, 600 mm wide for trenching/digging buckets, and 1200 mm wide for mud sieve or trim buckets.

The trench width should allow for at least a 100 mm space on either side of the pipe to allow for backfill material. Additional excavation maybe necessary at each joint to provide sufficient room around the pipe to properly make the joint and to ensure the pipe rests on its barrel and the socket does not bear any weight. After the joint is completed, the bedding and haunching should be compacted under the collar of the pipe to ensure it is supported when the backfill loads are placed on the installation.

When pipes are installed either perpendicular to or in line with a slope that exceeds 15°, a geotechnical engineer should be consulted to verify and/or recommend what should be done to ensure the stability of the slope. The base of the trench should be excavated so as not to create any area where water can accumulate, as water pooling may be a factor that could destabilise the slope and have a negative impact on the bearing capacity of the substrate.

Monitor the levels

There are numerous ways levels can be monitored, from the traditional boning rod technique to modern laser levels and sensors/receiver fitted to the excavation equipment. When starting, the work levels are calculated, the excavation is commenced and the instruments are set to the appropriate levels. These levels should then be double-checked at various intervals as the work progresses or at the end of a straight run to ensure the work is done to the specifications and the settings remain correct.

Grading of the main

The grade of the main may be determined by its use. A gravity main requires a set uniform grade/fall to enable it to work efficiently. Some mains may be laid through hilly or mountainous territory and therefore need to be installed through tunnels or at a grade that follows the topography of the terrain. Where possible, water mains should be laid with grade so as to prevent air build up and to ensure the velocity of its contents remain within specified limits. If possible, the main should be given a uniform grade of at least 2 or 3 metres per 1000 metres. This will help the air in the system rise to a point where it can be expelled.

Even if the pipeline is laid over undulating territory, too many changes in slope should be avoided, particularly in large water mains. If the pipeline has several highpoints, the grades for the downward slope should be a minimum of 4 to 6 metres per 1000 metres and the minimum upward grade should be 2 to 3 metres per 1000 metres. When pipelines are being laid in level territory, artificial high points should be incorporated into the installation. This is because, after some inevitable earth settlement, a level pipeline may form high points with no means of expelling the air.

Laying pipes in water-charged ground

Generally, because one litre of water weighs one kilogram, the substrate on which the water service is laid must have a sufficient bearing capacity to support the weight of the pipeline and the water it is carrying. Therefore, when water services are to be laid in water charged or filled ground, they should be supported by a means that is certified by a suitably qualified engineer.

Refer to AS/NZS 3500.1 for more information.

When the ground water is above the level of the base of the trench, the water must be lowered to enable the pipe-laying process to be carried out by installing a dewatering system.

Mains pipe system support mechanisms

When a mains pipe is in service, it may be subject to various loads and forces such as, but not limited to, those:

- angular, circumferential (Poisson's ratio) and linear forces, which act on the pipe from the head (pressure) and flow of the water
- water hammer and the way water acts within the pipe as it moves, changes direction and/or suddenly stops
- external forces from gravity; soil loads
- the weight of the pipe itself
- the weight of the water in the pipe.

Thrust/anchor blocking

Water mains services with unrestrained elastomeric rubber ring joints installed below ground require anchoring. The joints that fit into this category are mostly used when installing ductile/cast-iron, and the various PVC piping systems that use rubber ring jointing methods (see **Figure 2.11**).

When installing thrust blocks, they should be placed in accordance with the manufacturer's recommendations and in consultation with a qualified soil engineer. In general, thrust blocks are made of concrete, with one side bearing against a firm surface of the excavation. They should be designed by an appropriately qualified structural engineer.

Thrust blocks are engineered based on:

- the local soil stiffness or bearing capacity, and backfill materials
- installation conditions, such as the pressure or head the pipe is subjected to
- the possibility of water hammer and the variations in pressure exerted by the way the water acts within the pipe
- the material being used for the piping system.

Thrust blocks need to be installed at all:

- bends, tees and junctions
- termination points of piping, e.g. at end capping and hydrant points
- valves installed in the piping
- reducing fittings in the direction of the smaller pipe
- changes in direction greater than the normal allowable deflection of the pipe joints or in excess of 5°
- grades in excess of 1:5 or as specified by the manufacturer.

Restrained jointing systems

When certain situations arise, to avoid the need for thrust blocks and where approved by the water agency, a restrained joint piping system such as DICL with restrained elastomeric joints may be used. The use of restrained jointing systems can alleviate the need for thrust and anchor blocks, especially in common trench installations, that is, where drinkable water mains are laid in the same trench as recycled (non-drinking) water mains. The use of restrained joint pipeline systems also has the potential to reduce future difficulties that could be faced by maintenance personnel when large amounts of concrete may need to be removed to facilitate repairs.

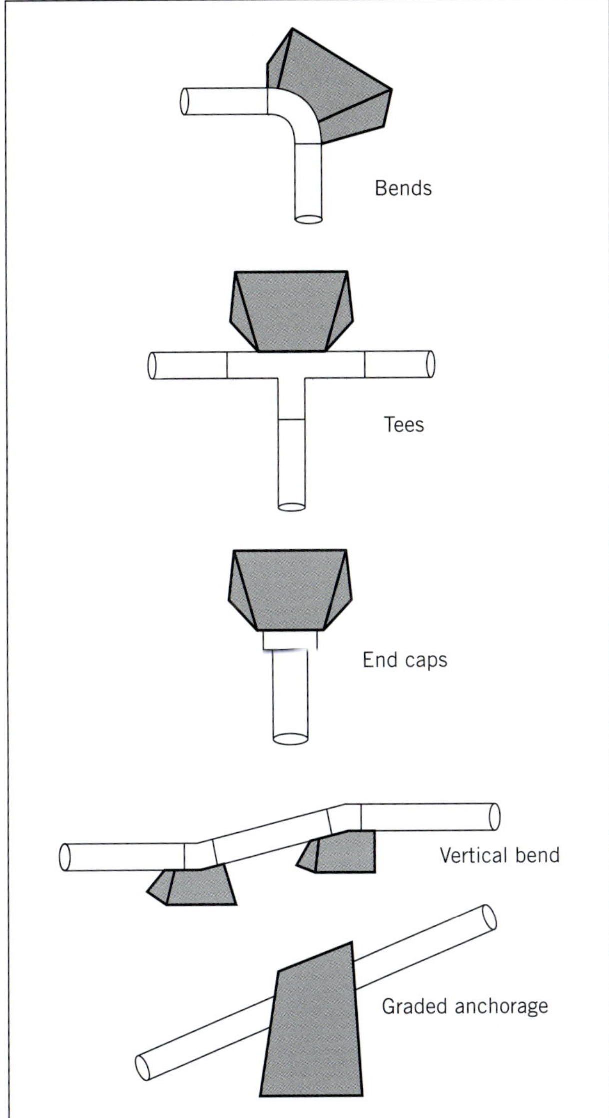

FIGURE 2.11 Typical thrust block position

Standards Australia/AS3500.1

Valves, fittings and flow control devices

Water mains systems require a number of in-line valves and ancillaries in order to operate correctly.

These valves and ancillaries are used for functions such as:

- isolating a portion of the supply or distribution system
- enabling slow release of accumulated air at high points in the pipeline to avoid interruptions to flow and pressure
- allowing air to enter, with the use of air and vacuum relief valves to avoid interruptions to flow and pressure
- clean out (scouring points) and drain water mains pipes.

Apart from the normal varieties of valves, such as gate, butterfly and ball valves, water mains require the installation of other specialised valves. To avoid problems associated with incorrect installation, it is imperative to understand the operation principles of these valves.

Air entrapment can cause a reduction in flow and pressure due to a loss in the cross-sectional area of the water occupying the pipe. It can also cause increased energy consumption in pumped installations, corrosion from trapped air by exposing metal piping to oxygen and therefore oxidisation. It can even create a situation that will result in blockages. Rapid inflow of water into a large air pocket in the pipeline can also cause water hammer. Negative pressure caused by draining, scouring or pipeline rupture can also result in collapse of water mains pipes. All of these problems can be extremely damaging to a pipeline and can ultimately affect its performance and longevity.

Kinetic air valves

Kinetic air valves (see Figure 2.12) are sometimes referred to as large orifice valves. The primary purpose of this type of valve is to release large volumes of air when the pipeline is being filled. The valve should be sized sufficiently to ensure the correct volume of air is allowed to leave the orifice while not exceeding the specified velocity.

Scouring points

Scouring points are installed in the line to facilitate draining for maintenance purposes and they are sometimes used as a flushing point when the main requires pigging or cleaning.

Spring hydrant valves

Spring hydrant valves are used to connect a hydrant tee or flanged hydrant riser. Sometimes these are used for flushing water mains and for temporary standpipes. Temporary standpipes should be fitted with a water meter and an appropriate backflow prevention device. The owner or operator of the standpipe must have authority to possess and use it.

FIGURE 2.12 A kinetic air valve installed on an above-ground water main

Hydrant bends

Hydrant bends (see Figure 2.13) are installed when a water main terminates at a hydrant point. The installation of a hydrant at this point facilitates the flushing of the dead leg of a main. This is to avoid creating what is termed black water or water that contains a large number of impurities, which have been transferred from the pipe material into solution such as ferrous oxide in cast-iron mains.

FIGURE 2.13 Hydrant bend

Duck foot bends

Duck foot bends are installed where there is a potential for high thrust or vertical loads.

Gibault joints

Gibault joints (see Figure 2.14) are used for maintenance and repairs to existing water mains installations. They are also used for joining spigot tees into a main when cutting a branch line, such as a tee and valve for a hydrant service. It is of utmost importance when installing these valves that the existing piping is clean and free of corrosion.

Note: Refer to relevant design manuals for installation details.

FIGURE 2.14 Gibault joint

Ball valve main taps

Whether they are installed under pressure or during construction, ball valve main taps (see Figure 2.15) are the most economical means of connecting house services to a water main. They can be used with a tapping band, pre-tapped fitting or be directly tapped into a ductile-iron main.

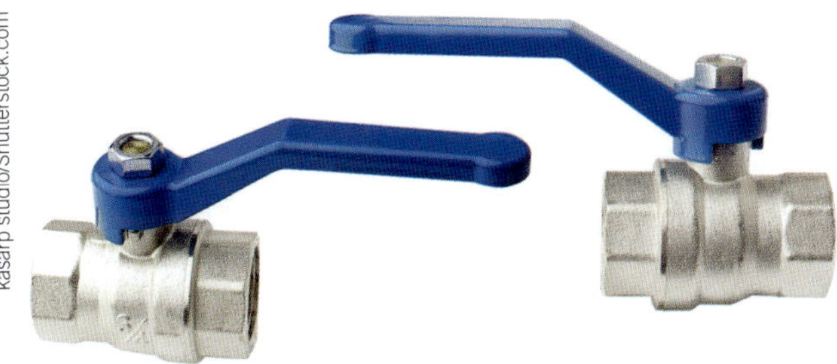

kasarp studio/Shutterstock.com

FIGURE 2.15 Ball valve main tap

Note: Direct tapping into a ductile-iron main can only be considered when the service connection is less than one-sixth of the nominal pipe size of the main to which it is being connected.

Tapping bands

Tapping bands (see Figure 2.16) are used as a pre-tapped saddle for the connection of house services.

FIGURE 2.16 Tapping band

Check valves

Swing check valves (see Figure 2.17) allow flow in one direction only. They utilise a hinged gate like disc that seals against a tilted seating face. The flow of the water keeps the hinged gate open, thus allowing flow to continue in the desired direction. The disc must seat by its own weight and stop backward flow as soon as forward flow ceases.

Krashenitsa Dmitrii/Shutterstock.com

FIGURE 2.17 Check valve

Resilient seated sluice valves

Resilient seated sluice valves (see Figure 2.18) are used to isolate sections and branch lines in pipelines or to manually operate discharge points, such as a scouring point. These valves are manually operated using a

FIGURE 2.18 Resilient seated sluice valve

removable key/bar or hand wheel. A cap is fitted to the top of the spindle and is removable for operation by the key. They are closed in a clockwise direction (opposite to the normal closing direction of most valves). Resilient seated sluice valves are sealed by a central metal gate, which has a rubber coating. The valve gate is raised and lowered by a screwed non-rising stem. The gate is aligned using an integral guiding system. To allow for full flow, when the valve is completely open the gate must raise clear of the valve's internal diameter.

Metal wedge sluice valves

The use and operation of the metal wedge sluice valve is similar to the resilient seated sluice valve, except the valve gate is uncoated and is sealed by a corrosion-resistant metal gate. These valves also must be installed with the valve spindle in the vertical position.

Pressure reducing valves, pressure ratio valves and automatic control devices

To help conserve water, pressure management is becoming a more prevalent issue for water supply authorities. More and more water authorities are taking steps to balance out the supply pressure in their reticulation systems. The balancing of pressure throughout the system will ultimately save water due to a reduction in burst mains piping and leakage.

Some areas in a reticulation system have higher pressures due to the vertical height between the service reservoir and the level of the property. Water pressure will often vary at different locations and is dependent on how far the property is from the service reservoir and its elevation in relation to the service reservoir (see Figure 2.19). Put simply, mains in low-lying areas are usually subject to higher pressures than those at a higher elevation.

In some situations, particularly in low- and high-flow periods, head/pressure may build up beyond the specified limits. To address this, the installation of pressure reduction, pressure ratio and/or automatic control devices may be required to keep pressures within the specified limits.

Some of the available options are as follows.

Pressure reduction valves

Basic pressure reducing valves automatically reduce a higher inlet pressure to a predetermined lower downstream pressure regardless of any changes in the upstream flow rate or pressure.

Pressure reduction valves work on the principle of when the downstream pressure exceeds the pre-set pressure of the control pilot, the main valve and the pilot valve closes to drip tight and thus holds the downstream pressure to a predetermined set point.

Anti-cavitation valves

Anti-cavitation valves are pressure reduction valves, which in their design reduce the effects of cavitation. Cavitation levels range from the relatively harmless levels often known as incipient cavitation, to the more significant levels of cavitation that can damage the water mains' services, valves and related pipework. Cavitation results when the velocity or speed of the water at the valve seat becomes excessive; this can then create a sudden severe reduction in pressure, which causes the water to become semi-vaporised. This results in the formation of millions of tiny air bubbles. A subsequent decrease in velocity and pressure occurs in the outlet of the valve seating area which, when re-pressurised, causes these vapour bubbles to collapse at a rate of many times per second. The resultant noise and vibration have the potential to damage piping and associated valves and equipment, not to mention the effect on the hearing of an operator through long-term exposure to excessive noise.

COMPLETE WORKSHEET 2

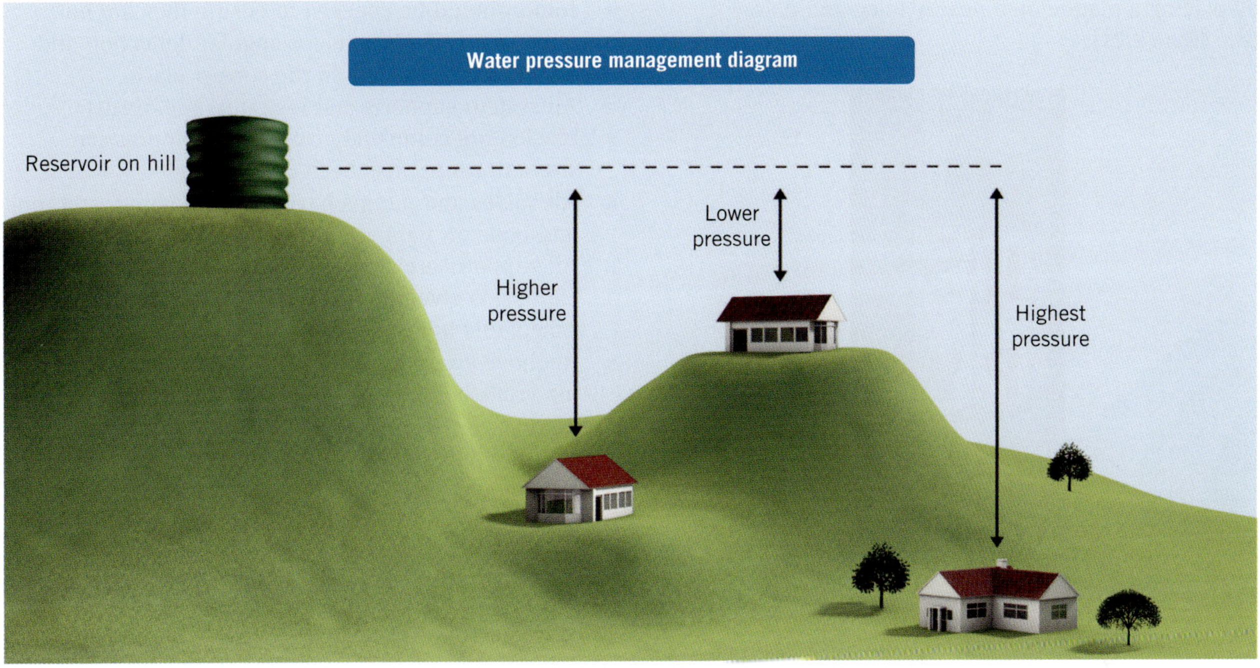

Water pressure management diagram

Reservoir on hill

Higher pressure

Lower pressure

Highest pressure

Source: © Sydney Water.

FIGURE 2.19 Water mains in low areas will have higher pressure than those in elevated areas

Constructing valve chambers, minor structures and thrust blocks

A basic consideration that needs to be allowed for when installing valves in a piping system is: will the valve need to be directly accessible, that is, installed in chambers – or can it be installed as a buried valve without the need for access chambers?

Generally, valves installed in smaller diameter mains piping systems are of the direct buried type (see Figure 2.20). This is the easiest and most cost-effective means of installing valves in smaller diameter pipe. However, there may be instances where a small diameter service valve needs to be accessible for servicing, such as in a basement of a building, and therefore it may need to be installed in a chamber.

FIGURE 2.20 Typical path boxes for buried stop valves

Buried valves should also be provided with a path box and conduit to allow for operation via a valve key.

To address thrust requirements, the valve may need to be encased in a reinforced concrete thrust block. The general assumption is that thrust works in one direction only. However, when a valve is closed, backpressure can create thrust loads in the opposite direction. Therefore, the structural support system should be designed to allow for this by considering the loads in either direction.

This method may also be used for larger diameter mains installations, depending on access requirements, the only limit being that of reasonable thrust block design for the space available.

Valve chambers should be constructed to allow access to operate, service and/or replace valves. The structures they are housed in must provide easy, safe entry for personnel with enough space to perform the appropriate task. Confined space entry may need to be considered when opening, entering and performing work in valve chambers.

Buried water mains markers

Valves that are buried need to be identified for two reasons. First, their purpose (i.e. hydrant or stop valve) and, second, so they can be located. Stop valves usually have a cover, which should be installed at 25 mm above the adjacent ground level when in grassed footpaths and flush with the surface (which must be graded away from the valve cover) when in roads and sealed footpaths. The cover must be located centrally above the valve spindle and able to be opened to allow the valve to be operated with a key.

The location of stop valves and hydrants are usually indicated by the installation of a post near the valve or by placing a plaque on a nearby telegraph pole (see **Figure 2.21**).

FIGURE 2.21 Hydrant valve and service valve indicators fixed to a telegraph pole

The hydrant/service valve identifying acronyms, as they would be placed on a nearby post, fence, telegraph pole or wall and the valve covers in the road or footpath, are set out in **Table 2.2**.

TABLE 2.2 Hydrant service valve identifying abbreviations

Abbreviation	Meaning
HR	Hydrant in road
HP	Hydrant in footpath
AV	Area valve
DV	Dividing valve
SV (or SVP)	Service valve in footpath
SV (or SVR)	Service valve in road

Test and commission the water mains installation

Prior to handing the installation over to the local water utility, the pipeline must be tested and accepted (passed off by the designated officer) in accordance with the plans and specifications, appropriate standards, manufacturers' recommendations and the local water utilities' requirements. Prior to this testing, there are some important procedures that must be performed.

Check the mains pipe system support structure

In preparation prior to filling and testing, the complete installation should be checked to ensure that all work has been completed as specified. Items of critical importance are, but not necessarily limited to, the following:

- All joints are assembled correctly and all bolts are tightened to specified torque settings.
- Joints should be checked to ensure they are not outside specified limits for angular deflection and all witness marks are in the correct place.
- All system supports and restraints (i.e. thrust blocks) are completely cured and compressive strength testing has been carried out.
- All valves and pumps have been anchored, and nuts and bolts on flanges and anchors are tightened.
- All supports are checked to ensure adequate support is given and pipe deflection is within acceptable limits.
- In above-ground installations and depending on materials being used, couplings/collars should be marked as per manufacturers' requirements as a future reference to measure any movement once the pipeline is pressurised.
- All gaskets and elastomeric rings should be checked to ensure they are correctly seated and the gap in the collar or coupling is free of concrete debris or foreign matter.
- The structural integrity of the support system should be checked, ensuring there are no visible failures or movement in structures prior to loading the pipe with fluid.
- Ensure all air has been expelled once the pipeline is loaded with fluid.

When the pipeline is filled and pressurised there are several items that should be checked. Items of critical importance are, but not necessarily limited to, the following:

- Inspect all joints for any signs of leakage.
- Check that joints and couplings have not rotated or moved more than the allowable tolerances relative to the marks placed prior to filling.
- Check there is no increase in pipe couplings' angular offsets and/or that joint deflections are still within allowable limits.
- Check the structural integrity of the pipe supports, ensuring the increase in pressure and load has not caused any settlement outside the set limitations. Use the prior-made marks to check if the pipe has moved relative to the anchors.

Dependent on the materials being used, the pipe should be checked for dark areas and weeping to ensure the integrity of the pipe barrels and that there was no damage to the pipes during installation. If pipe damage and weeping is detected, the installation should be drained and the appropriate repairs made prior to re-pressurising the line.

Fill the line with water

When filling the line with water, all valves should be opened to allow air to escape and to avoid pressure surges. The kinetic air valves should be checked to ensure they are functioning correctly and allowing any trapped air to escape. Once the pipeline has been filled and before any test pressure is applied, it should be checked and inspected as previously stated.

The line should be pressurised slowly. The rate of filling should be based on a maximum velocity of 0.05 m/s. It should be noted that considerable energy is contained in a pressurised pipeline and, to avoid any accidents, this should be done with caution.

The main should be left full and un-pressurised for between three and 24 hours to allow the test water temperature to stabilise and any dissolved air to vent from the system. Cement-lined pipes should be filled 24 hours prior to testing to allow for saturation of the lining.

Test the installation

Mostly, mains are tested in sections and so a maximum test length of 100 metres should be adopted.

Acceptance testing may be conducted in stages as the project progresses. Testing should be carried out as soon as the works are completed and where thrust restraint curing times are achieved. Where isolation is available, the water main may be progressively tested in sections of at least 100 metres or in its entirety if the main is less than 100 metres in length. Testing during wet weather should be avoided as visual detection of leaks may not be noticed.

Note: The work should not be covered in until it has been inspected by the water utility's representative. The joints in the pipeline should be left open so they can be inspected during testing. Test only the pipeline and fittings that are designed to withstand the test pressure and make sure any testing pressure remains within those design limits. The components that are not designed to withstand the test pressure should be isolated. All pipes and fittings should be securely anchored to prevent movement during the test. Prior to filling the installation with water and applying the test pressure, and to ensure the pipe is restrained from movement as required, backfill the trench adequately.

Note: Joints and fittings should be left exposed for visual inspection.

After the pipeline is pressurised, it should be allowed to stabilise for a short period of time. If after this time the pressure is still fluctuating, the effects of thermal change or entrapped air should be considered and ruled out as the cause. If the line pressure still fluctuates, then the following areas should be checked:

- flange and valve areas
- line tap locations
- kinetic valves are sealing correctly
- pipe joint soundness
- wet and dark patches on the pipes.

If this does not identify the leak, then the pipeline should be tested in smaller segments to help isolate the section containing the leak.

The pipeline is normally pressurised using a pump set up in a test manifold, which is made up specifically for the purpose of testing the pipe. It is usually fitted with two gauges, both of which are to be calibrated and have a current certificate to that effect. Each gauge should have a range of 0 to 2500 kPa. One of the gauges is to be used as a reference gauge while pressurising the installation and the other to ensure accuracy. Both gauges shall read within plus or minus 5% of the test head and within 5% of each other. The gauge recording the lowest pressure of the two should be read as the test pressure.

In preparation for testing, blank flanges or caps at each end of the test section should be installed. No testing should be done against closed valves unless they are fully restrained and it is possible to check for leakage past the valve seat. This is to ensure all potential leaks are visible and if for no other reason than to rule out the valve seat as an area of pressure loss/leakage. Mechanical ends that are not end load resistant to withstand the test pressures without movement need to be temporarily strutted/supported and anchored.

When recycled and drinking water mains are tested together, each ball valve on the property service should be closed. However, each property service valve can be temporarily opened to allow the escape of air, and in the case of recycled mains laid in conjunction with drinking water mains, to check for connection to the correct main. All personnel need to be clearly informed of the loading limits on temporary fittings and supports. Pressurise the line to 75% of the test pressure and leave for a minimum of 12 hours or as specified.

Unless tested in conjunction with the drinking water or non-drinking water main and to a pressure of 1500 kPa, each property service up to the meter valve must be separately pressure-tested to 1500 kPa (as per AS/NZS 3500.1).

The preliminary pressurisation is intended to:

- stabilise the water main by allowing most of the time-dependent movement to occur
- achieve saturation in absorbent materials such as the cement lining
- allow for pressure-dependent increases in volume of flexible pipes prior to the main test.

Provided there is no obvious leak in the water main, the pressure should be steadily raised until the specified test pressure is reached.

Maintain the test pressure for four hours. Measure and record, at half-hour intervals, the quantity of water added to maintain the pressure during the period of testing.

Visually inspect the line for leaks. If a leak is suspected but is not visible, use aural or electronic assistance.

Do not remove temporary supports until the test section has been depressurised.

Dispose of test water in accordance with the relevant environmental regulator and water agency requirements.

The test results should then be recorded as per the specifications. Records should be kept of the test results and any rectification work that was performed during the test period.

Disinfect the water main and make operational

During construction, utmost importance should be placed on always keeping the installation free of foreign material and/or storm or groundwater. If cleanliness during the construction phase is maintained, it will ultimately aid the final cleaning, filling, testing, sterilising and commissioning process for the installation.

Generally, it is written into specifications that the contractor has a responsibility to prevent the entry of foreign matter, storm and/or groundwater into the pipeline. It is also the responsibility of the contractor to remove all foreign matter from the interior of all pipes and fittings before they are placed into position. It is also mandatory that during any interruptions to the pipe laying, all open ends of the laid pipe be plugged or tightly sealed to prevent entry of foreign matter or groundwater.

After testing is carried out on a water main section, the section should be cleaned and flushed out. The method of cleansing the installation is usually set out in the job specification and is generally project specific.

One option for pipe cleaning prior to testing is by loading an appropriate/approved pipe cleaning pig into the line and forcing it along with hydraulic pressure. When forcing the pipe cleaning pig along the pipe, its velocity should not exceed 1.5 metres per second. The pipe cleaning pig should be carried out until all debris is removed. The water utilities representative should witness the pipe pigging process and sign off on its completion.

Pipe pigging may also be required in the future as part of a periodic maintenance schedule. This will help keep scale build up to a minimum by keeping the main internally clean and therefore in a condition where it will be able to perform to its full design capacity. There are a large range of pipe cleaning pigs available to perform this work without damaging the pipe and/or its lining. These pigs are also designed to fit through bends and full flow valves.

When the pigging process is complete, the main is then flushed with water until it flows clear.

Another option for pipe cleansing used for water mains of less than 300 mm is flushing the pipes with water at high velocity for as long as it is necessary to drain three times the volume of the pipe.

To satisfy local conditions, the disinfection and sterilisation procedures are normally set out in the job specification and are specific to each project.

Before an installation is placed into service, it must be sterilised. This can be achieved using a sterilising agent such as sodium hypochlorite in concentrations as specified.

Documentation

All the relevant documentation must be completed according to the local regulatory requirements. Any inspections of the work being carried out at different stages of the installation must be arranged and signed off as the local regulatory authority requires.

LEARNING TASK 2.3

1 In which jointing system are thrust blocks required?
2 Why is the water main given a uniform grade of at least 2 or 3 metres per 1000 metres?
3 Why are water mains in low-lying areas usually subject to higher pressures than those at a higher elevation?
4 What is the purpose of a scouring point?
5 In which direction are sluice valves closed?

Clean up

The final part, which is just as important as any other stage of the job, is the clean-up. At the completion of the testing, the clean-up process should be carried out. If the job has been managed correctly, this process should be relatively simple.

Clear the work area

The work area must be left clean and tidy, with all rubbish removed and surfaces reinstated.

The materials, which have been stockpiled during the project and not been removed as the work progressed, can be stored for the next project, sent to the recyclers or disposed of appropriately in accordance with the environmental management plan.

Clean tools and equipment

Plant and equipment are cleaned in readiness for transport to the next job or holding yard. Cleaning the plant of all residual soils, dirt and weeds is essential as this will prevent the spread of plants and noxious weeds.

Inspect all tools and equipment used during the work for any signs of damage. Any damage should be reported immediately to the supervisor, tagged as 'damaged' and taken out of circulation until the tool or equipment has been repaired.

Store all tools and equipment in a dry and secure place.

COMPLETE WORKSHEET 3

2

SUMMARY

The skills and knowledge required to install large water services/mains using various approved materials and techniques requires:

- understanding how to access, read and determine large water service installation requirements from job specifications, relevant Australian Standards, codes, manufacturers' instructions and jurisdictional requirements
- identifying and applying workplace, work health and safety (WHS) and environmental requirements
- an understanding of different piping materials and their limitations
- using correct jointing techniques with the appropriate valves and ancillaries as well as adequate bedding and backfill requirements for a compliant installation
- installing adequate support and restraints
- testing pipework according to standards and manufacturers' instructions
- pipework is thoroughly cleaned and flushed before use
- completing all documentation according to relevant regulatory requirements and submitted upon completion of work.

REFERENCES

AS/NZS 3500.1 Plumbing and Drainage: Water Services
WSA 03-2011 Water Supply Code of Australia Version 3.1.

GET IT RIGHT

1 When testing a large diameter PVC water supply piping system with elastomeric ring joints, what specific measures must be taken before testing begins?

WORKSHEET 1

Student name: ____________________

Enrolment year: ____________________

Class code: ____________________

To be completed by teachers	
Student competent	☐
Student not yet competent	☐

Unit competency code/title: CPCPWT3029 Install water pipe systems

Task: Review 'Identify installation requirements' and 'Prepare for work', and then answer the following questions.

1 Define the term 'water main'.

____

____

____

2 When considering the bedding and support of a water main, what factors should be taken into account?

____

____

____

3 Pipes are normally delivered to a site as a bulk delivery. List six factors that should be considered by the contractor onsite when storing and stacking these pipes.

____

____

____

____

____

____

4 List five factors that should be considered when selecting pipe materials for a water main.

____

____

____

____

____

WORKSHEETS 2

5 Identify three hazards associated with installing large water pipe systems.

6 Why is polyethylene pipe (PE) a popular choice for large water pipe installations?

WORKSHEET 2

To be completed by teachers	
Student competent	☐
Student not yet competent	☐

Student name: ______________________

Enrolment year: ______________________

Class code: ______________________

Unit competency code/title: CPCPWT3029 Install water pipe systems

Task: Review 'Install large water service', and then answer the following questions.

1 Calculate the number of cubic metres in a trench excavation 60 metres long, 450 mm wide and 600 mm deep. *Note:* All values are to be expressed in metres.

Formula: V = L × W × D

Where:

V =

L =

W =

D =

Answer: V =

2 When taking delivery of the pipe and fittings, what action should the contractor take to ensure that the delivery is complete?

____

____

____

3 Most modern water mains piping systems are joined using elastomeric, neoprene or rubber rings that are moulded to fit into a machined recess in the socket of the pipe. The seal is then compressed between the spigot and socket ends of each pipe length.

Why is grease or soap not considered suitable as a lubricant for these items?

____

4 Water mains services with unrestrained elastomeric rubber ring joints installed below ground require thrust/anchor blocking. The joints that fit into this category are mostly used when installing ductile/cast-iron and the various PVC piping systems that use rubber ring jointing methods.

List five situations where these thrust/anchor blocks need to be installed.

____

____

____

____

____

5 Water mains systems require a number of in-line valves and ancillaries in order to operate correctly.

List four functions of these valves and ancillaries:

6 What is the purpose of a scouring point?

7 Describe the function of a sluice valve.

8 Describe the two functions of a kinetic air valve.

9 Why can cavitation be an issue in large water pipe systems?

WORKSHEET 3

To be completed by teachers

Student competent ☐

Student not yet competent ☐

Student name: ______________________

Enrolment year: ______________________

Class code: ______________________

Unit competency code/title: CPCPWT3029 Install water pipe systems

Task: Review 'Install large water service' and 'Clean up', and then answer the following questions.

1 Water mains systems require testing. When should testing be carried out?

____

____

____

2 Prior to this testing, some important procedures must be performed. List nine items that must be checked.

____

____

____

____

____

____

____

____

____

3 How are water main stop valves and hydrants easily located?

____

4 List the five items that should be checked once the pipeline is filled and pressurised.

____

____

____

____

____

5 Once the pipeline has been filled and before any test pressure is applied, it should be checked and inspected. Describe the rate of filling, how the line should be pressurised and what should be remembered when pressurising the line.

6 Once filled with water, the main should be allowed to remain full for a certain amount of time. State the minimum and the maximum desirable timeframes the main should be left full and left unpressurised, and why?

7 State how long cement-lined pipes should be left full and unpressurised and why.

8 State when testing should be avoided and why.

9 What should be considered when covering the work in, prior to testing and inspection?

10 What is the name of the equipment used to clean the pipeline from debris prior to filling it with water?

INSTALL PROPERTY SERVICE

3

Chapter overview

This chapter addresses the skills and knowledge required to install a water service from the utility's water main, up to and including installation of a water meter (property service), using approved materials and techniques. Information is provided for the installation of recycled water services.

Learning objectives

Areas addressed in this chapter include:

- identify installation requirements
- prepare for work
- install, test and commission property service
- clean up.

Water for drinking (potable), irrigation and firefighting services, as well as recycled water in more recent years, is delivered to most large communities through a network of private and public water mains systems. A water main is a pipeline that conveys drinking or recycled water throughout the community and is owned and maintained by the water supply utility or the local council.

This chapter outlines the installation of a **property service (main to meter)**. The **installation of a property service** involves the use of approved pipework and fittings for the supply of water to a property from the water main, up to and including the meter assembly, or to the stop tap if there is no meter. A single connection to the main and a single property service for each water service type (drinking and **recycled water systems**) must be provided for each property.

Identify installation requirements

In some new housing estates, a separate drinking (potable) and non-drinking (if available) water property service is installed for each property (see **Figure 3.1**). These services must only be connected to the water utility's appropriate water main. The minimum size of a residential property service is usually DN 20; however, the utility may allow a DN 25 service when the total length of the water service exceeds 30 m. The local authority's website will provide details and procedures for a main tapping (see, for example, **Figures 3.2**, **3.3** and **3.4**).

GREEN TIP

Drinking water is more precious and scarcer, so using a recycled water supply for toilet flushing and irrigation is an excellent way to help address this problem.

The water utility may approve a larger-diameter property service to accommodate local requirements in remote areas. These services are called **trunk service lines**, trunk service or **joint service**. Individual properties connect to this service the same way that a connection is made to the utility's water main.

FROM EXPERIENCE

A recycled water main is always located closest to the property boundary and is purple/lilac in colour.

The water utility may require a split service to feed two properties where the water main is located on the other side of the road (a **long service**). The service

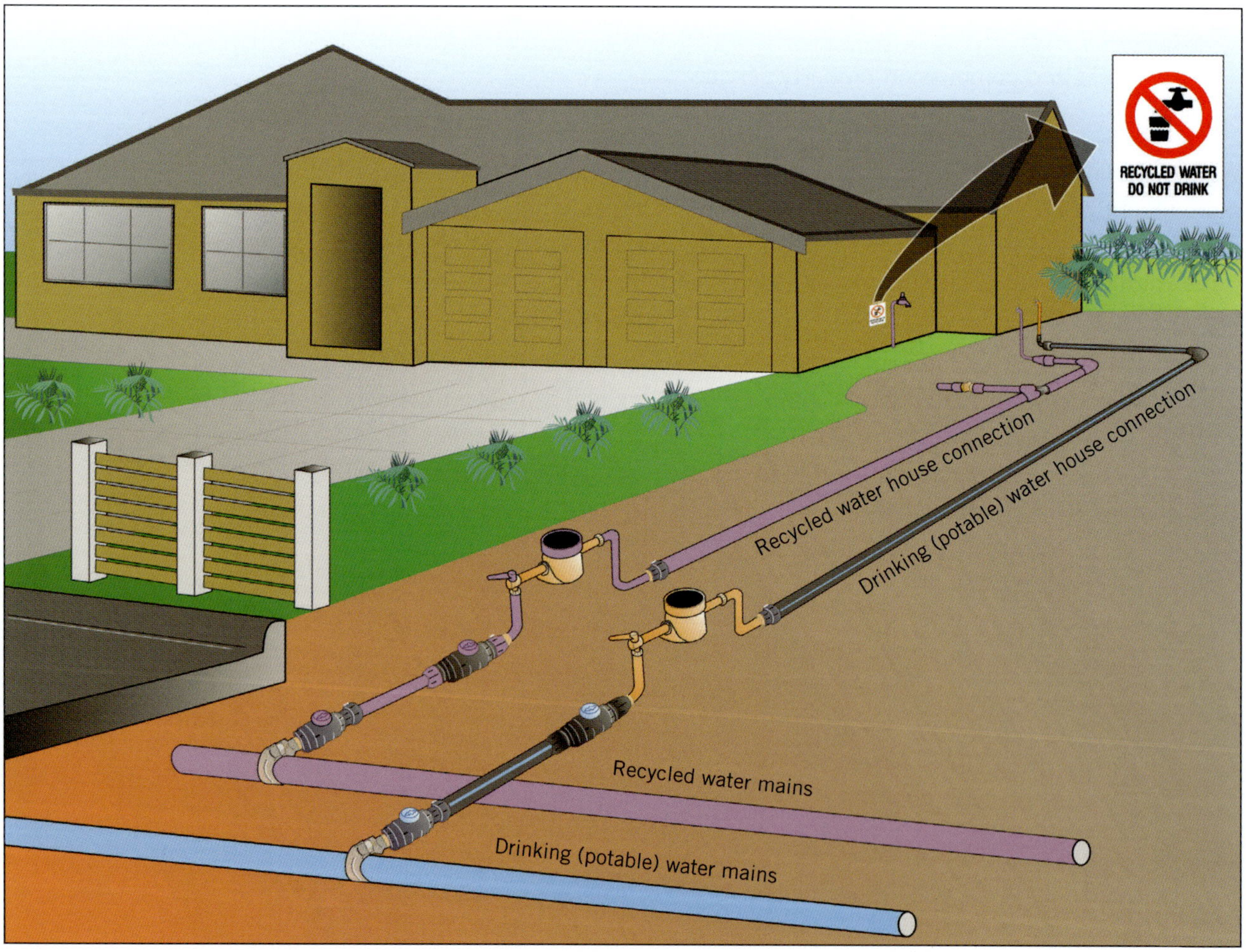

Source: © Sydney Water

FIGURE 3.1 Typical installation of a drinking and a non-drinking water service

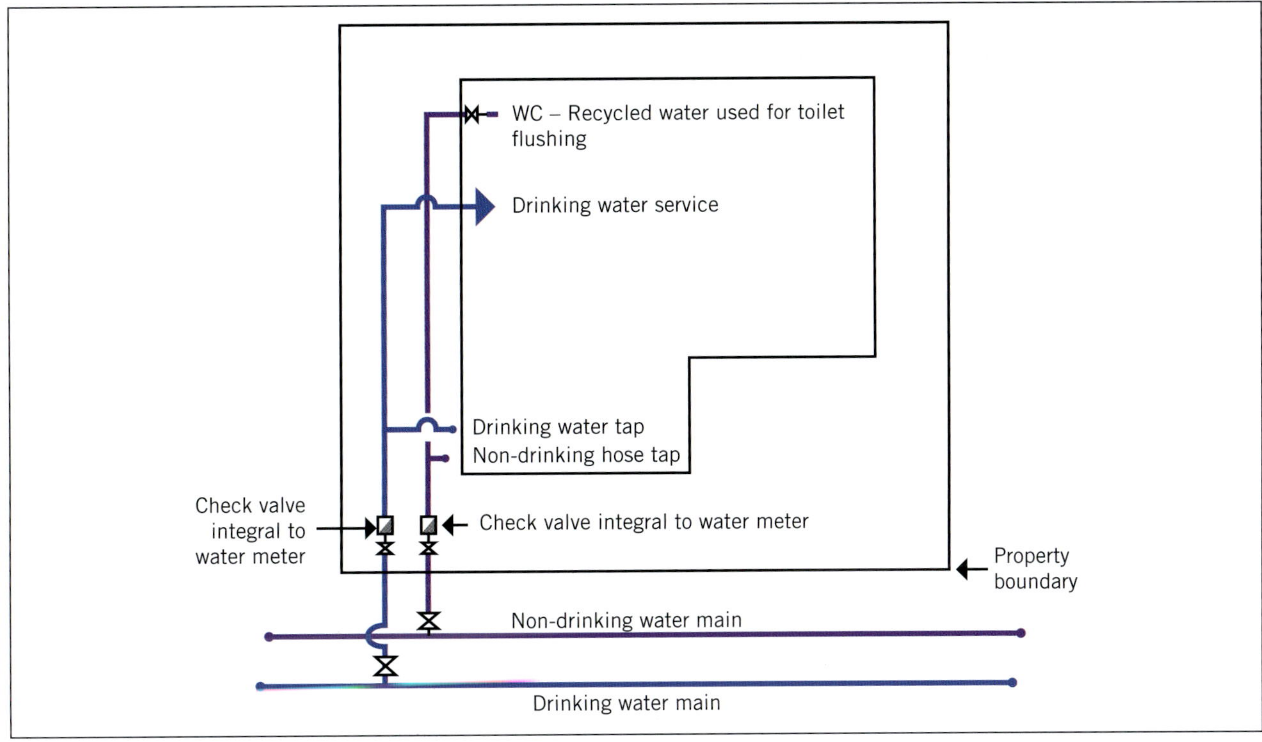

FIGURE 3.2 A recycled (non-drinking) water service and a potable (drinking) water service

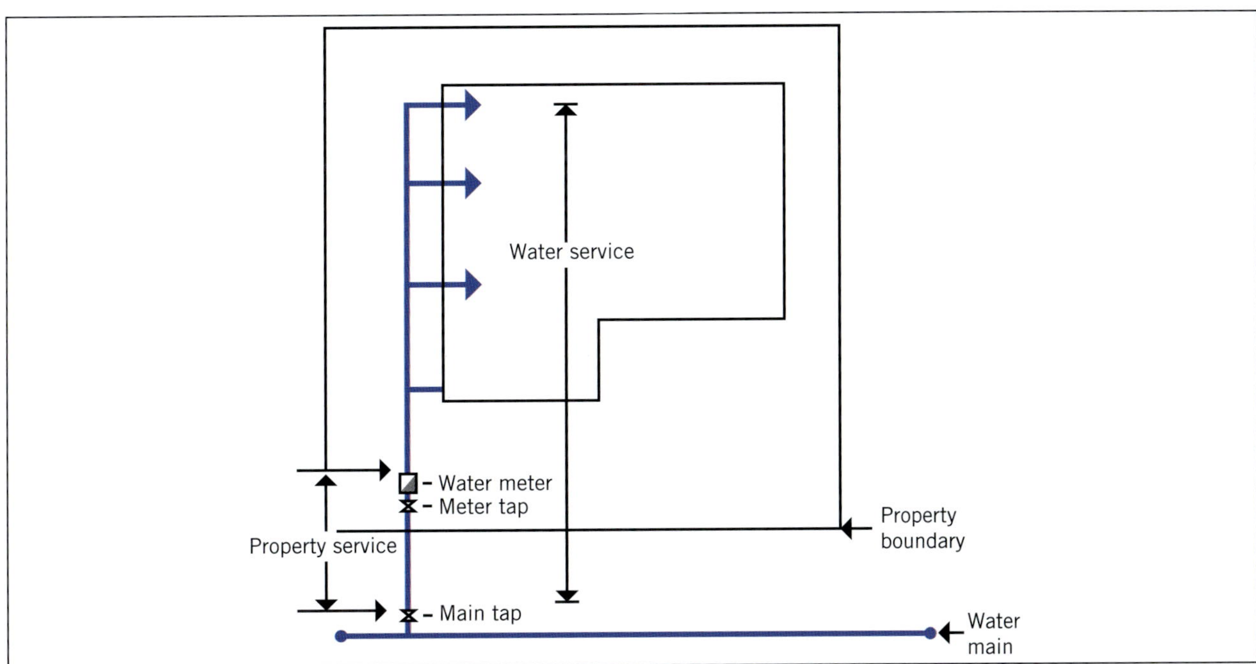

FIGURE 3.3 Property service and water service connected to water authority's water main

pipe then splits into individual services, feeding each property. A **short service** is where a property service is connected to a water main located on the same side of the street as the property. In this case, a split service may not be required.

LEARNING TASK 3.1

1 If a house property service exceeds 30 m, what diameter should it be?
2 Where should the water meter be located?

Single residential only application

(Office use only) Case no. ☐☐☐☐☐☐☐☐ P.S.P. no. ☐☐☐☐☐☐☐☐

Particulars of property

Lot no. ________ Street no. ________ Street name ________

Suburb ________ P/code ☐☐☐☐ Melways Ref. ________

An A4 scaled site plan is required to be submitted with the application detailing the building outline, and dimensions from the property foundations to the property boundaries and preferred sewer locations. **Note: If the application is not completed in full and/or a plan is not accompanying the application,** it may not be processed.

Particulars of owner

Name ________ Phone ________ Fax ________

Particulars of applicant

By lodging this application, the applicant warrants they are either the owner or occupier of the property, or they are authorised by the owner or occupier to make this application on their behalf.

Name ________ Phone ________ Fax ________

Address ________ P/code ☐☐☐☐

Signature ________ Date ________

Particulars of licensed plumber (Mandatory)

Name ________ Phone ________ Licence no. ________

Detail of work / fees

New connections (Must include potable water, recycled water and sewer when available) **Note:** Standard Connection is for 20mm only.

* ☐ Single dwelling sewer connection with dry tapping 20mm (Applies where recycled water is not available) Total $496.00
* ☐ Single dwelling sewer connection with wet tapping 20mm Total $458.00

New connections which include class 'A' recycled water

* ☐ Single dwelling sewer connection with dry tapping 20mm and class 'A' recycled water
  Dry tapping connection (If available) 20mm Total $1,270.40

Water tanks

* ☐ Does the development include below ground water tank interconnected with the drinking water supply? Yes ☐ No ☐

If 'yes', please specify ________

Dry tapping Is the site available for dry tapping installation? Yes ☐ No ☐

Note: Dry Tapping location must be accessible and clear of obstructions, if not a re-booking fee may apply.
Note: 20mm Dry Tappings are installed (Includes Potable and Class 'A' Recycled Water Connections). This includes the meter assembly and garden hose bib tap. Where Class 'A' Recycled Water is utilised, mandatory inspections are required. Inspection fees total $323.40 which is included in the fees. *GST of $29.40 is included in this charge.
Note: Contribution fees may apply to some backlog (Septic to Sewer) connections.

Other - sewer only

* ☐ Sewer only (Demolishing & rebuilding - retain water meter)	$45.00	* ☐ House extensions (Recycled)	$45.00
* ☐ Septic to sewer connection	$45.00	☐ Drain alteration / cut & seal	$45.00
* ☐ House extensions	$45.00	Details of work ________	

Water only **Note:** All application fees and charges are GST free. Vacant land contribution fees may be applicable.

* ☐ Dry tapping (20mm only) - sewer unavailable	$451.00	☐ Wet tapping (20mm only)	$413.00
* ☐ Plugging & re-tapping (Wet area & 20mm only)	$463.00	☐ Plugging only (20mm only)	$125.00

Method of payment

Cash ☐ Cheque ☐ Credit card ☐

Card type Mastercard ☐ Visa ☐ Card holder's name ________

Card no. ☐☐☐☐ ☐☐☐☐ ☐☐☐☐ ☐☐☐☐ Expiry date ____ / ____ / ________

Signature ________

Please notify me by

Mail ☐ Fax ☐ Email ☐ Email address ________

FIGURE 3.4 An example of a plumbing application form (although most applications are now completed online)

Access, read and determine property service installation requirements from job specifications, relevant Australian Standards, codes, manufacturers' instructions and jurisdictional requirements

It is the plumber's responsibility find out the local water authority's regulations as they can differ between areas. When working in an unfamiliar area, the plumber must contact the local water authority to understand the local regulator's property service installation requirements. It is also important to check job specifications and manufacturer's instructions for specific requirements. The minimum standard requirement is outlined in AS/NZS 3500.1 and this must not be compromised. Refer to Chapter 4 'Carry out interactive workplace communication' in *Basic Plumbing Skills* for further information on how to access relevant jurisdictional regulations and other standards.

Site inspection

A site inspection allows you to visualise any areas relevant to the task that may not be apparent from the plans and specifications, such as:

- access
- traffic conditions
- site conditions
- water main location
- suitable position of the meter (as required by the local water utility)
- obstructions that may impede the installation
- hazard identifications
- confirmation of the information obtained from the plans and specifications.

A risk analysis during the initial site inspection will identify all the apparent hazards. The risks can be assessed so the appropriate control measures can be planned and put in place.

Safety (WHS) and environmental requirements

This sequencing is an ideal opportunity to carry out a risk assessment and prepare a safe work method statement (SWMS) for any high-risk activity and, for every task, a job safety analysis (JSA) must be completed. Recording and reviewing these procedures will help to refine and enhance your company's quality assurance procedures. The sequencing and planning will also identify what safety equipment, personal protective equipment (PPE) and tools are needed for the task.

Identify cables, conduits, pipes or other services

Before carrying out any excavation within public property (road or footpath), it is essential that a check be made to identify and locate all underground services. This is particularly important when using mechanical equipment. It is advisable to contact Before You Dig Australia (http://www.byda.com.au; see **Figure 3.5**) to arrange the location information for services to be sent to you from the utilities, such as:

- electricity cables
- stormwater mains
- sewer mains
- water mains
- communication (telephone and television) cables
- gas mains.

Due to the danger to life and amenities, *no work should proceed* until this process has been thoroughly completed, the services identified and the area marked (see **Figure 3.5**).

iStock.com/Hailshadow

FIGURE 3.5 Footpath marked by a locating contractor in readiness for work

FROM EXPERIENCE

When excavating near existing services that are highly valued and easily damaged, such as optic fibre communication cables, 'non-destructive excavation' is recommended.

Traffic control and road opening permits

When working in an area where there is any type of traffic movement, properly trained personnel must be employed to arrange traffic management plans and control the flow of traffic in and around the area. This is a WHS requirement in the workplace. Failure to implement proper traffic control could be an infringement of the relevant WHS Act applicable to the workplace and result in prosecution or a fine. This is usually part of the company's quality assurance policy to ensure a safe work procedure.

When excavation works are to be carried out in a public area, all applications and permits should be obtained from the relevant authority, such as the local council or road and traffic authority. This usually requires the payment of a restoration fee. This may also affect the tender price, as each authority uses a different scale when calculating this fee. The rates can differ quite substantially in different areas. The costs are calculated on the type, load capacity and size of the road. For example, a main highway with a concrete surface would be very expensive to restore. Local emergency services, such as the police, fire brigade and ambulance services, may also need to be notified, especially if roads are to be closed, as this may affect the emergency routes to hospitals, fires and other emergencies.

The flowchart in Figure 3.6 illustrates the process for determining the planning for the installation of a property service.

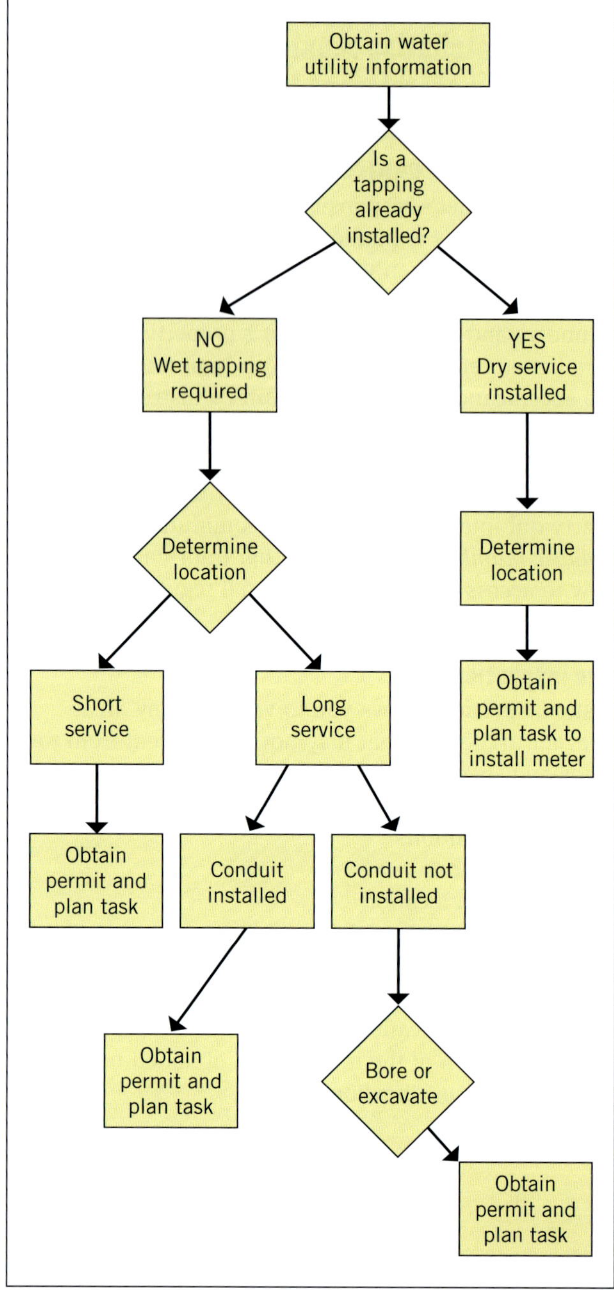

FIGURE 3.6 Installation flowchart

HOW TO

CARRY OUT PRELIMINARY WORK BEFORE THE PROPERTY SERVICE INSTALLATION BEGINS

The owner or licensed plumber must do the following:

1 Pay the appropriate fees and obtain a road opening permit from the local council. Familiarise yourself with the local council and water authority applications and requirements.
2 Obtain a copy of the plans, which must be approved by both the local council and the water utility.
3 Obtain appropriate consent forms (plumbing permit) and pay the required fees to the water utility.
4 Understand the local council or water utility's requirements as they may differ depending on the area.
5 Carry out a Before You Dig Australia enquiry.

Determine the location of property service and the main

Before either arranging or making a tapping in the main, licensed plumbers with a main drilling accreditation must find a hydrant or valve cover in either the footpath or the road. It is important to ensure that it is the correct main, as some streets may have more than one water main laid in them. A plate affixed to a wall, fence or telegraph pole identifies the location of the main (see Figures 3.7 and 3.8). This plate is marked in a series of different abbreviations that will help you determine the location (see Table 3.1). It also has stamped into it the size of the main and the distance from the plate to the main. Other local water utilities may use a cat's eye (reflector) affixed to the centre of the road to identify a hydrant location (see Figures 3.9 and 3.10).

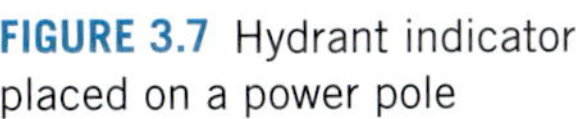

FIGURE 3.7 Hydrant indicator placed on a power pole

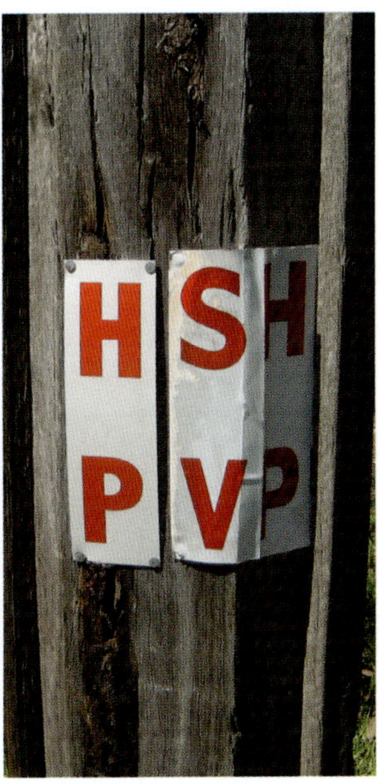

FIGURE 3.8 Hydrant valve indicators placed on a power pole

TABLE 3.1 Hydrant/service valve identifying abbreviations

Abbreviation	Meaning
HR	Hydrant in road
HP	Hydrant in footpath
AV	Area valve
DV	Dividing valve
SV (or SVP)	Service valve in footpath
SV (or SVR)	Service valve in road

In the planning and preparation stage, it is important that a JSA and a SWMS are carried out addressing the excavation method, whether trench support is required and the type of barricades needed.

To obtain the line of the main you need to repeat the above steps up or down the road. The line between the two hydrants or valves is the line of the utility's water main.

LEARNING TASK 3.2

1 Where does the information to locate underground services come from?
2 What type of excavation is advisable to expose optic fibre cables?
3 How do you precisely locate a water main?

Prepare for work

Being prepared and planning well will help to avoid costly problems and help the job to run more efficiently.

Source: Ray-O-Lite Aust Pty Ltd.

FIGURE 3.9 Drinking water hydrant indicator

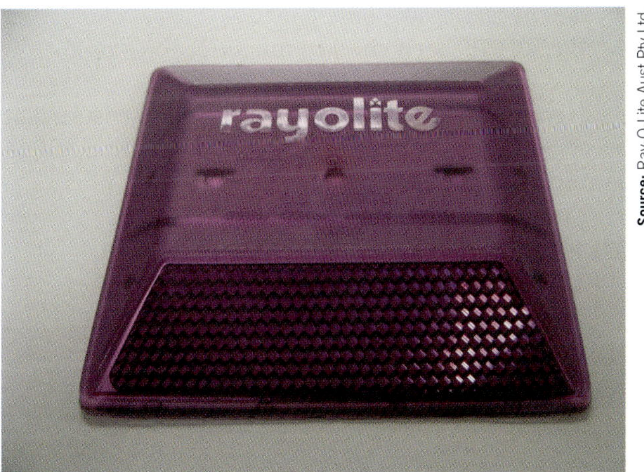

Source: Ray-O-Lite Aust Pty Ltd.

FIGURE 3.10 Non-drinking water hydrant indicator

FROM EXPERIENCE

Planning ahead and being well organised results in time and cost efficiency.

Prepare the work site

To protect the workers and the public, the work site must be clearly defined and secured to prevent unauthorised access. Special consideration should be given to vehicular traffic, pedestrian traffic and deliveries of material and equipment.

Planning the work around vehicular traffic requires specific requirements such as traffic control, signage, barricades and road plates.

GREEN TIP

Be prepared to place silt and sedimentation barriers in place prior to excavation to protect the environment from pollution. Heavy fines could be incurred otherwise.

Before starting work

Before commencing any installation work, a final check of the information gathered at the planning stage should be made. Some companies will have a checklist for this procedure as part of their quality assurance program. The checklist should include, but is not limited to, the following questions:

1 Are you familiar with the local authorities' regulations and requirements?
2 Do you have a copy of the approved plans and specifications?
3 Do you have the appropriate permits and have all the fees been paid?
4 Have the materials ordered been checked for compliance and quantity?
5 Are all the tools and equipment required available and in serviceable order?
6 Has all PPE been checked for correct operation and condition?
7 Have you made a final site inspection to ensure that there have been no alterations or obstructions since the planning inspection?

This is also your last chance to check the calculations used to determine pipe size and material compliance to standards, codes and regulations.

Job sequencing

Once all the required information is obtained, the job is planned by sequencing it into logical steps (see **Figure 3.11**).

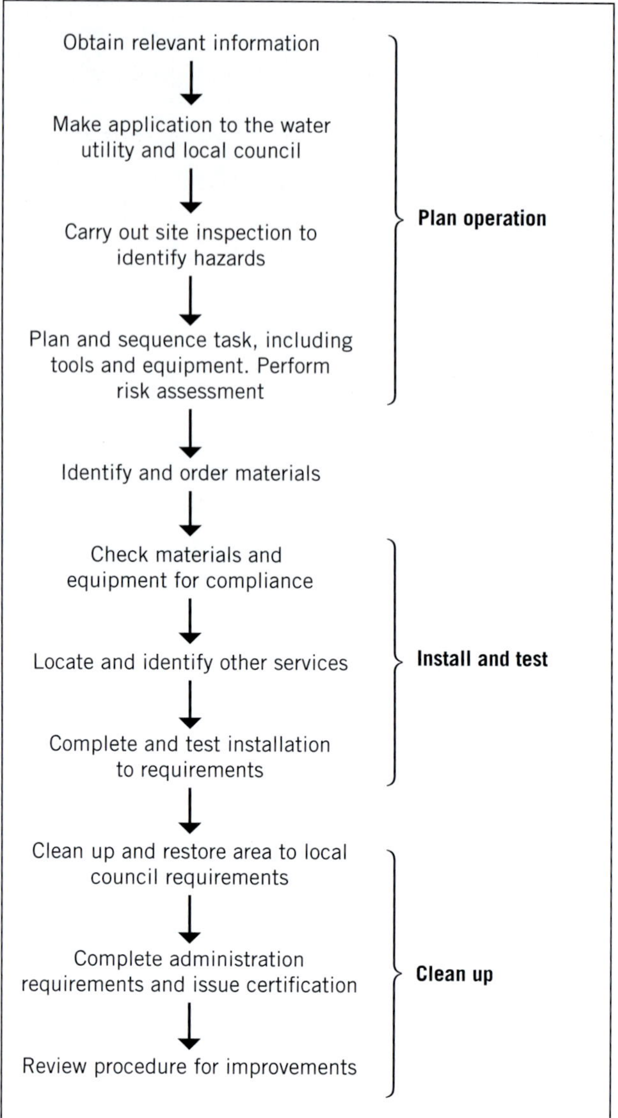

FIGURE 3.11 Installation sequence

Create a materials list and collect materials

It is important to take the time to create an accurate list of materials, so time is not wasted and the job is completed as planned.

When material is delivered, check the order is correct so any problems can be sorted straight away and prevent lost time.

Pipe material

Each material has its advantages and disadvantages. Being able to recognise the limitations of each type of material will help ensure the longevity and serviceability of the installation. While the materials listed in AS/NZS 3500.1 and covered in Chapter 1 of this book are approved for use in a property service, the local water utility may restrict the type of materials permitted to be used. For example, within Sydney Water's (NSW) area of operation, only copper (type A and B) and polyethylene (PE) material may be used. PE pipe and fittings must be manufactured to AS/NZS 4130 and rated to a minimum of PN 12.5.

LEARNING TASK 3.3

The job is to install a property service at a customer's house. The water main has already been tapped and the meter location is 12 metres away from the water main.

The property service will be installed in 25 mm polyethylene pipe, and a water meter and hose tap installed ready for the house to commence construction.

Create a materials, tool and PPE list to complete this task (teachers note: the list may vary slightly depending on the student's interpretation, and teacher's discretion should be used when marking).

Pipe equivalency tables are listed in AS/NZS 3500.1, showing the appropriate nominal diameter (DN), based on internal diameter for different pipe materials.

Learners should check with their supervisor/ instructor for the sizing requirements in their local area.

AS/NZS 3500.1 PLUMBING AND DRAINAGE: WATER SERVICES

AS/NZS 4130 POLYETHYLENE (PE) PIPES FOR PRESSURE APPLICATION

Copper (types A and B)

Copper is required if pressures are more than 1200 kPa. Copper may be used on most property services and is essential to use for risers to the meter assembly. Some water utilities prefer the use of PE, as it provides insulation and corrosion protection for metallic water mains. Types A (green) and B (blue) copper tubes are the only types of copper tube allowed for use between the main and meter (property service).

Polyethylene (PE)

PE has the following classification:

- allowable size range DN 20, 25, 32, 40, 50 and 63
- maximum pressure 120 m head (1200 kPa) for type PN 12.5.

Pipe identification for PE pipes is as follows:

- drinking water (non-dual water areas): black with blue stripes.

The following colours cannot be used outside of **dual water areas**:

- drinking water (dual water areas): blue
- recycled water: black with purple stripes.

PE pipe must not be used where it is subject to direct sunlight, as part of the water meter assembly or vertical riser, or as specified in AS/NZS 3500.1 Plumbing and Drainage: Water Services and AS/NZS 4130 Polyethylene (PE) Pipes for Pressure Application. It must be a single length of pipe and be free of joints or fittings between the main isolation valve and the service connection valve or meter riser.

AS/NZS 3500.1 PLUMBING AND DRAINAGE: WATER SERVICES

AS/NZS 4130 POLYETHYLENE (PE) PIPES FOR PRESSURE APPLICATION

Only copper pipes may be used as part of the meter riser; where copper pipes are used as part of a non-drinking water service, they must be sheathed with purple PE. The non-drinking water pipe must have a 75 mm purple identification tape installed on top of the pipe, running longitudinally, and fastened to the pipe at not more than 3 m intervals.

All pipes and fittings must display a standards mark **WaterMark** as required by AS/NZS 5200, as applicable.

Select appropriate tools and equipment including personal protective equipment (PPE)

PPE may include, but is not limited to, any of the in-depth list of PPE that can be found in Chapter 7 'Handle and store plumbing materials' in *Basic Plumbing Skills*.

Tools and equipment may include, but are not limited to, any of the in-depth list of tools that can be found in Chapter 8 'Use plumbing hand and power tools' in *Basic Plumbing Skills*.

FROM EXPERIENCE

Showing initiative by selecting the correct tools and equipment for the job is a valued skill.

LEARNING TASK 3.4

1 Name the two risk assessment documents to be completed before starting any high-risk activity.
2 State the Australian Standard that applies to a property water service.

COMPLETE WORKSHEET 1

Install, test and commission property service

This section explains how to prepare and carry out a main tapping, install the property service (main to meter), install a water meter, and test and commission the property service.

Mark out and excavate for installation of property service

Mark out the excavation and be sure that the meter is directly opposite and at a right angle (90°) to the main. This is important so the main tap can be easily located in the future. Where the excavation crosses any identified other services, they must be located by hand (excavation) and identified to prevent costly damage, delays and possible fatality.

The following requirements must be met:

1. Any excavation for connection to water mains must be in accordance with the water utility's requirements (for an example, refer to **Figure 3.12**). The excavation for drilling must be a minimum of 1 m × 1 m in size for a single drilling, with a clearance below the main of a minimum of 150 mm.
2. The excavation must be maintained in a dewatered condition.

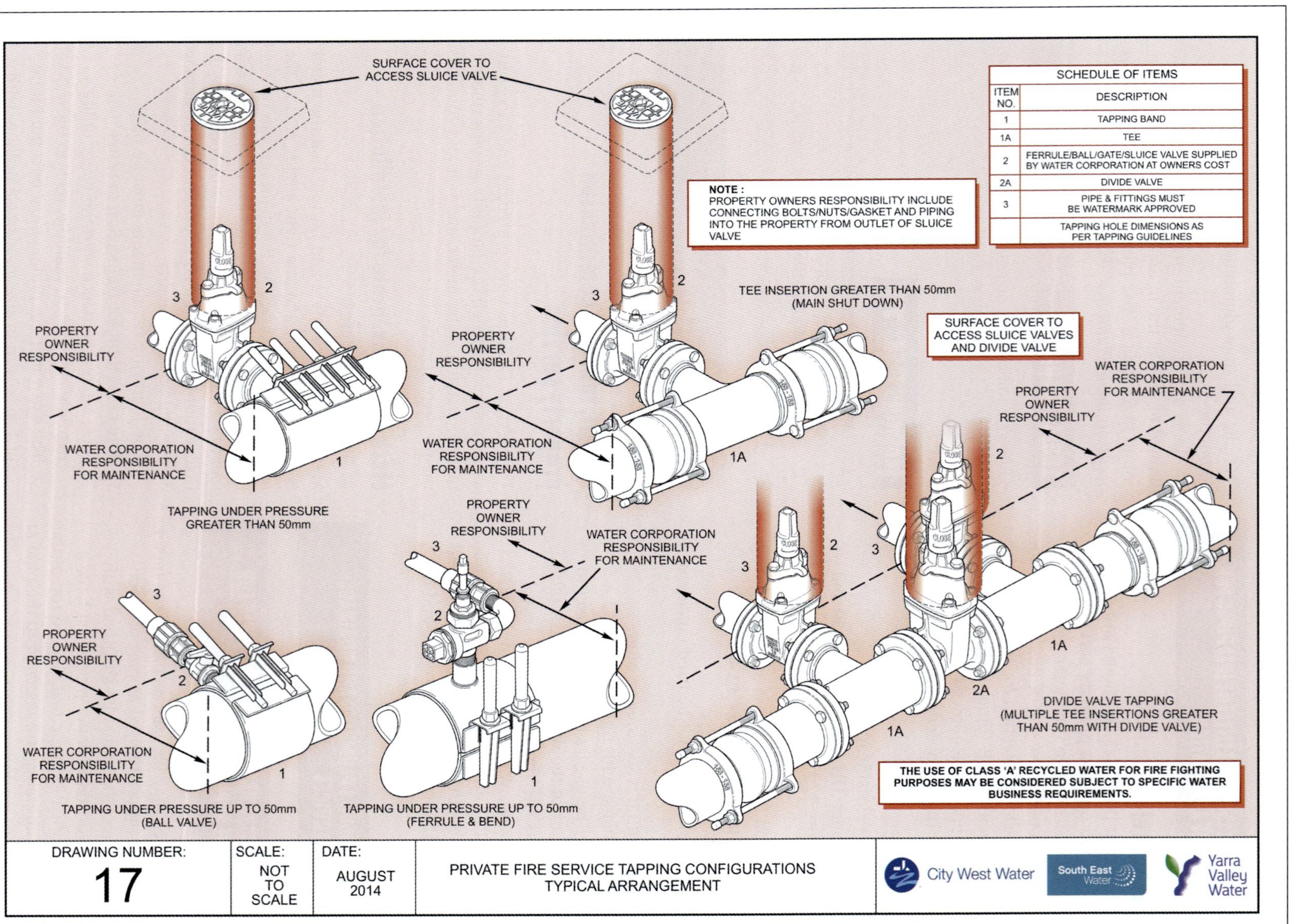

FIGURE 3.12 Installation of a tee and valve

Source: South East Water. This diagram is current at time of printing. It appears for training purposes only and should not be reproduced without the express permission of the owner, South East Water.

3 The excavation from the water main into the property can be an open trench if it is close by. However, if the water main is in or on the other side of the road, an under-road boring machine should be considered.

4 Drilling of the main or installation of the tee and valve must be carried out as required by the water utility (for an example, refer to **Figure 3.13**).

5 If a protective sleeve around a cast iron/ductile iron main is damaged, it must be repaired to the water utility's specifications by the licensee.

The size of the property service and meter

The diameter of a property service is sized to meet the demand and flow rate requirements of the user or users. This will be covered within the Certificate IV unit CPCPWT4011 Design and size heated and cold-water services and systems. Property service connections up to DN 65 are connected to the water utility's main via a single tapping or twin tappings (drillings). Larger services are connected by a tee and valve inserted in the main (see **Figures 3.12**, **3.13**, **3.14** and **3.15**).

Remember to check the local authority's guidelines prior to commencing any works.

FIGURE 3.13 Tee and valve main connection

FIGURE 3.14 Main tap connection on PVC pipe

Each state and territory may have supply tables that indicate the:

- drilling size requirements
- size of the water service
- piping material type
- total length of the water service for drilling size
- size of the drilling
- number of drillings for the property service size
- size of the water meter.

FIGURE 3.15 Twin main tap connection

Tapping (drilling) the main

The tapping (drilling) can be made as either a 'dry tapping' or a 'wet tapping'.

- **Dry tappings** are usually made during the development stage of a subdivision prior to pressurisation of the main before the roads are sealed. The property service is terminated as described earlier in this chapter.
- **Wet tappings** are carried out on pressurised water mains as directed by the water utility.

As a general rule, drinking water mains are typically made from polyvinyl chloride (PVC), high-density polyethylene (HDPE) and cast iron/ductile iron; however, other materials may have been used in the past. For example, while asbestos cement (AC) mains are no longer installed, there are many still in existence. When tapping AC mains, special precautions must be observed as older mains are likely to contain asbestos.

The proper PPE and hazardous materials handling must be used when working with asbestos. Refer to the safety data sheet (SDS) for asbestos.

The connection of a main isolation (ferrule) valve to a cast iron/ductile iron main is made via drilling and tapping a thread in the wall of the pipe; or more currently by a **tapping saddle** and then drilling the main, as shown in **Figure 3.16**. The connection of a property service to a PVC and PE main is made via a tapping saddle and drilling the main, as shown in **Figure 3.17**.

FIGURE 3.16 Connection to a ductile iron drinking water main

FIGURE 3.17 Connection to a PE recycled water main

HOW TO

INSTALL A TAPPING SADDLE

1 Prepare the area of pipe to be covered by the tapping saddle. Be sure it is clear of any joints or fittings.
2 Position the bottom half of the tapping saddle, making sure it is away from scored, pitted or damaged areas, as this will not provide a good seating area.
3 Fit the top of the tapping saddle around the pipe so that the bolts pass through the bolt holes in the top of the tapping saddle.
4 With one hand underneath the bottom half of the tapping saddle and pressing upwards, locate the bolt holes and screw on the nuts supplied with the other hand.
5 Finger-tighten the nuts so that the gap between the two halves of the tapping saddle is equal on both sides.
6 Tighten the nuts to 20 Nm (15 ft/lb) (see **Figure 3.18**).

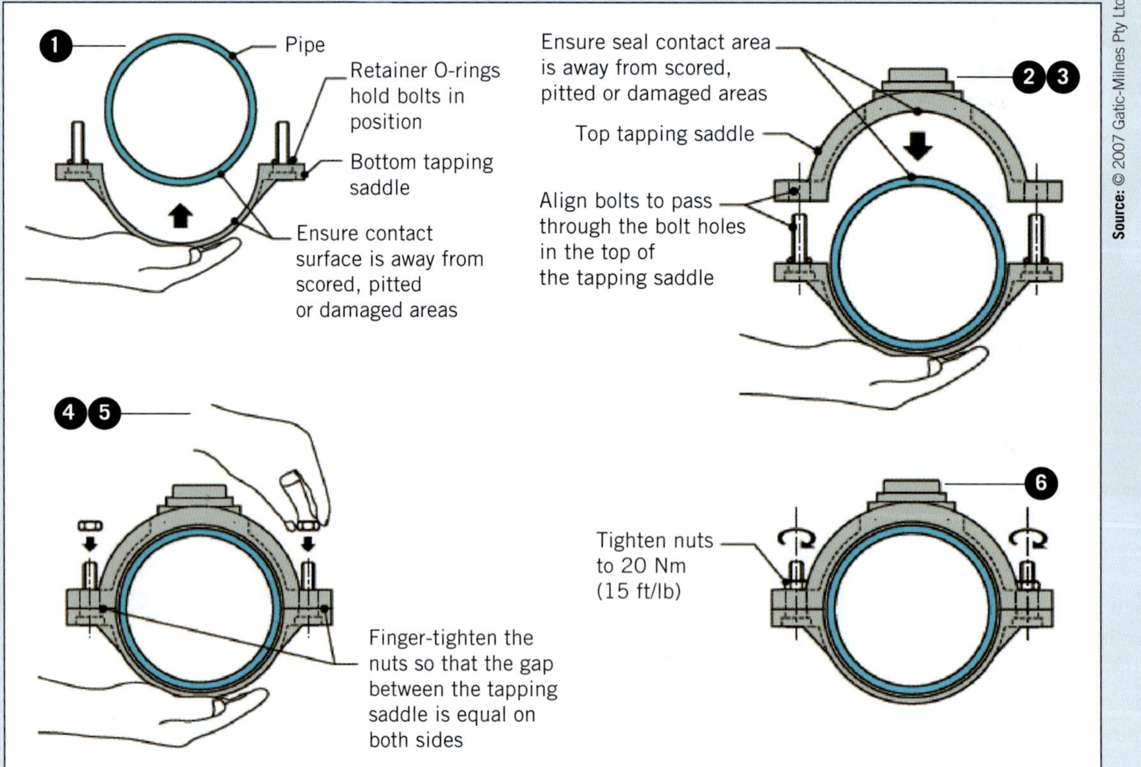

FIGURE 3.18 Installing a tapping saddle

Only licensed plumbers with the required qualifications and approved by the water utility are allowed to carry out either wet or dry tappings. Tapping a water main already under pressure allows services to be connected without interrupting the water supply for existing consumers.

Additional requirements for lip seals

When installing tapping saddles with lip seals, make sure the lip seal is in the correct position. The lip seal should be placed in the sealing groove with the sharp inner side of the lip seal facing out towards the pipe. A small moulding line running around the lip seal also indicates this should be facing down towards the pipe. If the lip seal is placed in the incorrect position, the joint may leak.

Seal installation

An O-ring is installed as shown in Figure 3.19.

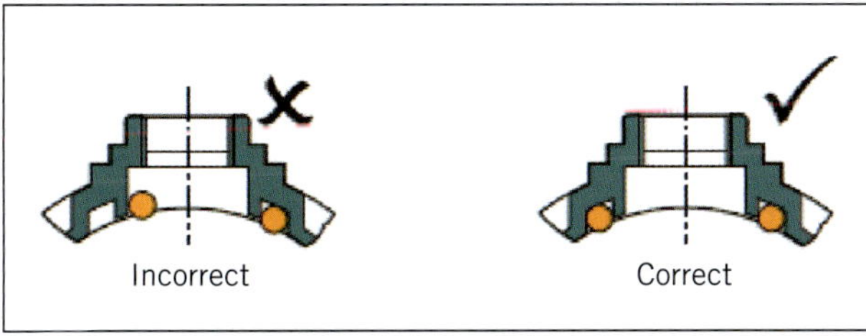

FIGURE 3.19 O-ring installation

Standard hand-operated under-pressure tapping machines allow tappings of 20 mm and 25 mm to be made in all pipe types without the need for electrical power in remote locations or on new construction sites.

Important operating instructions

Regardless of the machine used, always follow these important operating instructions:

- Operate by hand only; added leverage will damage the machine and the drill bit.
- Minimise sideways movement during operation; rocking of the machine will damage the drill bit.
- Always limit the feed rate to ensure a smooth cutting action; the feed nut may need to be restrained to prevent the drill tip from jamming.
- Do not use the upper handles to retract the feed nut; this may loosen the bearing cap, which, if removed under pressure, will cause the following results:
  - dislodgement of the circlip
  - damage to the bearing
  - injury to the operator.
- Keep the bearing cap screwed on tightly.
- Never place your head directly over a live tapping as it can blow off if bumped or damaged in the installation/excavation process.

Tapping small-diameter copper mains

The following procedure is the usual method used for connecting a property service to a copper trunk main:

1 Tapping to a small-diameter copper main is made by inserting a brass tee with a female iron (FI) centre outlet in the main, using a compression coupling, press-fit or by silver soldering.
2 The FI outlet and the ferrule should be kept in a vertical position to allow the service pipe to be connected as early as possible.

Insulating bush

All metallic water services *must* be insulated from the main using an insulating bush either:

1 fitted between the main tap and the property service, or
2 incorporated into the tapping saddle, into which is screwed the main tap.

The insulator is designed to prevent corrosion between the dissimilar metals (cast iron/ductile iron and copper) and to help prevent stray electrical currents passing from the property services to the main, thus reducing the life of the main and possibly causing it to become live, should an electrical fault exist.

HOW TO

DRILL TAPPING AND FIT MAIN COCK

1 Fit an appropriate tapping band to the pipe and screw in the selected under-pressure tapping ferrule.
2 Unscrew the bonnet assembly, remove the jumper valve and fully open the poly ferrule plug. (A ball valve may be used in lieu of the under-pressure tapping ferrule where acceptable.)
3 Screw the tapping machine into the top of the ferrule, using the correct adapter as required.
4 Rotate the feed nut until the drill tip makes contact with the pipe.
5 Apply pressure to the drill bit via the feed nut, turning the ratchet at the same time to cut the hole.
6 When drilling is completed, reverse the feed nut to retract the drill bit until clear of the poly ferrule plug.
7 Allow water escaping from the ferrule to flush the system, then close the poly ferrule plug. (A hose may be connected to the ferrule to divert the waste water.)
8 Close the poly ferrule plug, remove the tapping machine and replace the jumper valve and bonnet.
9 Connect the service pipe and open the poly ferrule plug to charge the service line (see Figure 3.20).

>>

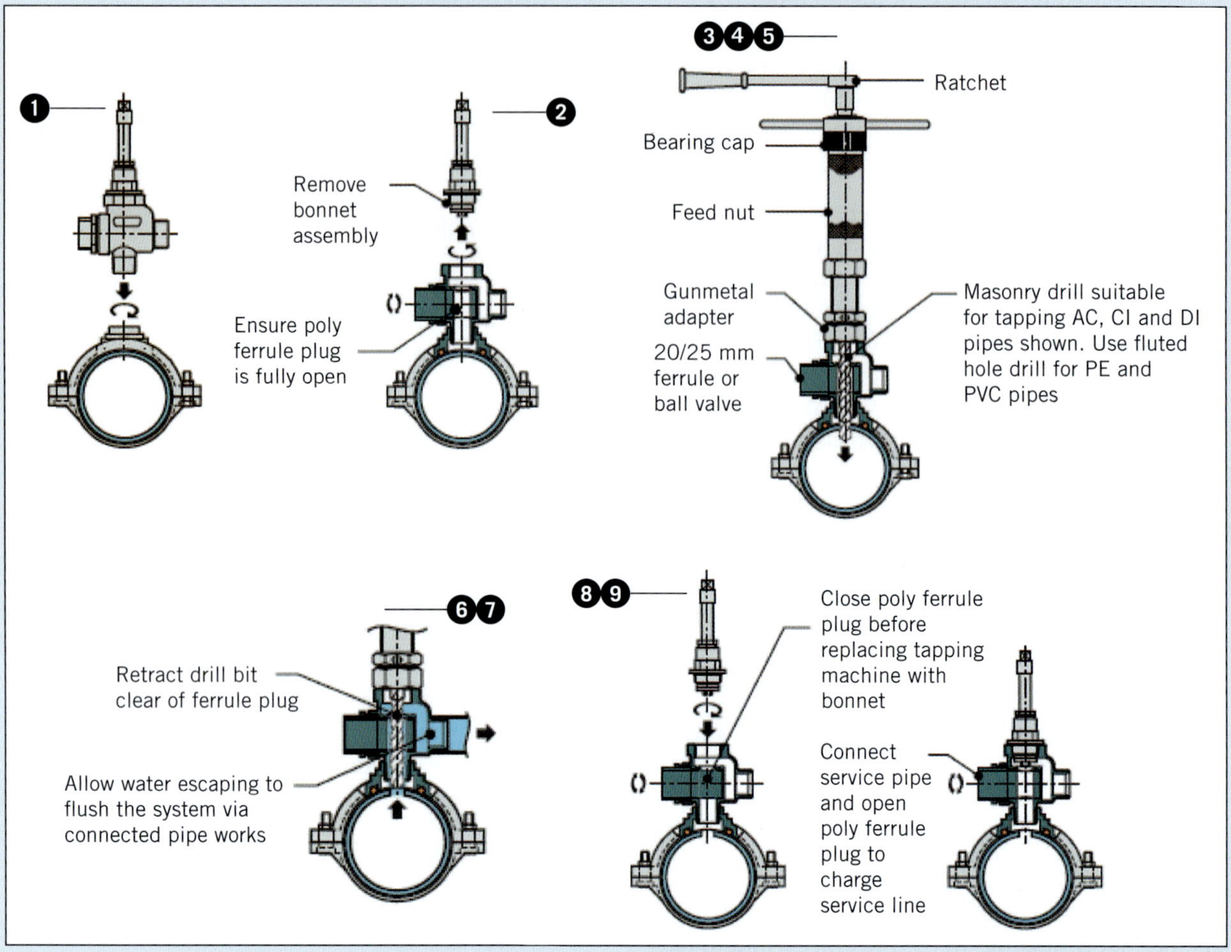

Ratchet

Bearing cap

Feed nut

Gunmetal adapter

20/25 mm ferrule or ball valve

Masonry drill suitable for tapping AC, CI and DI pipes shown. Use fluted hole drill for PE and PVC pipes

Remove bonnet assembly

Ensure poly ferrule plug is fully open

Retract drill bit clear of ferrule plug

Allow water escaping to flush the system via connected pipe works

Close poly ferrule plug before replacing tapping machine with bonnet

Connect service pipe and open poly ferrule plug to charge service line

Source: © 2007 Gatic-Milnes Pty Ltd

FIGURE 3.20 Tapping sequence

LEARNING TASK 3.5

1 Who is authorised to carry out main tappings?
2 How are water services larger than 65 mm connected to the water main?
3 What is the purpose of an insulating bush?

Determine the water meter location

Water meters that are used for billing purposes (see Figure 3.21) must be installed:

- within the property
- as near as practicable to the street alignment
- immediately downstream of the meter isolating valve
- directly opposite the connection and at a right angle (90°) to the main
- at the front of the property within easements.

Where the property service is required to be offset within a private easement or right-of-way, a below-ground isolating valve must be provided downstream of the offset:

- as required by the relevant water utility in other locations

FIGURE 3.21 Water meter installation

- horizontally
- so that it is protected from damage – if located in frost-sensitive areas, meters must be protected against damage caused by freezing of the water. This can be achieved by installing the water meter underground in an accessible box.

It must be readily accessible for reading, maintenance or removal, and be clear of obstacles; and meters DN 50 or larger must be supported independently of the piping.

FROM EXPERIENCE

Learners are reminded to check the local authority's guidelines prior to commencing any works. See Figures 3.22, 3.23 and 3.24 for water meter variances.

Some water utilities may have other specific requirements, such as:

- a minimum clearance of 300 mm is to be left under the meter (though water utilities in some areas may only require 150 mm clearance under the meter)
- a minimum clearance of 300 mm is required when installing containment backflow prevention devices (at the meter)
- water meters must be installed within 1 m of the front boundary and within 1 m and parallel to the side boundary.

AS/NZS 3500.1 PLUMBING AND DRAINAGE: WATER SERVICES

Recycled water meter installation requirements

Non-drinking (recycled) water meters must be:

- purple in colour
- a minimum of 300 mm from the drinking water meter
- installed with a minimum of 150 mm clearance between the underside of the meter and the finished ground level. Additional clearance (150 mm) is required if a testable backflow device is installed on the outlet side of the meter.

Check with the local water utility to ascertain the local installation requirements and whether special inspections are required in its recycled water area. Generally, in recycled water areas, mandatory inspections are required at three stages:

1. meter to dwelling
2. rough-in
3. commissioning and fit-off.

Inspections for meter to dwelling and rough-in may be combined.

AS/NZS 3500.1 PLUMBING AND DRAINAGE: WATER SERVICES

LEARNING TASK 3.6

1. How can water meters be protected in frost-prone areas?
2. What is the minimum clearance under a water meter?

COMPLETE WORKSHEET 2

Installing a property service

Once the main tapping and water meter location has been established, the property service (main to meter) is installed. This water service connects from the main tapping. Follow the 'How to' procedure.

AS/NZS 3500.1 PLUMBING AND DRAINAGE: WATER SERVICES

Corrosive areas

Corrosive areas are those that contain compounds consisting of magnesium oxychloride (magnesite) or its equivalent, coal wash, ash, sodium chloride (salt), ammonia or materials that may produce ammonia. A benign soil is typically considered to be a sandy, free-draining soil of low salt content that is non-corrosive. Anything else should be treated as a potentially aggressive/corrosive area. If copper tube is being used, it will need to be protected by insulating the pipe with a plastic sleeve (lagging) if installed in a corrosive area.

Depth of cover

Any water service located below ground on public and private property must have a minimum cover as indicated below as AS/NZS 3500.1 states, unless otherwise stipulated by the local water utility or council:

- 75 mm under concrete slabs and footings
- 300 mm not subject to vehicular loading (excluding fire services)
- 600 mm for fire services not subject to vehicular loading
- 450 mm subject to vehicular loading but not under a carriageway
- 600 mm subject to vehicular loading under a sealed carriageway
- 750 mm subject to vehicular loading under an unsealed carriageway
- 750 mm for pipes in embankments or subject to construction equipment loads.

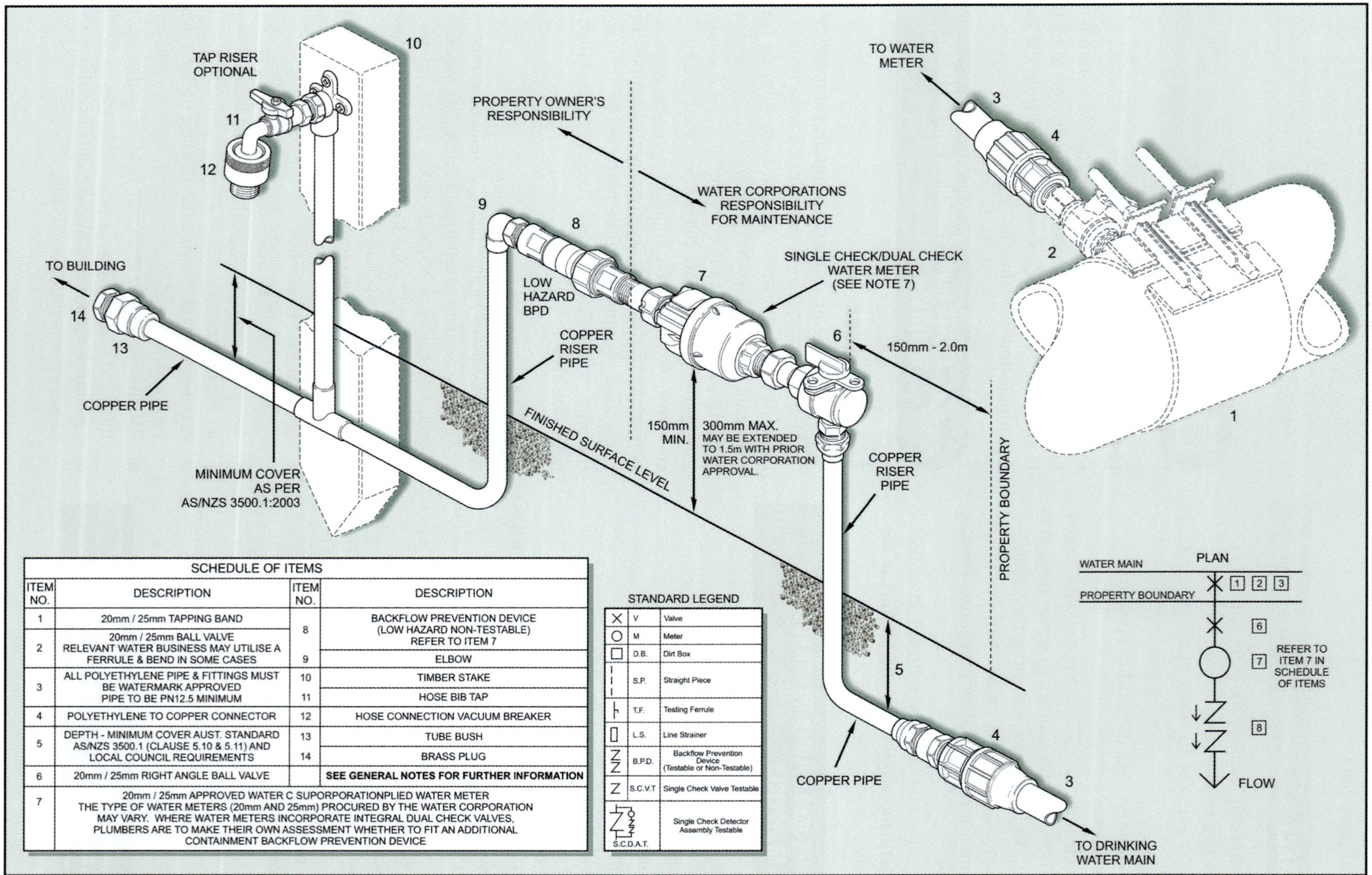

TAP RISER OPTIONAL
10
11
12
PROPERTY OWNER'S RESPONSIBILITY
TO WATER METER
3
4
WATER CORPORATIONS RESPONSIBILITY FOR MAINTENANCE
9
8
2
TO BUILDING
14
13
LOW HAZARD BPD
7
SINGLE CHECK/DUAL CHECK WATER METER (SEE NOTE 7)
COPPER RISER PIPE
6
150mm - 2.0m
COPPER PIPE
FINISHED SURFACE LEVEL
150mm MIN.
300mm MAX. MAY BE EXTENDED TO 1.5m WITH PRIOR WATER CORPORATION APPROVAL.
1
MINIMUM COVER AS PER AS/NZS 3500.1:2003
COPPER RISER PIPE
PROPERTY BOUNDARY
PLAN
WATER MAIN
PROPERTY BOUNDARY
1 2 3
6
7 REFER TO ITEM 7 IN SCHEDULE OF ITEMS
8
5
FLOW
4
COPPER PIPE
3
TO DRINKING WATER MAIN

SCHEDULE OF ITEMS

ITEM NO.	DESCRIPTION	ITEM NO.	DESCRIPTION
1	20mm / 25mm TAPPING BAND	8	BACKFLOW PREVENTION DEVICE (LOW HAZARD NON-TESTABLE) REFER TO ITEM 7
2	20mm / 25mm BALL VALVE RELEVANT WATER BUSINESS MAY UTILISE A FERRULE & BEND IN SOME CASES	9	ELBOW
3	ALL POLYETHYLENE PIPE & FITTINGS MUST BE WATERMARK APPROVED PIPE TO BE PN12.5 MINIMUM	10	TIMBER STAKE
		11	HOSE BIB TAP
4	POLYETHYLENE TO COPPER CONNECTOR	12	HOSE CONNECTION VACUUM BREAKER
5	DEPTH - MINIMUM COVER AUST. STANDARD AS/NZS 3500.1 (CLAUSE 5.10 & 5.11) AND LOCAL COUNCIL REQUIREMENTS	13	TUBE BUSH
		14	BRASS PLUG
6	20mm / 25mm RIGHT ANGLE BALL VALVE		**SEE GENERAL NOTES FOR FURTHER INFORMATION**
7	20mm / 25mm APPROVED WATER C SUPORPORATIONPLIED WATER METER THE TYPE OF WATER METERS (20mm AND 25mm) PROCURED BY THE WATER CORPORATION MAY VARY. WHERE WATER METERS INCORPORATE INTEGRAL DUAL CHECK VALVES, PLUMBERS ARE TO MAKE THEIR OWN ASSESSMENT WHETHER TO FIT AN ADDITIONAL CONTAINMENT BACKFLOW PREVENTION DEVICE		

STANDARD LEGEND

Symbol	Abbreviation	Description
X	V	Valve
O	M	Meter
□	D.B.	Dirt Box
¦	S.P.	Straight Piece
⊢	T.F.	Testing Ferrule
▯	L.S.	Line Strainer
ZZ	B.P.D.	Backflow Prevention Device (Testable or Non-Testable)
Z	S.C.V.T	Single Check Valve Testable
	S.C.D.A.T.	Single Check Detector Assembly Testable

DRAWING NUMBER:	SCALE:	DATE:	
1	NOT TO SCALE	AUGUST 2014	20 - 25mm GENERAL WATER SERVICE (WET TAPPING) LOW HAZARD RESIDENTIAL / COMMERCIAL / INDUSTRIAL TYPICAL ARRANGEMENT

City West Water South East Water Yarra Valley Water

FIGURE 3.22 Diagrammatic water meter installation

Source: South East Water. This diagram is current at time of printing. It appears for training purposes only and should not be reproduced without the express permission of the owner, South East Water.

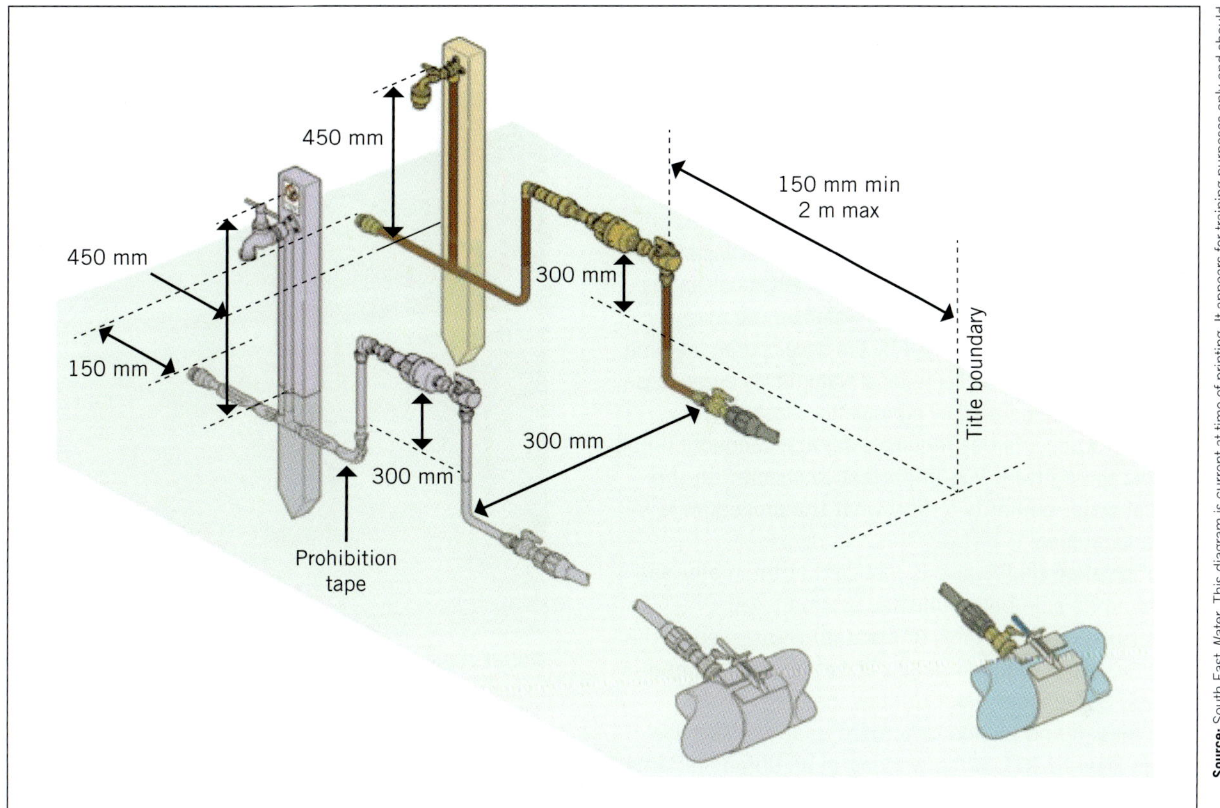

Source: South East Water. This diagram is current at time of printing. It appears for training purposes only and should not be reproduced without the express permission of the owner, South East Water.

FIGURE 3.23 Drinking and recycled water meter assembly

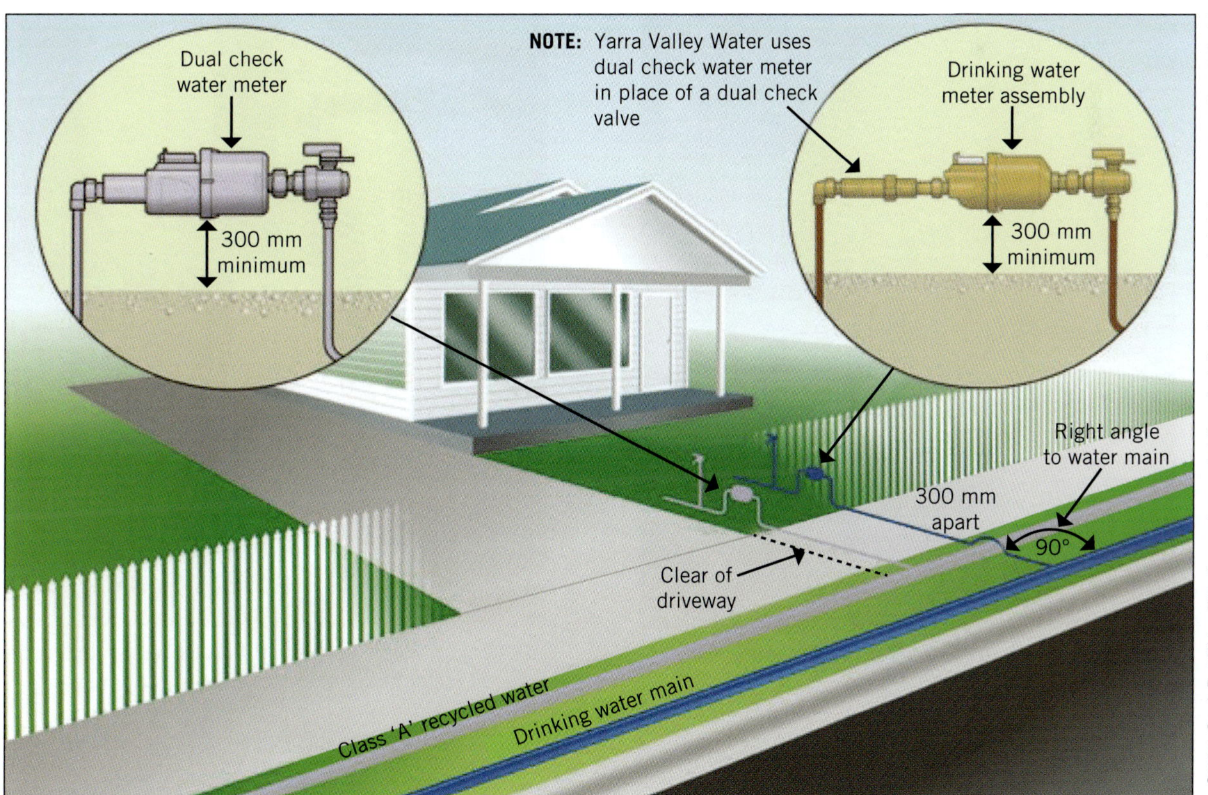

Source: South East Water. This diagram is current at time of printing. It appears for training purposes only and should not be reproduced without the express permission of the owner, South East Water.

FIGURE 3.24 Drinking and recycled water meter positioning

HOW TO

INSTALL A PROPERTY SERVICE AT THE TIME OF THE DEVELOPMENT OF A SUBDIVISION AND TERMINATE

1 A service connection ball valve (see Figures 3.25 and 3.26) is located within the property below ground level, to which a licensed plumber will connect once the property development has started, when they install the meter riser. The property service connection ball valves are located below ground with a water service marker tape brought to ground level by the developer at the time of installation. Within some local water utility areas these valves may be covered by a protective plastic cover.

2 Not all water utilities require a service connection ball valve to be fitted. It is important to check with the local water authority to find what the procedure is in the local area.

3 An isolation (ball) valve is installed at the meter and laid on its side below finished ground level.

4 An isolation (ball) valve (meter tap) is installed at the meter at the required height above finished ground level. Some local water utilities may require it to be tagged and pinned to prevent unauthorised use (see Figure 3.27). When working in an unfamiliar area, find out the local water authority requirements for a compliant installation.

FIGURE 3.25 Property service connection ball valve marking tap

FIGURE 3.26 Property service connection ball valve and meter riser

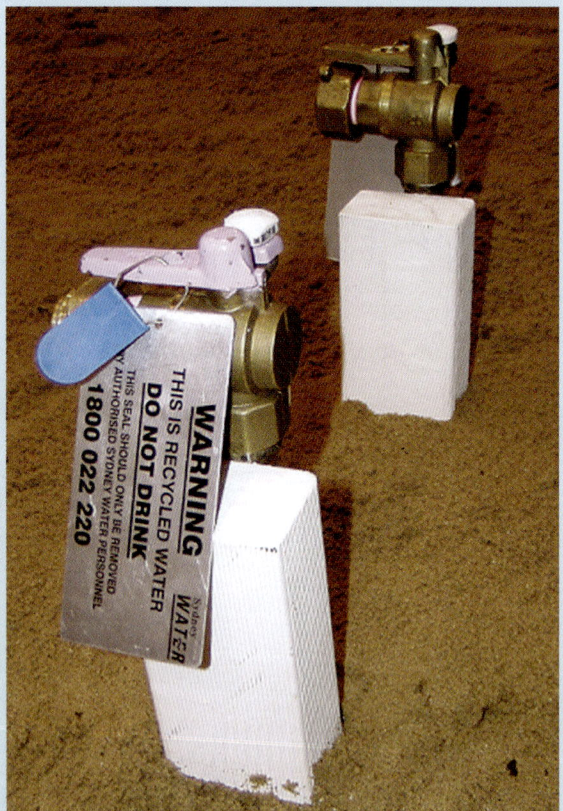

FIGURE 3.27 Meter isolation (ball) valve terminated at the required meter location

Bedding and backfill requirements

The bedding and backfill requirements for a property service are covered within AS/NZS 3500.1. Learners should refer to AS/NZS 3500.1 to gain a more in-depth understanding. Some of the main points are as follows:

- Water services must be surrounded with no less than 75 mm of compacted sand or fine-grained soil. Avoid contact with any hard-edged object such as rocks.
- Water services must be backfilled within the public property as required by the local council requirements. Learners should contact their local council to find out what the backfill requirements are within public property in the local area.
- Unless specified to the contrary by the local water utility or council, copper pipes may be installed in soil excavated from the trench in which they are to be installed, provided the soil is compatible and free from rock and rubble.

AS/NZS 3500.1 PLUMBING AND DRAINAGE: WATER SERVICES

Installing a long service

If the water main is on the other side of the street (long service) to the property, there are three alternative methods of installing the service:

1. Use the conduit supplied by the developer when the road was laid. Markers are usually located in the concrete kerb.
2. Arrange for an under-road boring machine to drill a small hole under the road through which the service can be pushed.
3. Use a water drill to drill a small hole under the road through which the service can be pushed.

Proximity to other services

A non-drinking water service must be separated from drinking water services as follows:

- Above-ground installations of non-drinking water services must not be installed within 100 mm of any parallel drinking water service except when installed in pipe duct or structurally separated.
- Below-ground installations of non-drinking water services must not be installed within 300 mm of any parallel drinking water supply.

Some local water utilities may require the non-drinking water service to be installed on the left-hand side (facing the property).

AS/NZS 3500.1 PLUMBING AND DRAINAGE: WATER SERVICES

Electrical safety

When carrying out any work on a metallic drinking or non-drinking water service connected to the water utility main, consideration needs to be given to electrical safety precautions and earthing to protect against potential electrical shock. Plumbers must always check for stray current with a neon tester before working on any metallic service. Further information may be found in AS/NZS 3500.5 and SafeWork NSW's website (under 'Electrical hazards for plumbers' at https://www.safework.nsw.gov.au). See also http://www.plumbingconnection.com.au/bonding-straps.

> Bonding straps must be used when disconnecting a meter or when cutting a metallic water service to prevent an electrical shock (see Figure 3.28). Plumbers have died from electrocution when not using bonding straps.

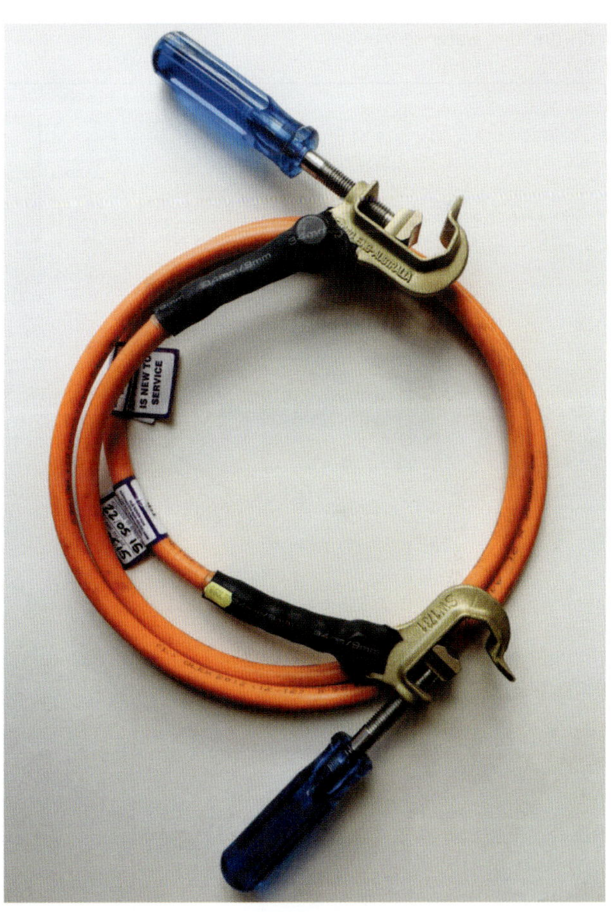

FIGURE 3.28 Bonding strap connection

Temporary interconnection between the drinking and non-drinking water services

While both the drinking and non-drinking water services need to be charged with water during construction, any interconnection between the drinking and non-drinking water services *is permitted at the meter assembly only* (see Figure 3.29). A temporary bypass is to be installed with a shut-off valve between the drinking and non-drinking water services at the water meter assembly only; a connection is not permitted to be made to the non-drinking water meter

FIGURE 3.29 Temporary interconnection between drinking and non-drinking water services

isolation valve. All meter assemblies and temporary bypasses may need to be inspected by a water utilities inspector.

Example of a property water service

Figure 3.30 shows an example of water services for a Class 2 building.

This example is based on:

- a minimum head of 30 m
- the highest fixture being 13 m above the main
- a 5 m minimum head being required at any fixture outlet.

The size of the service may change if any of the design requirements indicated above were to change. (The sizing of a water service is not part of this unit's requirements.)

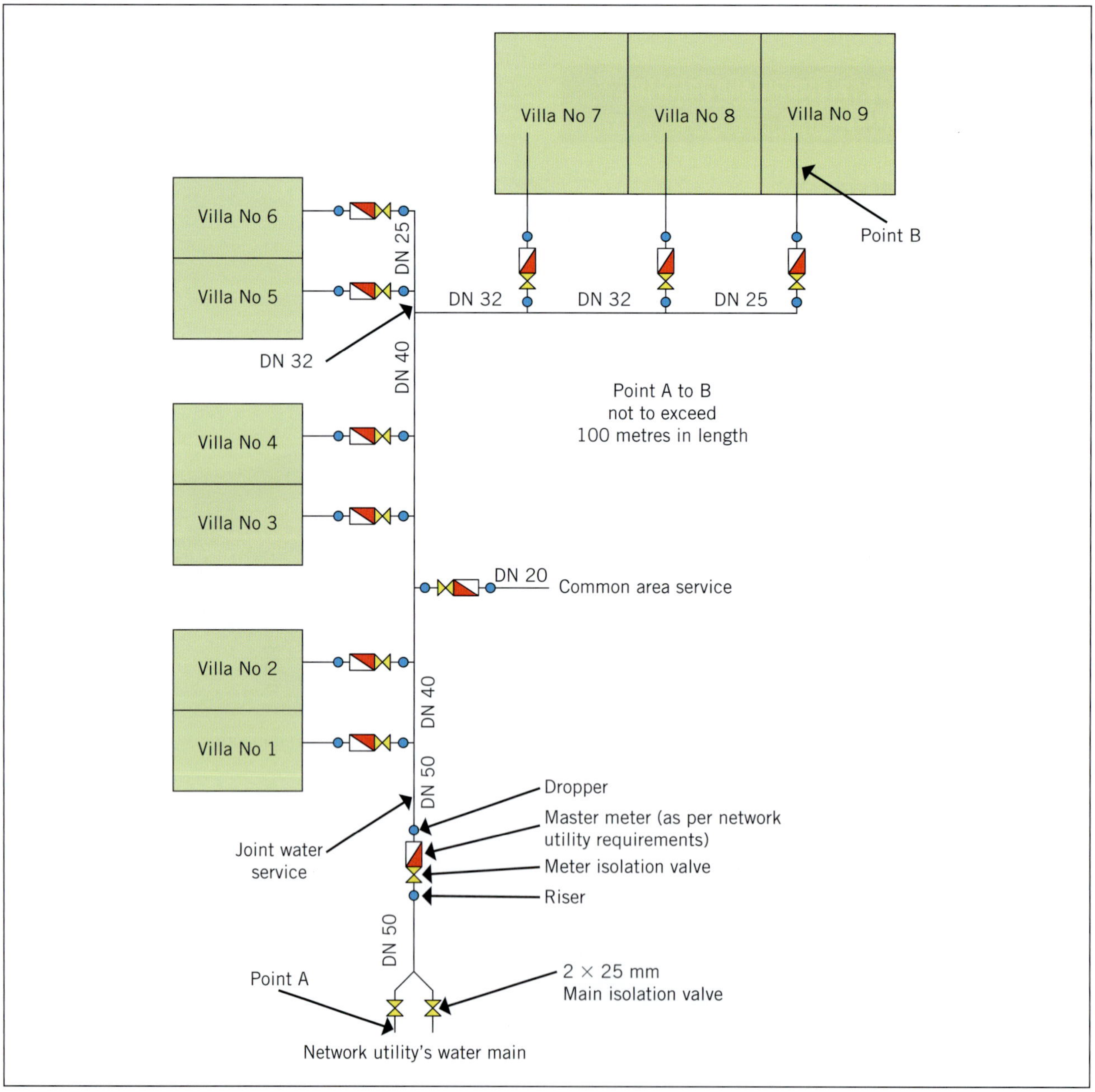

Villa No 7
Villa No 8
Villa No 9
Point B
Villa No 6
Villa No 5
DN 25
DN 32
DN 32
DN 25
DN 32
DN 40
Point A to B
not to exceed
100 metres in length
Villa No 4
Villa No 3
DN 20
Common area service
Villa No 2
Villa No 1
DN 40
DN 50
Dropper
Master meter (as per network utility requirements)
Meter isolation valve
Riser
Joint water service
DN 50
Point A
2 × 25 mm
Main isolation valve
Network utility's water main

FIGURE 3.30 Water services for a Class 2 building

Example of combined fire and domestic water services

The general layout for combined fire and domestic water services is shown in **Figure 3.31**. This is commonly known as a trident service because the one property service is feeding three services – the domestic water service, the fire sprinkler service and the fire hydrant service.

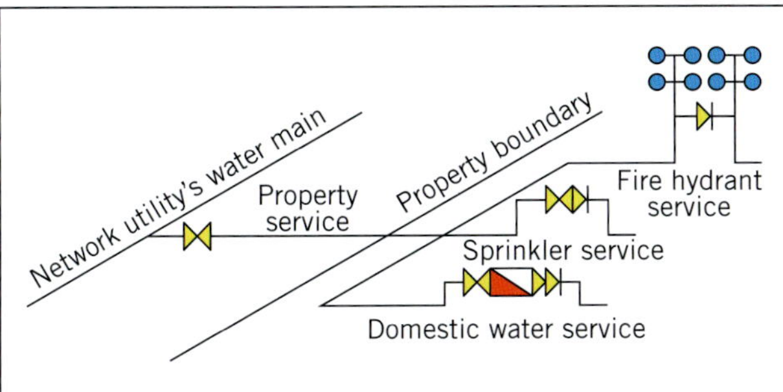

FIGURE 3.31 Combined fire and domestic water services

Test and commission installation

Before covering any joints in the property service, it must be tested to the requirements of AS/NZS 3500.1 and commissioned as stated below:

- The pipework must be flushed to remove any foreign matter until the water runs completely clear.
- The service must be subjected to a hydrostatic pressure test of 1500 kPa for a minimum of 30 minutes.
- The service must show no signs of leakage during the test.
- Check the water main valve and meter valve are operational – fully open and close.
- Check the water main valve is fully open before backfilling.

When carrying out a test, it is important to ensure that all pipework is isolated from any fixture or appliance that may be damaged by the test.

AS/NZS 3500.1 PLUMBING AND DRAINAGE: WATER SERVICES

LEARNING TASK 3.7

1 Who is authorised to carry out main tappings?
2 What are the test requirements of a property service?
3 Describe a 'long service'.
4 How is stray current tested?

Clean up

Remember that the job is not complete until the work area is cleaned up and the rubbish removed.

Clear the work area

The work area should always be left clean and tidy by removing the rubbish to the appropriate bins and sweeping the area clean.

Be sure to reduce waste by storing any material left over that can be reused. Scrap pieces of material, for example copper, should be recycled and reused when possible, in accordance with company procedures.

Clean tools and equipment

Clean all tools and equipment when packing up and keep them lubricated and maintained. Inspect tools for damage; if damage is found, ensure this is reported to the supervisor and the tool is tagged and put out of service until repaired.

LEARNING TASK 3.8

1 Why is it important to flush pipes before testing?
2 Why must tapware and fixtures be disconnected when testing the water service?
3 How are tools lubricated?

COMPLETE WORKSHEET 3

SUMMARY

- The service from the water main up to and including the water meter is called the 'property service'.
- Check local authority requirements, specifications and manufacturers' instructions before starting work
- Make sure all permits and fees are completed before starting work on the site.
- Make sure a thorough search for underground services is carried out.
- The property service is to be installed at 90° to the water main.
- Copper or polyethylene are generally the piping materials used for property services.
- Water meters must be installed within 1 m from the front property boundary.
- Extra inspections are usually required for recycled water service installations.
- Licensed plumbers must be accredited to install main tappings.
- Always check for stray electrical current and use bonding straps when disconnecting a service.
- Test the water service to 1500 kPa for 30 minutes as per AS/NZS 3500.1.

REFERENCES

AS/NZS 3500.1 Plumbing and Drainage: Water Services

Plumbing Code of Australia (PCA): **https://ncc.abcb.gov.au/**

Plumbing Connection: **http://www.plumbingconnection.com.au/bonding-straps**

Safe Work NSW: **https://www.safework.nsw.gov.au**

South East Water: **http://www.southeastwater.com.au**

GET IT RIGHT

1 Which water meter is incorrectly installed?

2 Why is this an issue?

3 What is the correct procedure?

WORKSHEET 1

To be completed by teachers	
Student competent	☐
Student not yet competent	☐

Student name: ______________________________

Enrolment year: ______________________________

Class code: ______________________________

Unit competency code/title: CPCPWT3028 Install property service

Task: Review 'Identify installation requirements' and 'Prepare for work' and answer the following questions.

1 In your own words, define a water main.

____

____

____

2 Describe a dual water system.

____

____

____

3 What is the minimum size of a property service?

____

4 What colour is used to identify a recycled water service?

____

5 In your own words, describe three tasks the licensed plumber must complete before installing a property service.

____

____

____

6 Before You Dig Australia is a service provided to help locate various public utilities services. List four of these utilities services.

____

____

____

____

7 The water utility may require a DN 25 service if the total length of service exceeds how many metres?

____

8 What do the letters 'HP' on signs fixed to telegraph poles mean?

____

9 Give four reasons why it is recommended to carry out a site inspection before starting work on a water service.

____

____

____

____

10 Is the potable (drinking) water main or the recycled water main closer to the property boundary?

____

11 Identify the following symbols in relation to hydrant and service valves for water mains:

HR

____

HP

____

SVP

____

SVR

____

12 Which classes of copper are allowed to be used for property services?

____

13 Name two limitations of using polyethylene pipe.

____

14 Copper must be used on a service that exceeds ______________ kPa.

WORKSHEET 2

To be completed by teachers	
Student competent	☐
Student not yet competent	☐

Student name: ______________________

Enrolment year: ______________________

Class code: ______________________

Unit competency code/title: CPCPWT3028 Install property service

Task: Review 'Install, test and commission property service' and answer the following questions.

1 Up to what diameter is a single or twin tapping used?

____

2 How would a 100 mm water service connect to a water main?

____

3 Define the term 'dry tapping' in relation to drilling a water main.

____

____

4 What are the minimum dimensions of an excavation for a main tapping?

____

5 What provides the seal between a tapping band and a PVC water main?

____

____

6 What is the purpose of an insulating bush on a cast iron main tapping?

____

____

7 What is the approved colour of a recycled water meter?

____

8 Where should a water meter be located in relation to the water main and why?

____

____

9 What is the minimum clearance below a water meter?

____

10 What support requirements are needed for water meters 50 mm and larger?

____

____

11 What is the minimum height a water meter can be installed with a backflow prevention device?

____

12 Name the backflow prevention valve incorporated in most domestic water meters.

____

WORKSHEET 3

To be completed by teachers	
Student competent	☐
Student not yet competent	☐

Student name: ____________________

Enrolment year: ____________________

Class code: ____________________

Unit competency code/title: CPCPWT3028 Install property service

Task: Review 'Install, test and commission property service' and 'Clean up' and answer the following questions.

1 Underground drinking water services should have ____________ mm clearance from a non-drinking water service.

2 An above-ground recycled water service must not be installed within ____________ mm of a potable water service.

3 List three requirements of a non-drinking supply property service.

____

____

____

4 What is a long service?

____

____

5 What is the minimum pressure and time a water service is tested for, as per AS/NZS 3500.1?

____

6 When should bonding straps be used, and why?

____

____

____

7 Why is it important to check the local authority's requirements before commencing work?

____

____

____

8 At what three stages must a recycled water installation be inspected?

____

____

____

9 What is the recommended minimum cover for a water service located below ground in public or private property as per AS/NZS 3500.1?

a Under a concrete slab or footing: __________ mm

b Not subject to vehicular loading (excluding fire services): __________ mm

c Fire services not subject to vehicular loading: __________ mm

d Subject to vehicular loading but no carriageway: __________ mm

e Subject to vehicular loading under a sealed carriageway: __________ mm

f Subject to vehicular loading under an unsealed carriageway: __________ mm

10 When should the property service be tested?

____

11 What is the procedure to commission a property service?

____

12 How is copper tube protected in soil that has a high content of ash?

____

SET OUT AND INSTALL WATER SERVICES

4

Chapter overview

This chapter focuses on the installation requirements of a water service from the outlet of a meter up to the connection points of fixtures, appliances and outlets, including drinking and non-drinking water services and hot water services.

Once the water supply has been connected to the property from the authority's main, it is metered and then connected to the points of discharge/use. Appropriate material selection and the correct setting out of the installation are important, as much of this work will be concealed and so will be hard to access later if mistakes are made at this point. Correct planning and preparation are essential before the walls are sheeted or rendered.

Learning objectives

Areas addressed in this chapter include:

- identify installation requirements
- prepare for work
- install and test pipe system
- clean up.

Identify installation requirements

To be organised to set out and install a water service, it is important to do the pre-planning work:

1. Obtain the required permits from the water utility and any other authority that has control over the work and that may have specific local requirements.
2. Check the plans for any specific installation requirements such as fixture set-outs and the type of tapware.
3. Check AS/NZS 3500 and Plumbing Code of Australia (PCA) for installation requirements.
4. Select the material type to be used (copper or polymer).
5. Carry out a site inspection.
6. Ensure quality assurance and work health and safety (WHS) issues are addressed.
7. Arrange any special installation requirements such as backfill material and depth of cover.
8. Organise the appropriate tools and equipment, including personal protective equipment (PPE).
9. Write a detailed material list of pipe and fittings required.

Access water service installation codes, standards and local regulations

It is the plumber's responsibility to find out the local water authority's regulations, as they can differ between areas. When working in an unfamiliar area, the plumber must contact the local water authority and find out the local regulator's installation requirements as well as what inspections are required.

It is also important to check job specifications and manufacturer's instructions for specific requirements of piping material, fittings, fixtures and tapware. This ensures the set-out of pipework and connection points are accurate to enable a straightforward fit-off.

The Plumbing Code of Australia (PCA), volume three in the National Construction Code (NCC), is the overarching guideline and refers to AS/NZS 3500.1 Plumbing and Drainage Part 1: Water Services for the minimum standard requirements. Refer to Chapter 4 'Carry out interactive workplace communication' in *Basic Plumbing Skills* for further information on how to access relevant jurisdictional regulations and other standards.

AS/NZS 3500.1 PLUMBING AND DRAINAGE: WATER SERVICES

Work health and safety (WHS)

When preparing for work it is important to complete a risk assessment and identify potential hazards so suitable controls can be implemented.

Electrocution hazards associated with cutting metallic piping are significant risks when working with water services and precautions should always be taken as described in previous units.

When working at heights and running water pipes through roofs, it is important that a fit-for-purpose platform with a handrail is used whenever possible to reduce the risks of falls.

Be sure the hazards of working under buildings are identified, such as adequate ventilation, access and hot works.

Always check for stray current and use bonding straps when working on a metallic water service.

Design considerations

It is important to plan the installation so the job runs smoothly and is kept to a high standard. Designing the most effective way to supply water to fixtures and outlets by choosing the appropriate route, material, fittings and fixings will help achieve a professional job.

Drinking and non-drinking services

When designing drinking and non-drinking water services, you need to consider:

- the most direct route from the water meters to the fixtures, appliances and outlets
- the location and type of the hot water system – it should be installed central to and as close as possible to the most frequently used outlet, which is usually the kitchen sink. It is a good rule to locate the hot water heater so *no more than 2 L of cold water is drawn off from a hot tap* before hot water is present. If this cannot be achieved, then consideration should be given to the installation of a flow and return hot water system, or the installation of more than one hot water heater, and to the type and location of temperature control devices
- running the hot and cold water services close to each other; if installed horizontally, the hot water service should be installed above the cold water pipe
- ensuring hot water branches are as short as is practicable.

GREEN TIP

It is important to plan the hot water system location well so that less energy and water draw-off are achieved with lower running costs.

Non-drinking water usage

Non-drinking (recycled) water may only be used for:

- toilet flushing
- garden watering/irrigation

- washing cars
- filling ornamental ponds
- firefighting (check with the local authority if it may be used in your local area)
- construction and industrial purposes (only with special approval of the water utility)
- clothes washing (check with your supervisor and the local water authority if it may be used in your local area).

Non-drinking (recycled) water must *not* be used for:

- drinking
- cooking or other kitchen purposes
- personal washing, such as bathing
- evaporative coolers (although some state/territory health authorities have given approval for highly treated recycled water to be used for these purposes in some areas)
- household cleaning
- internal wash-down outlets (taps)
- swimming pools
- recreation involving water contact (such as children playing under sprinklers)
- the irrigation of fruit trees and crops that are eaten raw or unprocessed.

GREEN TIP

Using recycled water is very important to help conserve precious drinking water, which will become even more scarce in the future.

Thermal insulation and frost protection

If a water service is to be installed in an area subject to regularly low temperatures (below 0°C), the service must be protected to prevent the water from freezing. This can be achieved by installing in-ground pipe to a minimum depth of 300 mm and providing the pipe with a protective waterproof insulation and a trace heating system.

A water service located on a metal roof must be designed to prevent it from coming into contact with the roof material because dissimilar metals in contact will cause electrolysis. Consideration needs to be given to installing pipes in other locations. Where this is not possible, pipes need to be insulated with the required minimum thickness thermal insulation for that area and provided with a waterproof sheathing.

FROM EXPERIENCE

Showing an awareness of the environment that you are working in by using correct applications and materials increases customer confidence.

Insulation of hot water pipes

Thermal insulation of hot water pipes must meet the minimum thermal insulation R-value for that area. This information may be found in AS/NZS 3500.4.

Thermal insulation must be provided for a domestic hot water service as follows:

- on the cold water inlet pipe between the isolating valve and the hot water heater
- for a minimum of the first 500 mm of pipe from the outlet of the hot water heater or 150 mm down the first vertical leg of a heat trap (if fitted)
- on multiple installations, on the hot water manifold to a point 500 mm past the last hot water branch (see Figure 4.1).

FIGURE 4.1 Insulated pipework

LEARNING TASK 4.1

1 Which Australian Standard regulates the installation of water services?
2 Name two design considerations for a water service.
3 Where is pipe insulation required on hot water services?
4 What is a solution for a hot water outlet far from the hot water heater (more than 2 L draw-off till hot water is present)?

Prepare for work

Unless otherwise stated in the plans or specifications, the types of materials that may be used are covered within AS/NZS 3500 and include:

- copper
- polybutylene (PB)
- cross-linked polyethylene (PE-X)
- polypropylene (PP)
- polyethylene (PE)
- unplasticised polyvinyl chloride (PVC-U)
- modified polyvinyl chloride (PVC-M; used on large water services)
- oriented polyvinyl chloride (PVC-O; used on large water services).

When selecting material, you need to consider the limitations, such as:

- what the pipework is going to be used for
- water quality and its temperature
- the type of ground (for considerations such as corrosion and leaching)
- the possibility of chemical attack from the environment
- the compatibility of material and products
- frost protection
- the water pressure within the water utility's supply system
- any special material installation requirements, such as:
  - some polymer pipes and fittings may not be installed in direct sunlight
  - polymer pipes may not be used between the cold water isolation valve and the hot water heater
  - polymer pipes may not be used as part of a temperature pressure relief valve
  - polymer pipes and fittings may not be used within 1 m of the outlet of a hot water heater
  - polymer pipes and fittings may be installed on the outlet side of a temperature control valve.

Installation requirements for materials should always be checked for the minimum requirements, which are stated in AS/NZ 3500.1. When selecting a material other than copper, it is important to refer to the Australian equivalent pipe size guide in AS/NZS 3500.1 (see Table 4.1 for an example).

Apart from the installation requirements of AS/NZS 3500.1 and AS/NZS 3500.4, the installation of each material is also covered by other standards such as AS 1432 Copper Tubes for Plumbing, Gasfitting and Drainage Application, AS 2492 (Cross-linked Polyethylene (PE-X) Pipe for Hot and Cold Water Applications and AS/NZS 2642 (Polybutylene Pipe Systems.

TABLE 4.1 Example from the Australian equivalent pipe size guide

Material size (DN)	Acceptable equivalent size according to material type				
	Copper	PVC	PB	PE-X	PP
20	20 (15 mm internal bore)	20 (15 mm internal bore)	22	25 (20.5 mm internal bore)	25 (20.5 mm internal bore)

Create a material list and collect materials

A material list for a hot and cold water piping installation (rough-in) for a typical domestic house may look similar to this:

- 1 × 50 m coil 20 mm PE-X tube (black)
- 1 × 20 m coil 16 mm PE-X tube (black)
- 1 × 50 m coil 16 mm PE-X tube (red)
- 1 × 20 m coil 20 mm PE-X tube (green)
- 1 × 20 m coil 15 mm PE-X tube (green)
- 1 × 6 m length 20 mm copper tube (type B)
- 18 × 16 mm no. 19s BP (PE-X)
- 2 × shower breeches (PE-X)
- 1 × bath breech (PE-X)
- 8 × 20 mm tees (PE-X)
- 12 × 16 mm tees (PE-X)
- 10 × 20 × 16 × 20 mm tees (PE-X)
- 6 × 20 × 16 × 16 mm tees (PE-X)
- 4 × 20 × 20 × 16 mm tees (PE-X)
- 10 × 20 mm elbows (PE-X)
- 10 × 16 mm elbows (PE-X)
- 4 × 20 mm CU elbows (press fit)
- 2 × 20 mm CU no. 2s (press fit)
- 25 × 15 mm breech plug and caps
- 6 × 20 mm CU – PE-X adaptors
- 1 × box 20 mm PE-X saddles
- 1 × box 16 mm PE-X saddles
- 1 × box 25 mm × 8 g chipboard screws.

Select and check serviceability of tools, equipment and PPE

A variety of tools, equipment and PPE necessary to set out and install a hot and cold water piping installation (rough-in) generally include:

- tape measure (8 m)
- marking pen/pencil
- PE-X cutters
- PE-X crimp tool
- copper press fit tool
- copper tube cutters
- copper tube benders (15 mm and 20 mm)
- circular saw
- impact driver
- drill with speedbore/spade drill bit (20 mm and 25 mm)
- stud/hole puncher
- test bucket
- hacksaw
- retractable knife
- safety glasses
- earmuffs/ear plugs
- steelcap boots
- high-visibility shirt.

COMPLETE WORKSHEET 1

LEARNING TASK 4.2

1. Name four types of piping material used on hot and cold water services.
2. Name three limitations to consider when choosing water service material.
3. What is the purpose of the equivalent pipe size guide?

Install and test pipe system

The hot and cold water piping system installation usually takes place at 'rough-in' stage. This must be well planned in the construction timetable so that different trades can be notified and booked in advance for efficient job progression.

Rough-in

'**Rough-in**' is the term used for the stage at which the pipework is installed inside walls, ceilings and under floors before the wall sheeting or rendering and ceiling sheeting takes place.

It is important to have clear communication with the builder/client and to be aware of the construction stage. It is far easier to install the pipework before any walls are rendered or sheeted. This helps to prevent clashes with other trades. The installation of the water service from the meter to the building is called the **front run** and can also be considered part of the rough-in.

The installation of the pipework at the relevant time will help you to:

- avoid damaging:
  - brickwork
  - wall sheeting
  - ceilings
  - floors
  - other services
- provide access for the correct supporting of pipework
- provide clear access for the installation, saving time.

It is important to remember the special installation requirements when installing water services underground in the proximity of other services, such as:

- a minimum of 300 mm separation between any drinking and non-drinking water service
- a minimum of 100 mm separation between any water service not greater than DN 65 and any electrical supply (provided the electrical cable has an orange marking tape along its length), and a consumer gas pipe with marker tape
- a minimum of 300 mm separation between any water service greater than DN 65 and any electrical supply (provided the electrical cable has an orange marking tape along its length), and a consumer gas pipe with marker tape
- a minimum of 600 mm separation between any water service and any electrical supply or consumer gas pipe where the electrical supply or consumer gas pipe has no marking tape or protection
- a minimum of 100 mm vertical separation and 50 mm horizontal separation from any stormwater or sanitary drain
- a minimum of 100 mm separation from a communication service.

AS/NZS 3500.1 states that water service pipes *shall not be* embedded or cast into concrete structures. Where a pipe passes through a concrete slab, it must be at right angles to the slab and sleeved with an impervious plastic conduit or impervious insulation with a minimum thickness of 6 mm.

Depth of cover

Any water service located below ground within private property must have a minimum cover as per AS/NZS 3500.1, as indicated below:

- 450 mm minimum when subject to vehicular traffic
- 75 mm when located under a concrete slab or house
- 300 mm in all other locations not subjected to vehicular load.

Bedding and backfill

As AS/NZS 3500.1 states, water services in trenches must be surrounded (underlay and overlay bedding) with a minimum of 75 mm compacted sand or fine-grained soil, with no hard or sharp objects touching any pipe or fittings.

Backfill material above the overlay must be free of rock or hard matter, with no soil lumps larger than 75 mm in diameter.

Any metallic pipe in corrosive soil must be protected with a polymer sleeve or appropriate wrapping.

GREEN TIP

Remember to contact Before You Dig Australia (http://www.1100.com.au or telephone 1100) to locate existing services before excavating.

Before installation

It is important to consult with the builder/client to establish the finished wall. This will have a direct impact on the depth that the fittings are recessed into the wall and the type of tapware to be installed. It will also determine the type of connection point required.

Termination methods

When installing hot and cold water outlets horizontally (such as recess sets or washing machines), the hot water outlet is to be on the left and the cold water outlet on the right (see Figures 4.2, 4.3 and 4.4). If they are installed vertically, the cold water outlet must be installed in the lower location and the hot water outlet in the upper location. These set-out methods are used at rough-in stage for the following fixtures:

- showers
- baths
- basins
- sinks
- dishwasher isolation valves
- washing machine isolation valves.

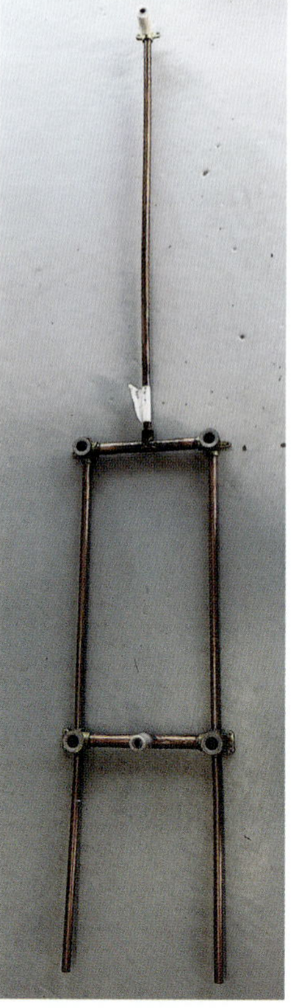

FIGURE 4.2 Combined shower/bath recess tee

FIGURE 4.3 Shower recess tee

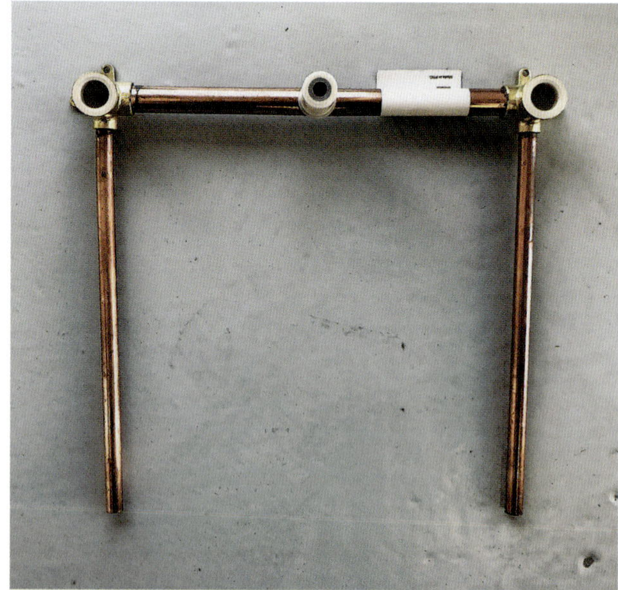

FIGURE 4.4 Bath recess tee

Setting out pipework

It is very important to have all the correct information at rough-in stage, such as the type of tapware to set out for. It is common practice to install single lever mixer taps in the wall for showers and baths, and they must be installed at rough-in stage. Also, the type of shower rose chosen will determine where the outlet will be positioned, such as a fixed arm shower rose or a hand shower on a sliding rail. Starting the rough-in without all the correct information may result in the use of the wrong type of fitting, or the incorrect tapware or valves.

As an example, when installing recess bodies for tap sets, the depth in the wall at which they are positioned may depend on where the wall finishes and the dress plates are to be fitted. This adjustable depth is dependent on whether the dress plates are screwed directly onto the spindles or held in position by a spring. If the recess bodies are installed too deep, the dress plates may not screw on. Alternatively, if they are not installed deep enough, the dress plates may not rest closely on the tiles or wall and there will be an unsightly gap. The depth requirements may also be different depending on whether the spindles are half-turn ceramic disc or just ordinary jumper valve sets. Each set of tapware has its own requirements regarding the depth of recess bodies (generally, the noggin is set back 35 mm from the face of the stud). Therefore, setting the recess at an inappropriate depth at the rough-in stage may result in having to chop out tiles to move them deeper into the wall, or purchase expensive recess adapters or extended spindles to fit the proper depth so that the dress plates and tap handles can be fitted correctly.

FROM EXPERIENCE

Clear communication with the builder/client will help avoid costly mistakes at the rough-in stage.

Installation of noggins

When setting out the installation, you may need to install support noggins for lugged/back plate elbows and recess sets (see Figures 4.5 and 4.6). If they are to be supported and fixed to noggins, it is important that the noggins are set back to the required depth for the type of tapware, as stated earlier. Care should also be taken to ensure that the noggins are installed with equal setback on both sides and that they are level and plumb. Also, noggins must be installed to support the fixing of wall hung basins, wall hung vanity units, cisterns, soap holders, toilet roll holders and shower screens. Poorly fixed noggins and lugged fittings can lead to fittings and fixtures becoming loose over time.

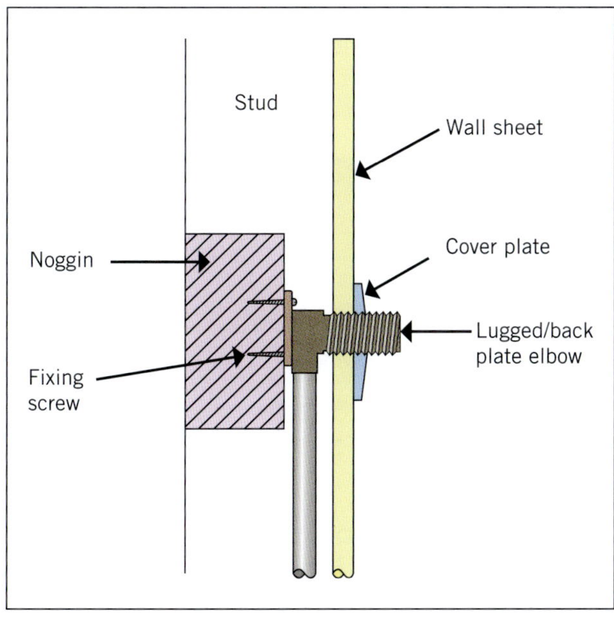

FIGURE 4.5 Lugged/back plate elbow fixed to noggin

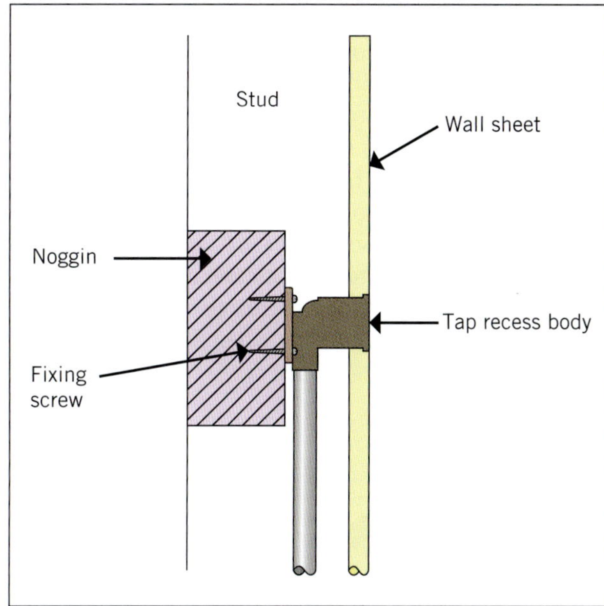

FIGURE 4.6 Tap recess body fixed to noggin

Over time, timber noggins tend to dry and shrink, so if nails are used for fixing, they may become loose. It is better to use screws as the fixing method for clips, hangers, lugged/back plate elbows and recess sets. Where possible, install the fitting in the centre of the noggin; if screws or nails are used on the edge of noggins, they may split the timber.

Pipe supports

Australian Standards require that all pipework installed has some form of support to hold the pipes in place. There are several factors that may influence the plumber's choice of pipe support method for a given installation. These factors include:

- the ability of the pipe to hold itself between supports (e.g. 15 mm copper tube may sit straight between two timbers placed 1.5 m apart, whereas 15 mm plastic tube in the same situation would sag between the timbers – it is supported at a maximum spacing of 600 mm) as per AS/NZS 3500.1
- the weight of the pipe and its contents
- compatibility (e.g. steel saddles or brackets would be unsuitable for use with copper tube without an insulator being placed between the two different metals)
- expansion and contraction
- appearance
- corrosion resistance – supports used outside need to be more corrosion-resistant than those inside (out of the weather), and areas near the coast may also require more corrosion-resistant materials
- the method used to fix the supports to the building.

Copper pipes

- DN 15 and DN 20 copper pipes must be supported at a maximum distance between supports of 1.5 m vertically and horizontally.
- DN 25 copper pipes must be supported at a maximum distance between supports of 2 m vertically and horizontally.

Saddles and clips (see Figures 4.7 and 4.8) used for supporting pipes need to be of a compatible material or coated with an inert, non-corrosive coating. Screws should be used instead of nails when fixing saddles and clips to timber.

FIGURE 4.7 Saddles and stand-off clips for copper pipes

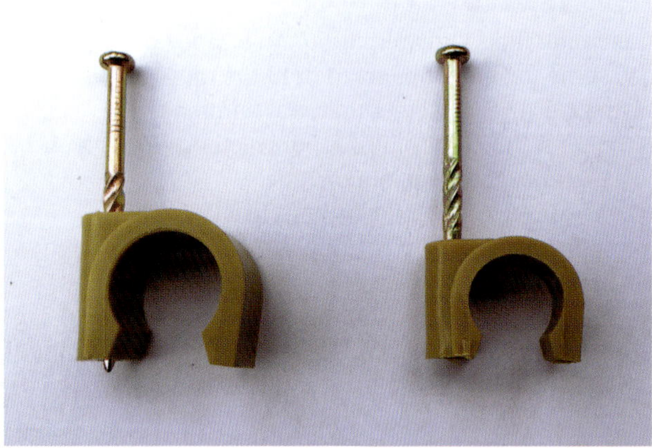

FIGURE 4.8 Polymer saddles

Polymer pipes

- DN 15 pipes need to be supported at a maximum distance between supports of 600 mm when installed horizontally and 1200 mm when installed vertically.
- DN 20 pipes need to be supported at a maximum distance between supports of 700 mm when installed horizontally and 1400 mm when installed vertically.
- DN 25 pipes need to be supported at a maximum distance between supports of 750 mm when installed horizontally and 1500 mm when installed vertically.

Polymer pipes conveying hot water may require a lesser distance between supports to prevent the pipe from sagging.

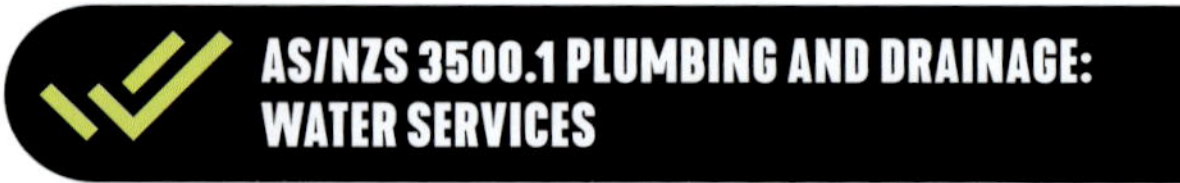

Saddles

Saddles (see Figure 4.9) hold the pipes tightly against the building structure to which they are fixed. Saddles may be made from copper, stainless steel, galvanised mild steel, plastic, or plastic and nylon-coated mild steel.

FIGURE 4.9 Copper tube fixed to timber with a nylon-coated saddle

Stand-off brackets

Stand-off brackets (see Figure 4.10) hold the pipes a set distance clear of the building structure and may be made from stainless steel, galvanised mild steel, plastic, or plastic and nylon-coated mild steel. They are used so the pipe is not in contact with the structure and so the pipe can clear other services.

FIGURE 4.10 Stand-off bracket used to support copper tube

Hanging brackets

Hanging brackets (see Figure 4.11) are used to support pipes hanging below the building structure. They may be made from galvanised mild steel or painted mild steel and nylon-coated mild steel.

FIGURE 4.11 Copper tube supported by a threaded rod hanger

Bracket systems

There are several bracketing systems on the market. They may incorporate saddles, stand-off brackets and hangers in the one system. A metal channel may hold a number of pipes of various sizes and material types (see Figure 4.12).

FIGURE 4.12 Copper pipework supported using channel and plastic brackets

Installing pipework in timber and metal wall frames

When installing water service pipes and fittings in or fixed to timber wall frames, you may be required to drill or notch the timber stud (see Figure 4.13). It is good practice to install horizontal pipe runs at a height between 300 mm and 450 mm from floor level. Wall sheets are not usually fixed in that area, so nails and screws into pipes can be avoided. It is also easy to locate pipework if alterations are made at a later stage.

- A hole of a maximum of 25 mm diameter may be drilled through timber studs for pipes to pass through.
- A notch of a maximum of 25 mm deep may be cut in the depth of a timber stud to install pipes.
- A notch of a maximum of 25 mm deep can be cut into a timber stud for a bath installation.

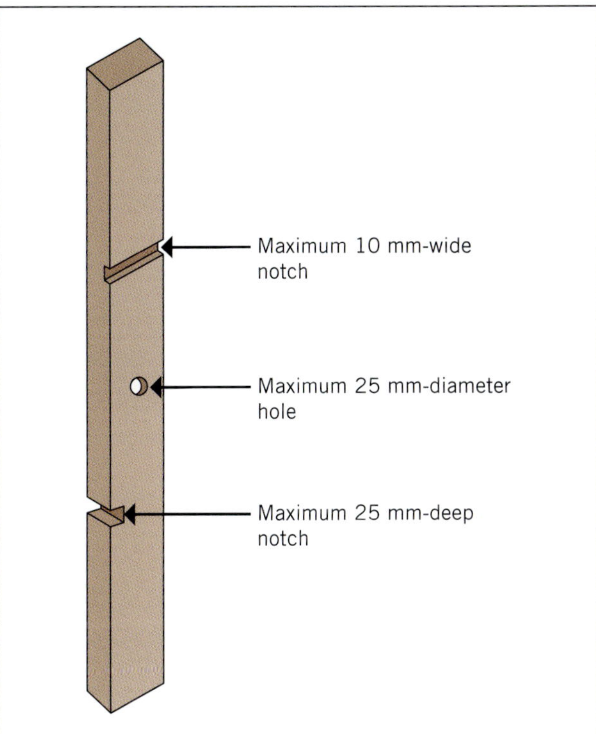

FIGURE 4.13 Notching timber wall studs

- A notch of a maximum of 10 mm wide may be cut in the breadth of a timber stud to install pipes.
- Holes and notches may not be closer than three times the hole or notch width.

Remember that it is better to use screws as the fixing method in timber studs. It is also best to use an insulator between any pipe, timber and clip; this will allow for expansion and contraction, so preventing a creaking noise.

Drilled holes in timber studs

Where pipes pass through wooden studs, the hole drilled should be larger than the pipe to allow a clear space for movement of both the finished wall and the pipe. The space around the pipe may be filled with silicon that, when set, will allow movement and keep the pipe secure. A side-fix clip may also be used to hold the pipe.

Holes in metal studs

Metal wall frames come with pre-drilled holes. The manufacturer also supplies polymer insulation grommets of different sizes. These are designed to hold the pipe firm, ensuring there is no direct contact between the pipe and the frame but still allowing expansion and contraction. It also protects polymer from the sharp edge of the metal frame and stops electronic corrosion of metal pipes (see Figure 4.14). If there is no pre-drilled hole, a hole should be drilled or punched in the metal frame, ensuring that the size of the hole is accurate (see Figure 4.15). Under no circumstances should the hole be hacked into the metal frame, as shown in Figure 4.16.

FIGURE 4.14 Plastic tube passing through a metal stud with side-fix clip in place

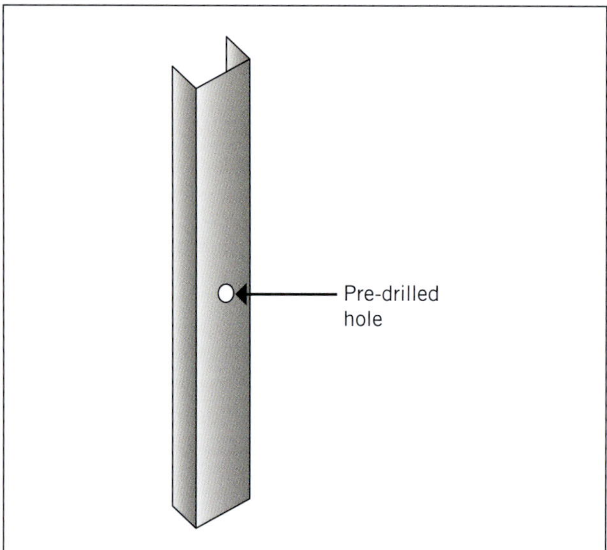

FIGURE 4.15 Metal stud with pre-drilled hole

FIGURE 4.16 Incorrectly installed water pipe in a metal stud

Installing pipework in masonry walls

When installing water service pipes and fittings in a brick or masonry wall, it is important to lag all pipework with a plastic sheath to allow for expansion and contraction because the pipework is encased in cement render (see **Figure 4.17**).

FIGURE 4.17 Installation of a mixer tap in a brick wall (pipe not completely lagged)

When carrying out any work using cement or concrete, always use the correct PPE to avoid inhaling cement dust, which contains silica [causing lung disease [silicosis]], and avoid skin contact, as this can cause dermatitis.

HOW TO

INSTALL HOT AND COLD WATER IN A MASONRY WALL

1. Mark out where the pipes and outlets are to be located on the brick wall with crayon or spray paint.
2. Wear appropriate PPE: clear safety goggles, respirator, earmuffs and steelcap boots.
3. Cut the chase (trench) into the wall 25 mm deep with a diamond-blade saw (water-cooled double blade).
4. Chop the chase out to a consistent depth.
5. Install the hot and cold water piping and terminate at the required fixtures.
6. Make sure all pipework is lagged to allow for expansion and contraction.
7. Use a spirit level to check all outlets are level and plumb.
8. Cap off and test pipework and fittings to the required test pressure and time.
9. Cement render pipes and fittings so all chases are finished flush with the wall.

Winged connectors (see Figure 4.18) may be cemented into the brickwork; they are designed to be held in place by their brass wings. They are sometimes preferred as connection points for hot, cold and tempered water connection for hot water heaters, and for external hose taps on brick walls.

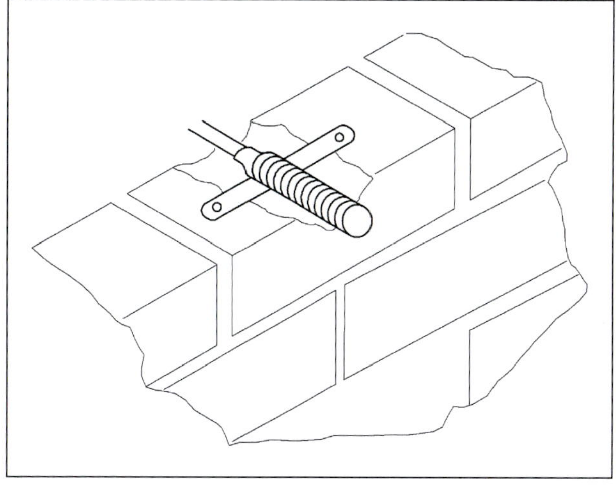

FIGURE 4.18 Winged connector

Tap heights

The heights and fittings shown in Table 4.2 should be used only as a guide; you will need to check with the builder/client if they have other height requirements and tap specifications.

COMPLETE WORKSHEET 2

TABLE 4.2 Guide for tap and outlet heights

Fixture	Outlet height	Centres	Termination fitting
Sink (hod/wall-mounted tap set)	1000 mm	Centre of sink	Recess tee
Sink (bench-mounted tap set)	600 mm	200 mm offset from centre	Lugged/back plate elbow – no. 19 (see Figure 4.19)
Laundry trough/tubs (wall-mounted tap set)	1000 mm	Centre of trough	Recess tee
Laundry trough/tubs (bench-mounted tap set)	500 mm	200 mm offset from centre	Lugged/back plate elbow – no. 19 (see Figure 4.19)
Clothes washing machine	1350 mm	200 mm offset from centre (may be installed vertically or horizontally)	Lugged/back plate elbow – no. 19 (see Figure 4.19)
Dishwashing machine	300 mm	In sink cupboard adjacent to dishwasher	Lugged/back plate elbow (no. 19) for each connection point; usually only cold water connection
Basin	600 mm	200 mm offset from centre	Lugged/back plate elbow – no. 19 (see Figure 4.19)
Shower (taps)	950 mm	200 mm offset from centre	Recess bodies (may be installed vertically or horizontally)
Shower rose	1850 mm	Centre of shower recess	Lugged/back plate elbow – no. 19 (see Figure 4.19)
Bath	600 mm (or 150 mm above bath rim) or to suit tiles	May be installed on the wall at the plug end or the centre of the side wall	Recess tee
Water closet (close coupled)	150 mm	200 mm right of centre	Lugged/back plate elbow – no. 19 (see Figure 4.19)
Water closet (concealed inlet)	Manufacturer's specification	Top rear of cistern	Lugged/back plate elbow – no. 19 (see Figure 4.19)

Source: Reece Group.

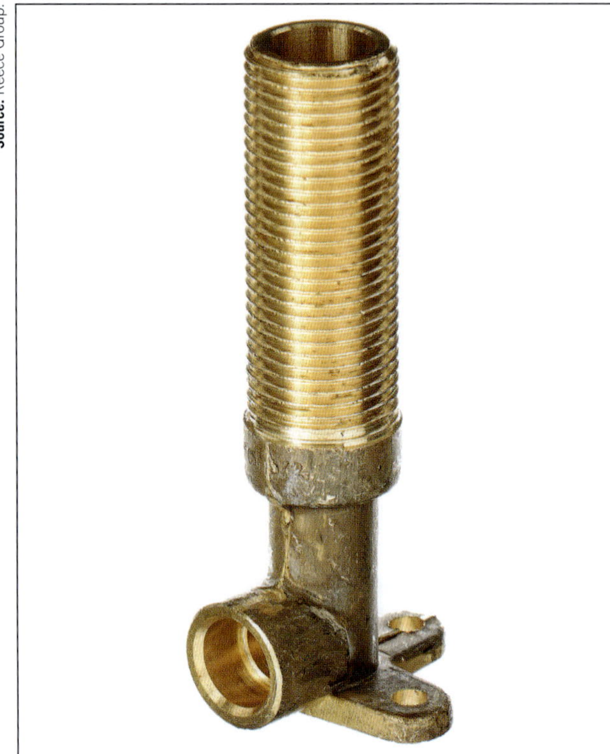

FIGURE 4.19 Lugged/back plate elbow (19 BP)

LEARNING TASK 4.3

1. What information is needed from the client before starting the hot and cold water installation?
2. Describe the term 'rough-in'.
3. What does the term 'front run' mean?
4. What is the maximum size hole that can be drilled in a timber stud?
5. What is a requirement for hot and cold water pipes chased in a brick or masonry wall?
6. Research the manufacturer's instructions for Auspex pipe and provide information about how the pipe needs to be installed into a metal frame.

Water properties

It is important to understand that water freezes at 0° Celsius and boils at 100° Celsius. It is necessary to protect piping with adequate insulation when exposed to extreme temperatures. Minimum insulation thicknesses are listed in AS/NZS 3500.1.

Also, water is most dense at 4° Celsius and becomes less dense and less viscous as it increases in temperature. So, this means hotter water will flow more easily due to less density and less viscosity.

Water is a virtually non-compressible liquid. This characteristic can have a direct impact on the water supply piping and may result in an effect called water hammer (a distinct loud hammering sound that can occur in a poorly designed or poorly installed piping system; this is discussed in Chapter 5). The kinetic energy contained in the moving water (which has a mass of 1 kilogram per litre) can create a shock wave that transmits as a violent vibration into the surrounding structures due to water's inability to absorb the impact of this change in velocity when the flow of water is suddenly halted.

To help alleviate this problem and overcome the noises transmitted by the transfer of water through the piping system to the point of use, AS/NZS 3500.1 stipulates maximum vertical and horizontal support spacings and places a maximum pressure limit of 500 kPa in a building and a maximum velocity of 3 m per second. The velocity can be reduced by increasing the pipe size.

AS/NZS 3500.1 PLUMBING AND DRAINAGE: WATER SERVICES

Sizing a water service

A water service must be sized to meet the flow rate and fixture requirements of a particular building, and this information is covered in depth in the Certificate IV unit CPCPWT4011 Design and size heated and cold-water services and systems'. Below is some basic information about flow rates and sizing of water services. The minimum size water service to the branch offtakes for a single dwelling is DN 20.

The flow of water through pipes

While static pressure is a guide to the volume of water that might be expected to flow through a piping system, other factors determine the volume that will actually flow and create pressure loss. The pipe sizing tables contained in AS/NZS 3500.1 satisfy the requirements of flow and friction loss. Correct installation is still required to ensure that the piping system performs as required. The most important factors are:

1. *The diameter of the pipe*. The larger the bore of the pipe, the more water it can deliver.
2. *The length of the pipe*. The longer the pipe, the greater the friction that occurs between the water and the internal surface of the pipe; therefore, a greater pressure loss occurs.
3. *The condition of the pipe*. Different pipes have different internal bores, which cause more friction and eddies through the rough surface and restrict the water flow.
4. *Changes in direction*. Sharp bends can create eddies, which in turn cause friction and restriction of the water flow. Frictional loss varies with the angle of the bend and the radius of curvature. The sharper the angle and the shorter the radius, the greater frictional losses will be. A radius of five times the pipe bore on the centre line has the same amount of friction loss as the same straight length of pipe.

5 *The quality of jointing*. Burred pipe ends, excess sealing compound and gaps between pipe ends restrict the flow of water.
6 *Passage through valves*. When water passes through a valve, it is restricted and is usually forced to change direction once or twice, depending on the design of the valve.
7 *Enlargement or contraction of size*. To gain maximum flow, enlargements and contractions of pipe diameter should be made as gradual as possible.

Flow rates

Each tap, valve and outlet is required to meet a minimum flow (discharge) rate as per AS/NZS 3500.1; for example:

- water closet (cistern), basin (standard outlet), shower, sink (aerated outlet): 6 L per minute
- bath: 18 L per minute
- laundry tub: 7 L per minute.
- washing machine/dishwasher: 12 L per minute
- 20 mm hose tap: 18 L per minute
- 15 mm hose tap: 12 L per minute.

Minimum size cold water branches

The following restrictions apply to cold water branch lines.

- A branch with a minimum internal diameter of 12.5 mm (DN 18 copper) must not exceed 6 m in length and may supply water to one fixture or outlet and in addition:
  - a flushing device.
  - a make-up tank supplying a gravity-fed hot water system.
- A branch with a minimum internal diameter of 10 mm (DN 15 copper) must not exceed 3 m and may supply a combination bath and shower unit, or laundry trough and washing machine, or a kitchen and dishwasher.

Supply to a hose tap (standpipe) has a minimum pipe size of DN 15 and must be installed a minimum of 450 mm above finished ground level.

Minimum size hot water branches

For a storage water heater in excess of 170 kPa, the minimum sizes are:

- a DN 18 service from the heater to the first branch
- a DN 15 branch to a kitchen sink or basin
- a DN 15 branch to a sink and laundry
- a DN 15 branch to a bathroom and one other room
- a DN 15 branch picking up all fixtures within a bathroom.

The pipe sizes indicated above and the others outlined in AS/NZS 3500.4 are a guide only – a larger pipe diameter may be required to meet the minimum flow rates.

Other considerations

As per AS/NZS 3500.1, there are minimum requirements for separation between different services and colour identification.

Proximity to other above-ground services

A drinking water service requires:

- a minimum of 100 mm separation from any parallel non-drinking water service
- a minimum of 25 mm from any gas service or electrical cable, wire or conduit.

Pipe identification

Pipe identification for PE-X pipe is as follows:

- drinking water – black/blue
- hot water – red
- rainwater – green
- recycled water – purple.

Any copper tube used as part of a non-drinking water service must be sheathed with purple PE (see Figures 4.20 and 4.21).

Source: Reliance Worldwide Australia.

FIGURE 4.20 Cross-linked PE pipe – the pipe for recycled water is purple and is clearly marked 'recycled/reclaimed'

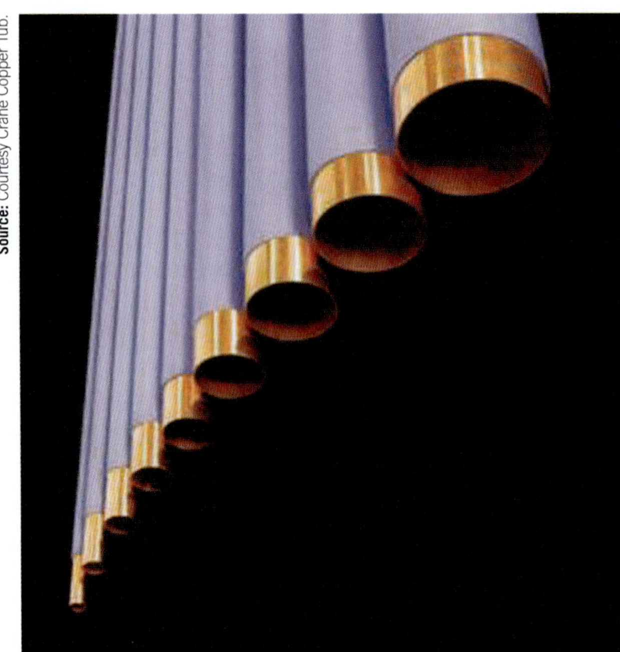

Source: Courtesy Crane Copper Tub.

FIGURE 4.21 Copper pipe with purple PE sheathing for recycled water

LEARNING TASK 4.4

1. What can cause pressure loss in a water service?
2. What can cause 'water hammer'?
3. What is the maximum length of a DN 15 cold water copper branch?
4. What is the minimum size pipe to connect to a hot water heater outlet?
5. How can velocity be reduced in a water service without reducing the pressure?

Testing installed pipework

All water services need to be tested before being covered or concealed from view. Prior to any testing, the services need to be flushed to remove all air and any foreign matter. The testing needs to be carried out with fixtures and appliances isolated (not subject to the pressure test). If the service can be charged (filled) with water prior to applying pressure, it will save considerable time.

AS/NZS 3500.1 requires all water services to hold a test pressure of 1500 kPa (15 bar) for a minimum of 30 minutes. A test bucket is the most common method used for testing a water service.

AS/NZS 3500.1 PLUMBING AND DRAINAGE: WATER SERVICES

A temporary interconnection is generally made between the hot and cold services by removing the jumper valves (tap washers) in a recess set (bath or shower) and sealing the outlets with approved caps. The test bucket (full of water) is connected to a convenient point within the system and pressure applied until the required pressure of 1500 kPa is reached (see Figure 4.22). As required by the standard, the system is left pressurised for a minimum of 30 minutes. The system is visually checked for leaks. If there are no leaks, and the test gauge on the test bucket indicates it is holding at 1500 kPa, this system may be backfilled or covered. The test bucket can be removed and the male thread sealed with an approved plug.

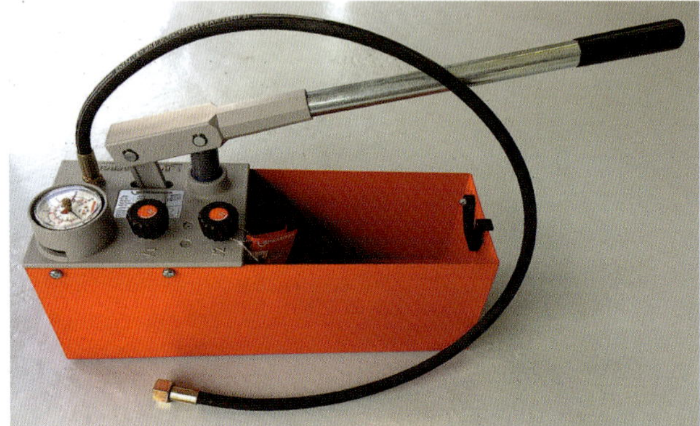

FIGURE 4.22 Test bucket with gauge fitted to hose for testing water pipework

If there is a non-drinking water service, it needs to be tested in a similar manner.

A check should be made with the local water utility if an inspection is required of the installation of the hot and cold water services and of any non-drinking water services.

It is recommended that the water service remain charged with water during construction. If the service is damaged, it will leak, requiring it to be reported and fixed. If the construction is going to take some time, the service should be regularly flushed with clean water.

LEARNING TASK 4.5

1. Why is it important to flush pipework before testing?
2. What piece of equipment is used to pump the water service to test pressure?
3. Why are tapware and fixtures not connected to the water service pressure test?

Alternative water supply: rainwater

Authorities are now encouraging property owners to install rainwater storage tanks (for new residential developments it is compulsory in most areas), with some state and territory governments offering a rebate for their installation based on the size and end usage. See *Roof Plumbing*, Chapter 18 'Collect and store roof water' for more information (see HB 230 and AS/NZS 3500.1 for 'Above-ground rainwater tank installation with charged (wet) system for all downpipes – cross-section', 'Above-ground tank with pit pump installation – uncharged (dry) system' and 'Flexible rainwater storage device – bladder under floor construction'). A rainwater collection system can provide water for several uses, including:

- toilet/urinal flushing
- clothes washing machines
- hot water systems
- garden irrigation
- car washing and similar outdoor use
- filling ornamental ponds
- filling swimming pools and spas
- firefighting (subject to the requirements of AS 2419.1, AS 2118 and AS 2441).

Properly maintained rainwater tanks can provide good-quality drinking water. If rainwater tanks are to supply water for cooking or drinking, the property owner should have in place an adequate system of filtering, cleaning and maintenance. It is important to be aware of potential risks associated with microbial and chemical contamination. Rainwater tanks in urban areas can be contaminated with airborne contaminants from heavy traffic, smelters and heavy industry. Rainwater tanks can also be contaminated from roof or plumbing materials or with bacteria from bird or animal droppings. Water from rainwater tanks is not recommended for drinking or cooking unless it has been properly filtered, in line with local authority guidelines.

A first-flush device (see 'Example of pre-storage filters for rainwater tanks', in HB 230–2008) should be installed on any rainwater collection system where the water is to be used for drinking or cooking, or connected to an appliance. The first-flush devices should be cleaned (emptied) after rainfall. Tanks must be installed with mosquito proofing on the inlet and overflow. The mosquito proofing installed on the inlet must be above the spill level of the tank. A rainwater tank overflow must discharge to a stormwater legal point of discharge (see **Figure 4.23**).

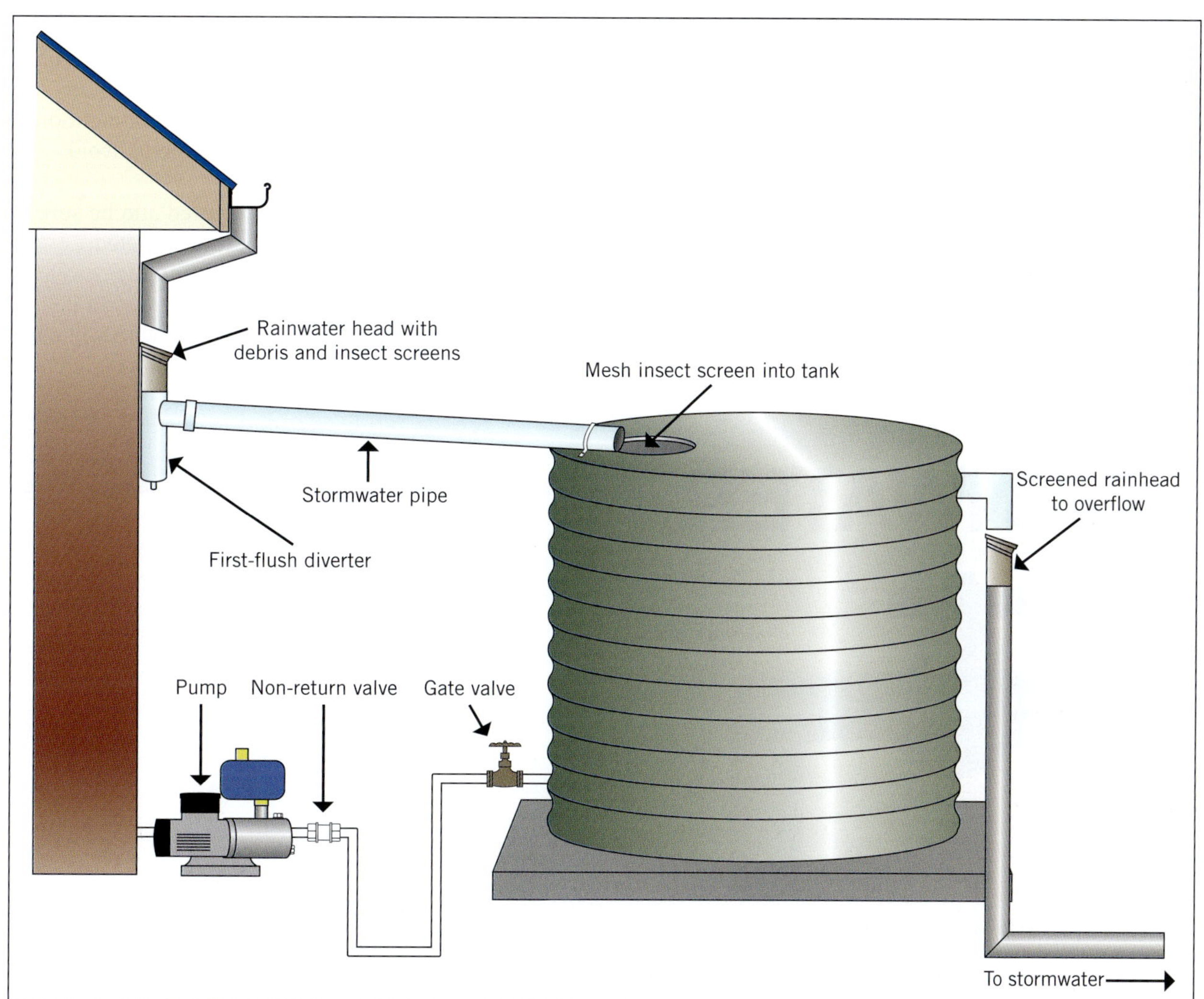

FIGURE 4.23 Above-ground rainwater tank with rainwater head and debris screen, first-flush divertor and insect screen

Installing an adequate filtering system that removes parasites and bacteria will help to mitigate health risks.

If a fixture or outlet is connected to a water supply from a rainwater tank, the tank will need to have a top-up system connected to the utility's water supply (see 'Above-ground rainwater tank installation with mains water top-up and rainwater supplied to appliances in the household' in HB 230–2008). A drinking water service interconnected with a rainwater service requires an approved backflow prevention device installed at the point of interconnection to prevent cross connection, and a further device installed for property containment on the outlet of the meter (discussed in Chapter 9). Rainwater piping is coloured green for easy identification to help prevent cross-connection with drinking water services.

GREEN TIP

Rainwater harvesting is now compulsory for most new developments to supply water for flushing toilets and an outside hose tap for washing and irrigation.

Other points to bear in mind when using rainwater tank supply include:

- All rainwater tanks and outlets are required to be identified with an approved sign (see the current version of AS 1319) (see also Figure 4.24).
- The catchment area and storage capacity need to be considered (toilet flushing quickly depletes stored water).
- Pressure requirements need to be checked; for example, will a pump be required to meet minimum pressure standards?

Source: © 421 Environmental Products.

FIGURE 4.24 Rainwater identification sign

LEARNING TASK 4.6

1 What can rainwater be used for?
2 What is a first-flush device?
3 What device must be installed when a rainwater tank has a mains water service top-up?

It is required to label all rainwater outlets with an approved sign stating 'rainwater'. Learners can refer to HB 230–2008.

Clean up

After the rough-in stage is complete, the work area should be left clean and tidy by sweeping and removing all debris and rubbish. Any leftover material should be stored and reused on another job to prevent waste and to save costs. All packaging waste should be placed in its respective bin. Paper and cardboard waste should be placed in the recycling bin.

All tools and equipment should be cleaned, lubricated and stored safely for reuse. Discard and replace any cutting tools or blades that are blunt, so they are ready to use next time. Tag out and report any tools that are faulty. Any tools and equipment on hire must be returned as soon as possible to avoid extra cost.

Complete all documentation required and be sure all mandatory inspections have been completed.

Leaving the job clean and tidy adds to the reputation of a well-run organised business, and it leaves a lasting impression on the client, who will pass on valuable referrals.

COMPLETE WORKSHEET 3

SUMMARY

- Gather all information from the latest plans and specifications and pay all fees.
- Confirm with the client the type and location of tapware and fixtures to be installed, and order in-wall single lever mixer units and in-wall cisterns if required.
- Mark out rough-in accurately from all the information gathered.
- Choose the appropriate piping material and check specifications.
- Choose the most direct path for the hot and cold water service, reducing the amount of bends for less friction loss.
- Consider the most convenient location for the hot water heater, minimising the longest hot water pipe run.
- Use appropriate guidelines when installing a recycled water service and check for extra inspections.
- Notch and drill appropriate sizes in timber stud walls.
- Correctly size pipework (refer to AS/NZS 3500.1).
- Be sure pipework is appropriately clipped and additional noggins to fix fixtures are installed.
- Flush pipework and test water service to 1500 kPa for 30 minutes.
- Clean and tidy work area and pack away all tools and equipment.

4

REFERENCES

AS/NZS 3500.1 Plumbing and Drainage: Water Services

AS/NZS 3500.4 Plumbing and Drainage: Heated Water Services

HB 230–2008 – Rainwater Tank Design and Installation Handbook

NCC – Part 3 Plumbing Code of Australia (PCA)

GET IT RIGHT

1 What is the main difference between the two rough-ins shown in the photographs?

____

____

____

2 What are the advantages and disadvantages of each one?

____

____

____

WORKSHEET 1

To be completed by teachers

Student competent ☐

Student not yet competent ☐

Student name: ____________________

Enrolment year: ____________________

Class code: ____________________

Unit competency code/title: CPCPWT3021 Set out and install water services

Task: Review 'Identify installation requirements' and 'Prepare for work', then answer the following questions.

1 Name four important points to consider when planning and preparing to set out and install a water service from the meter outlet to the fixture connection point.

____

____

____

____

2 Name four important considerations when designing a service.

____

____

____

____

3 List four situations in which non-drinking (recycled) water may be used.

____

____

____

____

4 List five situations in which non-drinking (recycled) water cannot be used.

____

____

____

____

____

5 What can be done to prevent water from freezing in a water service?

6 How can a hot water pipe be insulated, and why should it be insulated?

7 Materials for use in water services are often referred to by their abbreviation. Give the full name of the following five polymer products:

a PB

b PE-X

c PP

d PE

e PVC-U

8 What is the most important consideration when choosing a hot water heater location?

9 What does the term 'polymer pipe' mean?

10 Name an important consideration when installing a water service on a metal roof.

11 State the equivalent pipe sizes for DN 20 pipe in copper and PE-X.

WORKSHEETS 4

12 Name six tools/equipment needed to carry out a hot and cold water rough-in.

13 Create a list of materials required to carry out a hot and cold water pipe installation (rough-in) for a bathroom with a bath, shower, basin and toilet in a timber frame with standard recess tees.

WORKSHEET 2

To be completed by teachers	
Student competent	☐
Student not yet competent	☐

Student name: ______________________

Enrolment year: ______________________

Class code: ______________________

Unit competency code/title: CPCPWT3021 Set out and install water services

Task: Review 'Install and test pipe system' and answer the following questions.

1 What is the common name given to the task of installing a hot and cold water service inside the walls and ceiling of a building?

____

2 State the recommended minimum cover for a water service located below ground in private property in the following conditions:

a When subjected to vehicular traffic: ______________ mm

b When located under a concrete slab or house: ______________ mm

c In all other locations: ______________ mm

3 When hot and cold water outlets are installed horizontally, on which side is the hot outlet installed?

____

4 When hot and cold water outlets are installed vertically, is the hot water outlet installed in the higher or lower location?

____

5 What is the standard setback for a noggin supporting recess bodies and why is this depth very important?

____

____

____

____

6 What is the maximum size hole to be drilled for a DN 20 pipe through a timber stud?

____

7 When installing water pipes in metal wall frames, the frame comes with pre-drilled holes. Polymer insulation grommets are also provided. Why are these grommets provided?

____

____

____

8 Sometimes water service pipes are installed in masonry walls. Explain why these pipes must be lagged.

9 List five factors that must be considered when selecting pipe supports.

10 Where would a plumber find out the minimum allowable spacing for water service pipe supports?

11 What is the maximum spacing of supports for DN 15 and DN 20 copper pipes?

12 What is the maximum spacing of supports for DN 20 PE-X pipe?

13 What is the standard set-out height of a shower tap set and a shower rose?

14 What are the standard set-out measurements for a bottom inlet water connection to a low-level WC?

15 What is the common name for the section of cold water service from the water meter to the building?

16 Why is it better to use screws instead of nails when fixing noggins, recess tees, lugged elbows and saddles?

17 What is a reason to use stand-off brackets?

18 What is the appropriate PPE when cutting a chase into a masonry wall?

WORKSHEET 3

To be completed by teachers	
Student competent	☐
Student not yet competent	☐

Student name: ______________________

Enrolment year: ______________________

Class code: ______________________

Unit competency code/title: CPCPWT3021 Set out and install water services

Task: Review 'Install and test pipe system' and 'Clean up' and answer the following questions.

1 Why is it important to understand that water freezes at 0° Celsius and boils at 100° Celsius?

____

____

2 Describe in your own words the potential problem associated with the fact that water is a virtually non-compressible liquid.

____

____

____

____

3 Why does hotter water flow more easily than cold water?

____

____

4 State the recommended maximum pressure and the maximum velocity for water in a piping system as per current Australian Standards.

____

5 What is the minimum flow rate for a shower?

____

6 Name five factors that affect pressure loss in a water service.

____

____

____

____

____

7 When installing both hot and cold water pipes for a water service, some pipes are limited in both length and what they provide. Complete the following statements by filling in the missing information.

a A DN ______________________ branch may not exceed ______________________ m and may supply water to one fixture or outlet as well as:

- a make-up tank supplying a gravity-fed hot water system
- a flushing device.

b A DN ______________________ branch may not exceed 3 m and can supply water to a single fixture only as well as a combination bath and shower unit, or a laundry trough and washing machine, or a kitchen sink and dishwasher.

c A standpipe (hose tap) must not be smaller than DN ______________________ and must be installed a minimum of ______________________ mm above finished ground level.

d For a storage water heater in excess of 170 kPa, the minimum sizes are:

- a DN ______________________ service from the heater to the first branch
- a DN 15 branch to a ______________________ or basin
- a DN ______________________ branch to a sink and laundry
- a DN ______________________ branch to a bathroom and one other room
- a DN ______________________ branch picking up all fixtures within a ______________________.

8 What are the identification colours for different water services that PE-X pipe provide?

____

____

____

____

9 What are the most common uses for stored rainwater in an urban area?

____

____

____

____

____

____

____

WORKSHEETS 4

10 What is the purpose of a first-flush device when storing rainwater?

11 What is the test pressure and test time for water services?

12 What are three conditions that should be observed when testing a water service, after completing the work and prior to concealment?

13 How is the hot water pipework temporarily connected for testing?

14 How far must a drinking water service be apart from a non-drinking water service above ground?

15 How far must a drinking water service be apart from a non-drinking water service below ground?

16 Why is it good practice to keep the hot and cold water installation charged during the construction process?

17 How is cross connection between a rainwater supply and a mains water supply prevented?

18 Why should left over material be kept and reused?

5

INSTALL AND COMMISSION WATER HEATING SYSTEMS AND ADJUST CONTROLS AND DEVICES

Chapter overview

The aim of this chapter is to address the skills and knowledge required to install and commission water heating systems with the associated valves and control devices. The chapter covers the installation requirements for low pressure storage water heaters, mains pressure storage water heaters, continuous flow water heaters and solar water heaters, as well as the installation requirements for multiple water heater installation (manifold installation). The chapter also describes the different types of water controls and devices (valves), their functions and adjustment installed for water heating and plumbing systems.

Learning objectives

Areas addressed in this chapter include:

- identify installation requirements
- prepare for work
- install water heating system and service control device
- clean up.

Identify installation requirements

Understanding the right choice of water heating system and the appropriate valves for different buildings and situations is vital for plumbers to help clients make an informed decision for the most efficient and economical choice.

A **hot water installation** is classed as an installation of one or more water heaters and the required hot and cold piping system to supply hot water to several fixtures, appliances and outlets. Water heaters are generally divided into four classes:

1. continuous flow water heaters
2. storage water heaters
3. heat exchanger–coil heaters/calorifiers
4. commercial water heaters (boilers).

A **continuous (instantaneous) flow water heater** is designed to heat water only at the time it is being used and supplies hot water continuously while the hot tap is turned on. Continuous flow water heaters are generally connected to a mains pressure water supply.

A **storage water heater** is designed to hold a designated quantity of hot water in an insulated container ready for use as required. These units can be designed to both store and supply hot water at mains pressure (above 350 kPa), or to store water at atmospheric pressure and deliver it by gravity.

Heat exchanger–coil heaters/calorifiers and commercial water heaters (boilers) are not part of the learning for this unit and will be covered later in your training.

Storage water heaters are further classified by storage pressure:

- **Mains pressure units**. A mains pressure unit is designed to store and deliver water at mains pressure (recommended to be above 350 kPa); this provides the hot and cold water at the same outlet pressure.
- **Reduced-pressure unit**. A pressure-reducing valve (see the 'Valves' section) or an overhead storage (feed) tank connected to the cold water connection of a water heater reduces the delivery pressure to below that in the utility's water main. This may deliver a different pressure in the cold and hot water supply, making end-use balancing or mixing of hot and cold water more difficult.
- **Gravity unit (low pressure)**. A gravity feed hot water system has a cold water feed (cistern) tank fitted to the storage tank. It is designed to store water at atmospheric pressure and deliver it via gravity to the required hot water outlets. These units are no longer commonly installed due to problems with balancing the supply pressures on single-lever mixing valves, thermostatic mixing valves and tempering valves incorporated into the warm water supply. Also, there is a reliance on high usage of electricity as the fuel medium for heating.

GREEN TIP

Using electricity to heat water with traditional 1.8 kW, 2.4 kW, 3.6 kW or 4.8 kW elements consumes a high amount of greenhouse-gas-intensive energy.

Water heaters are further classified by their delivery method:

- **Single-point unit**. These units are designed to supply water to one tap/outlet only; they may be continuous flow or storage design.
- **Multipoint unit**. These units are designed with sufficient water flow capacity and thermal input to provide a consistent supply of hot water to several outlets at the same time. They may be of a storage or continuous flow type.
- **Push-through unit**. These units are also known as free outlet under-sink water heaters. They store a small quantity of water at atmospheric pressure; when the hot tap is opened, water is delivered at mains pressure. This will be covered later in this chapter.

FROM EXPERIENCE

It is necessary as a plumber to have the knowledge to select the correct water heater for the right application.

Access codes, standards and local regulations

It is the plumber's responsibility find out the local water authority's regulations as they can differ between areas. When working in an unfamiliar area, the plumber must contact the local water authority and find out the local regulator's installation requirements as well as what inspections are required.

It is also important to check job specifications and manufacturer's instructions for specific requirements of piping material, fittings, fixtures, valves and tapware. This ensures the correct installation of valves and water heaters complies with the requirements. Always check that the material, fittings, valves and fixtures are approved with the Watermark rating for compliance.

The Plumbing Code of Australia (PCA), which is volume three in the National Construction Code (NCC), is the overarching guideline and refers to AS/NZS 3500.1 Plumbing and Drainage: Water Services and AS/NZS 3500.4 Plumbing and drainage: Heated Water Services for the minimum Australian Standard requirements. Refer to Chapter 4 'Carry out interactive workplace communication' in *Basic Plumbing Skills* for further information on how to access relevant jurisdictional regulations and other standards.

All plumbing installations must have a permit, and local authorities must be advised before work is started. In most instances, an application for a permit must be submitted at least two working days before the work is begun. When carrying out work on some water service controls and devices, such as the installation of backflow prevention devices and pumps, notification and approval must be sought from the relevant water utilities before the work is carried out.

Pressure, velocity and flow rates

As per AS/NZS 3500.1, the maximum water pressure in a building other than a fire service is 500 kPa and the minimum is 50 kPa.

The maximum velocity in a building (the speed at which the water travels) other than a fire service is 3 m/second. This helps avoid excessive noise in the piping system.

Each tap, valve and outlet is required to meet a minimum flow (discharge) rate (see **Table 5.1**). A more in-depth list may be found in AS/NZS 3500.1. Note that no pipe sizes are indicated in **Table 5.1**. Pipe diameter may vary to meet the minimum flow rates and consumer needs.

TABLE 5.1 Minimum outlet flow rate

Outlet	Minimum flow rate
Basin (standard outlet) Shower Sink (aerated outlet)	6 L/min
Bath	18 L/min
Laundry tub	7 L/min

Water service controls and mixing devices

It is important to understand the valves required and their function for hot water heater installations as well as for whole plumbing systems. They perform a variety of functions, ranging from the simple task of isolating or controlling the water flow to a fixture or pipework section to more complex functions, such as automatically controlling the flow and pressure within a water mains system or high-rise building. Valves are also installed to ensure the safe operation of a system, such as pressure relief valves for hot water heaters, backflow prevention devices to protect the community's drinking water supply and thermostatic mixing valves to ensure that infants, children, older people and those with underlying health conditions are protected from the potential of scalding injuries from hot water.

AS/NZS 3500.0 defines a **valve** as a device for controlling the flow of fluid, with an aperture that can be wholly or partially closed by the movement relative to the seating of a component in the form of a plate or disc, door or gate, piston, plug, ball or flexing diaphragm. AS/NZS 3500.0 defines a **tap** as a valve with an outlet used as a draw-off or delivery point.

With Australia's increasing population and the threat of drought, there is more emphasis being placed on the conservation of resources. This has created a need for more environmentally friendly regulations, such as the Building Sustainability Index (BASIX) certificate and the Water Efficiency Labelling and Standards (WELS) star rating for new and renovated projects and products. These initiatives reduce water use, placing greater importance on the amount of water that is delivered through fixture taps and valves with the use of flow restrictors and non-drinking water services.

The BASIX model

The **Building Sustainability Index (BASIX)** is a model introduced in NSW to ensure that homes are designed to use less drinking (potable) water and reduce greenhouse gas emissions by setting energy and water reduction targets for homes and units. BASIX is a sustainable planning measure to deliver equitable and effective water and greenhouse gas reductions.

WELS rating

The **Water Efficiency Labelling and Standards (WELS) scheme** operates just like the star ratings on electrical appliances. It allows the designer, consumer or plumber to assess at a glance the water efficiency of the tapware being installed: the higher the water consumption, the fewer the number of stars on the label (see **Figure 5.1**).

Source: Images provided by DSEWPaC.

FIGURE 5.1 WELS ratings labels

GREEN TIP

The introduction of BASIX and WELS schemes has resulted in a large reduction in water and energy usage.

Identify and apply work health and safety (WHS) and environmental requirements

Work, health and safety (WHS) procedures involve hazard and risk management with the correct documentation. The work and the tasks associated with the installation of hot water heaters and valves must be properly assessed and the risk levels determined. These risks must then be eliminated or controlled to prevent or reduce the likelihood of accidents and injuries occurring during the performance of the work.

Some hazards may include:

- manual handling
- working at heights
- working in confined spaces
- hot works
- electricity
- hot water
- dust suppression.

A job safety analysis (JSA) should be completed before any tasks commence and a safe work method statement (SWMS) completed for any high-risk activity.

Electrical safety

When carrying out any work on a metallic water service connected to the water utility main, consideration needs to be given to electrical safety precautions and earthing to protect against potential electrical shock. Refer to the References at the end of this chapter for further information.

Always test for stray current and use bonding straps when disconnecting a metallic service to avoid electrocution.

Isolating valves

The flow from any water main to any water service pipe must be controlled by isolating valves such as stop taps, stop valves and non-return valves. All isolating valves on water services must be opened and closed slowly to prevent shock waves in the piping, which can cause leaks and burst pipes. This is because water is non-compressible. Isolating valve components are made from a wide range of materials, including brass, cast iron, steel, stainless steel, various polymers, rubbers and urethanes, and are available in a variety of designs to meet various requirements. The main types are:

FIGURE 5.2 Butterfly valve

- ball valves
- gate valves
- butterfly valves (these valves are used predominantly on large services and allow for full or partial closure of a service [see **Figure 5.2**]; they are not further elaborated on within this section)
- screw-down stop taps.

Isolating valves must be installed at the following water service locations:

- at the water main (at either a tapping or a tee insertion)
- at the water meter or within 1 m of the property boundary if no water meter is fitted
- at each flushing device/tank
- at each appliance
- at each backflow prevention device
- at each thermostatic mixing valve
- at each pressure-limiting valve
- at each commercial/industrial appliance or apparatus
- at each pumping apparatus
- at each storage tank inlet
- at each storage tank outlet (where capacity exceeds 50 L).

Isolating valves must be installed in multiple building and multistorey construction, as follows:

- at each branch serving individual buildings
- at each branch serving each floor in buildings of two or more storeys

- at each group of fixtures
- at each standpipe.

AS/NZS 3500.1 covers the function of isolating valves.

Isolating valves must be installed on fire services, as follows:

- at each water main connection
- at or near the property boundary
- at each hose reel
- at each pumping apparatus.

Ball valves

Ball valves are part of the rotary movement family of valves, which are often called quarter-turn valves and include butterfly valves.

Ball valves are opened and closed by turning a handle attached to a ball inside the valve body. The ball has a hole, or port, through the middle so that when the port is in line with both ends of the valve, fluid flow will occur. When the valve is closed, the hole is perpendicular to or across the ends of the valve, and the flow is blocked. When the valve is open, the handle or lever is usually in line with the port; its position therefore indicates whether the valve is open or closed (see **Figure 5.3**).

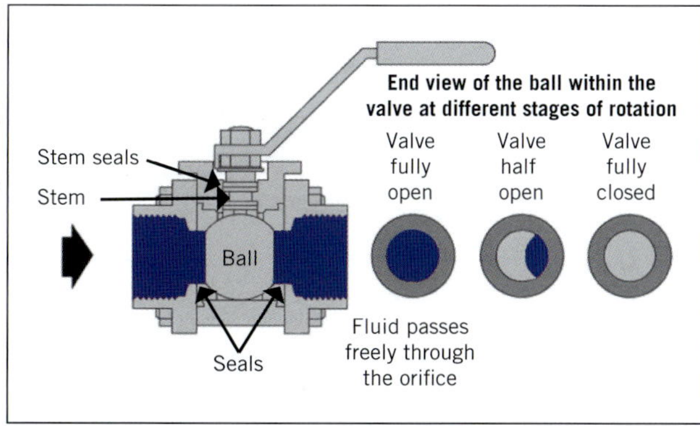

Source: Spirax Sarco. Illustrations and text are copyright, remain the intellectual property of Spirax Sarco and have been used with their kind permission; http://www.spiraxsarco.com/resources/steam-engineering-tutorials.aspx.

FIGURE 5.3 Cross-section of reduced bore ball valve

There are two basic designs of ball valves:

1. the floating ball design, which relies on the valve seats to support the ball
2. the trunnion-mounted ball, which uses a trunnion to support the ball. Trunnion mounting is used on larger valves, as it can reduce the operating effort required to open and close the valve to about two-thirds of that provided by a floating ball.

Ball valves are available as either reduced bore or full bore. **Full bore valves** have an orifice that is the same size as the diameter of the pipe, whereas in **reduced bore valves** the orifice diameter is less than that of the pipe. Full bore valves provide full flow, so there is no pressure loss across the valve.

Ball valves are durable and usually work to achieve perfect shut-off even after years of non-use. However, they do not offer the fine control that may be necessary in throttling applications, nor do they offer any backflow protection.

The body of ball valves may be made of metal and metal alloys such as brass, plastic or metal with a ceramic centre. The ball is often chrome plated or manufactured from stainless steel to make it more durable.

Gate valves

Gate valves (see **Figure 5.4**) are widely used valves in industrial applications (although it is not suitable for use as a throttling valve that regulates flow). They are mainly used as stop valves; however, their primary purpose is to fully shut off or fully turn on flow slowly and provide full flow, such as when fitted to a storage tank outlet. Fluid flow moves in a straight line and with minimal resistance when the wedge is fully raised.

FIGURE 5.4 Gate valve

Seating and operation is at right angles to the line of flow where the fluid flow is met head on. Seat types include a metal seat and/or a resilient seat.

A gate valve usually requires more turns to open it fully. Larger-diameter gate valves, or **sluice valves** as they are often called, open and close in the opposite direction to conventional screw-down valves. The direction for opening and closing is usually marked with an arrow on the valve stem or hand wheel.

Gate valves can be opened and closed via either a rising stem or non-rising stem. For the non-rising stem gate valve, the stem is threaded on the lower end into the gate.

Screw-down stop taps and valves

A common type of valve used in a water service as an isolating valve or stop tap is the screw-down type. This has the simple tap washer (loose jumper valve), which slowly lowers onto the body seat by a spindle to shut the flow of water. This slow-closing valve helps eliminate water hammer. When replacing jumper valves (tap washers), be sure the spindle is screwed out (open) so the washer is not damaged. Care should be taken to ensure that they are rated at the appropriate temperature. Some plastic jumper valves may fail under hot water conditions and result in injury.

Screw-down type valves are commonly installed as stop taps. They may be used as isolating valves or stop taps for hot water heaters, or installed as recess taps, pillar taps, **bib taps** (see **Figure 5.5**), hose taps, meter taps, main taps, cistern taps and washing machine taps. They do not provide full flow. The stop valve (**Figure 5.5**) is fitted inline and has a threaded inlet and outlet for connection to pipes.

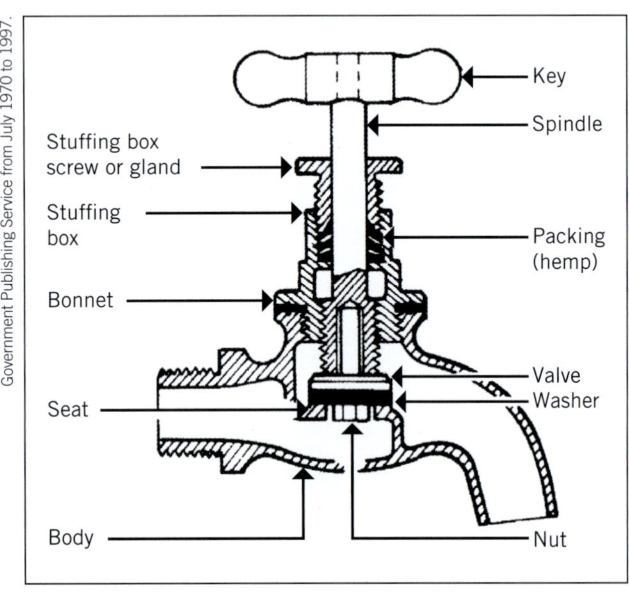

Source: *Basic Training Manual 11–2 Water Supply 3*, Australian Government Publishing Service from July 1970 to 1997.

FIGURE 5.5 Bib tap with valve closed

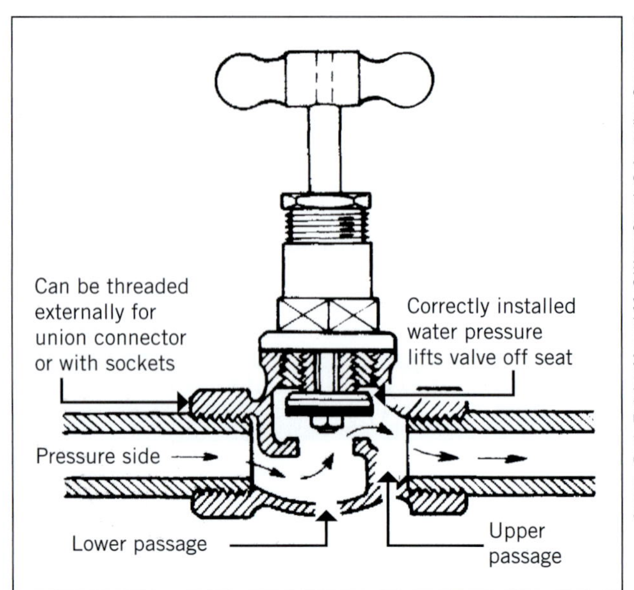

Source: *Basic Training Manual 11–2 Water Supply 3*, Australian Government Publishing Service from July 1970 to 1998.

FIGURE 5.6 Stop valve (open)

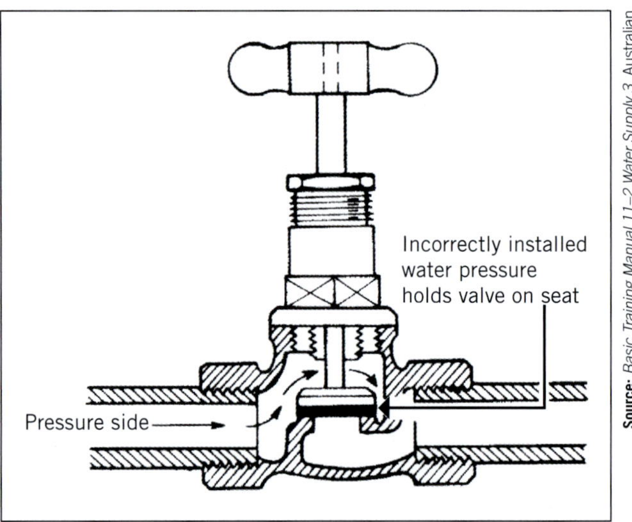

Source: *Basic Training Manual 11–2 Water Supply 3*, Australian Government Publishing Service from July 1970 to 1999.

FIGURE 5.7 Stop valve (closed)

Care is needed when installing a loose jumper valve, as flow can be in only one direction (see **Figure 5.6**). Installing the valve the wrong way will cause water pressure to hold the valve on its seat when the tap is opened, resulting in no flow through the valve (see **Figure 5.7**). Most stop valves have the direction of water flow marked on the body by an arrow to assist in correct installation (see **Figure 5.8**).

Older stop valves have a stuffing box screw, or gland, which screws into the stuffing box to compress the packing gland around the spindle to prevent water leaking past it (see **Figure 5.9**). Most modern taps are made with an O-ring, which has replaced the packing gland (see **Figure 5.10**); however, some stuffing box taps are still available, such as main taps.

FIGURE 5.8 Stop valve showing flow direction on the valve body

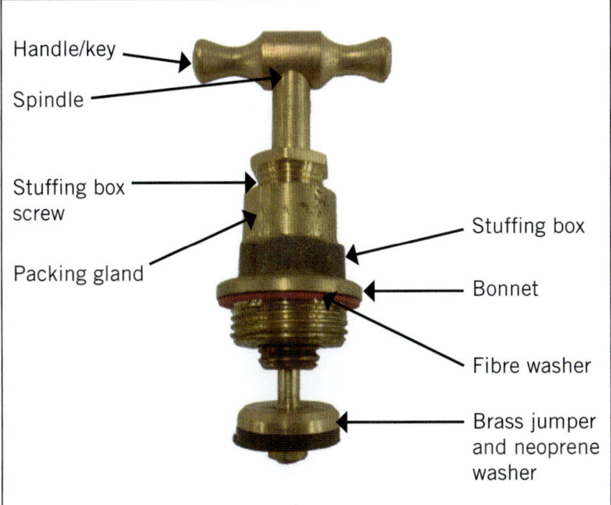

FIGURE 5.9 Stuffing box assembly and jumper valve

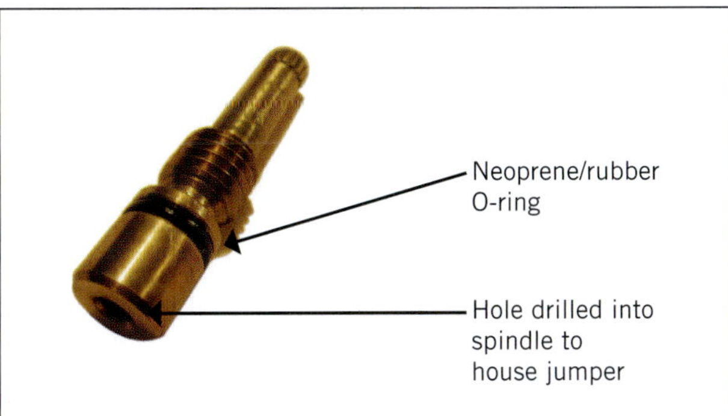

FIGURE 5.10 Spindle and O-ring

The bonnet holds the working parts and screws into the body. Water leakage between the bonnet and the valve body is prevented by inserting a fibre washer between them (see **Figure 5.11**). The bonnet has an internal screw thread into which the threaded part of the spindle engages. The lower part of the spindle is bored to house the valve stem and act as a guide.

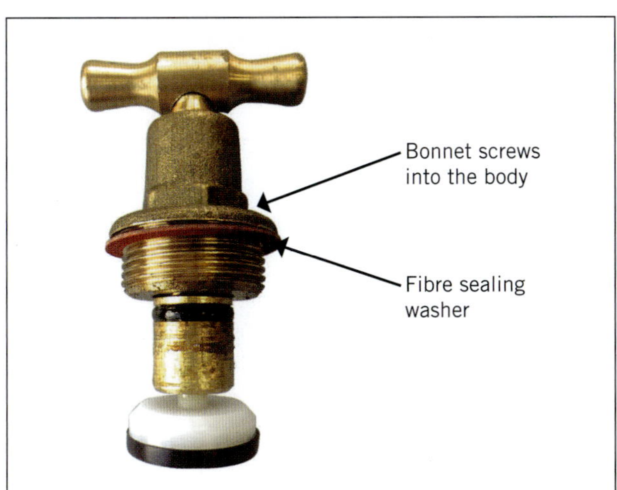

FIGURE 5.11 A typical jumper valve (tap washer) with its stem housed in the bored-out spindle

Ceramic disc technology

Ceramic disc taps use ceramic discs instead of conventional washers/jumper valves to give a quick action and are typically used in single-lever or half-turn taps. Ceramic washers are more resistant to wear and tear and therefore have a longer life. They are also less likely to leak and are not affected by any lime or solids in the water (see **Figures 5.12**, **5.13**, **5.14**). These types of taps are easier to operate and suitable for people with disabilities that affect their motor skills, such as arthritis, as they require less effort to turn the tap handle and usually require only a half turn of the handle to fully open or close the tap. When the tap is turned on or off, the lower of the two discs remains stationary while the upper disc turns with the movement of the tap handle. Water will flow only when the upper and lower slots match up in the open position. The quick-acting shut-off can cause water hammer. Consequently, most ceramic disc taps now have a half-turn action instead of a quarter-turn action to help eliminate water hammer.

Care must be taken not to overtighten the spindles when installing ceramic disc tapware so the ceramic discs are not damaged. Also, be sure to flush the pipes of debris before installing new tapware.

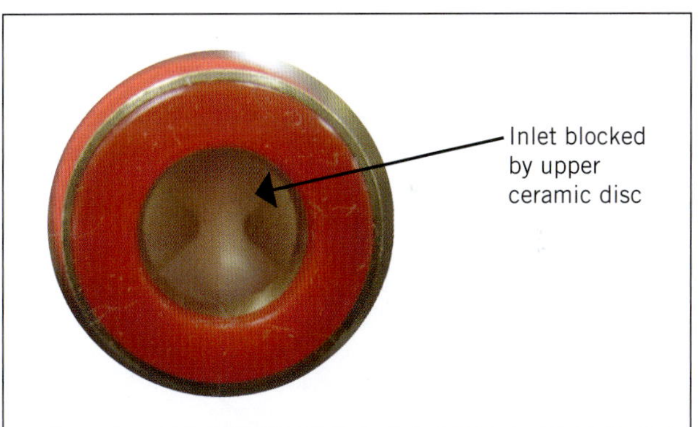

FIGURE 5.12 Ceramic disc inlet closed

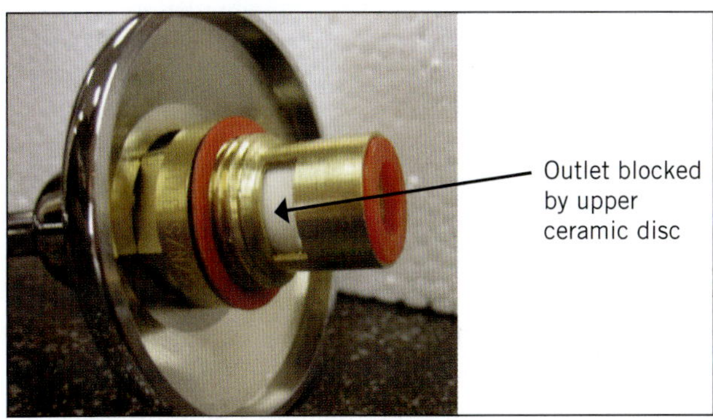

FIGURE 5.13 Ceramic disc cartridge outlet closed

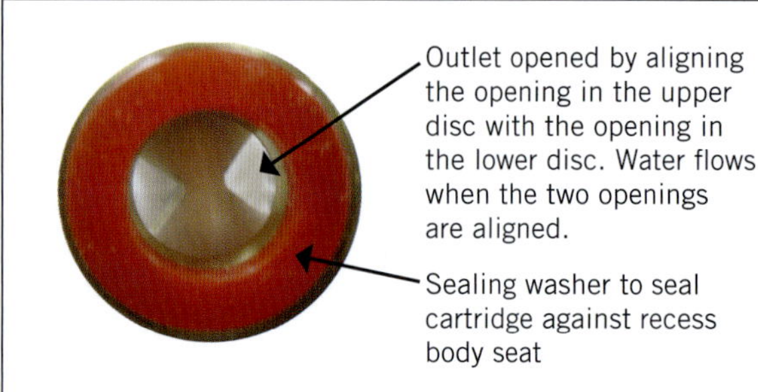

FIGURE 5.14 Ceramic disc open

Other types of control valves

There are many types of valves in the plumbing industry that are used for different applications. This section will explain the function and operation of regulating valves, non-return (check) valves, pressure reduction valves, line strainers and solenoid valves.

Globe valves

Globe valves are designed specifically for regulating or throttling water flow in a pipeline. They consist of a movable disc-type element and a stationary ring seat and are named because of the spherical appearance of their body. (see Figure 5.15).

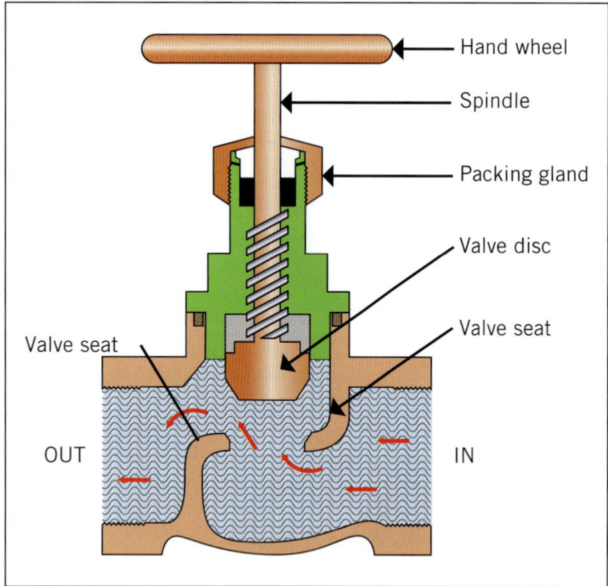

FIGURE 5.15 Cross-section of a globe valve

Globe valves are often installed in flow and return hot water services to balance the flow rates between the flow and return lines in the system. They are also installed in steam and other specialist fluid applications and can be controlled either manually or automatically.

The valve should be installed in the upright position with the inlet below the valve seat in a similar manner to screw-down-type valves.

Duo and trio valves

A **duo valve**, or **combined non-return/isolating valve** (see Figure 5.16), offers the dual functions of being an isolating valve and a non-return valve. These valves are used on installations of pressurised storage water heaters and serve as a compact, easy-to-install device.

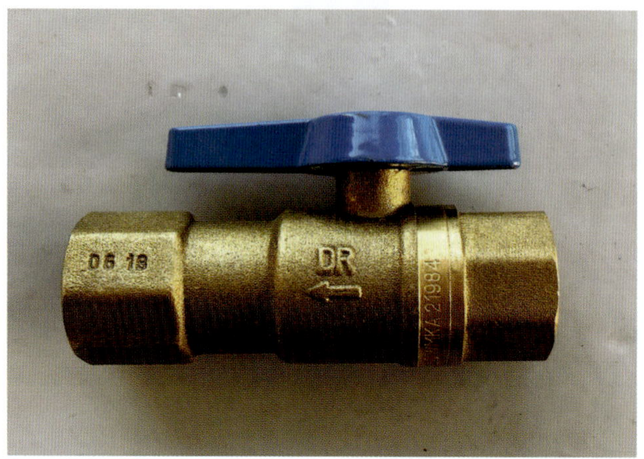

FIGURE 5.16 Combined stop non-return (duo) valve

Another versatile valve is the **trio valve** (see Figure 5.17), which has the combined functions of non-return, isolator and line strainer. In the trio valve, the line strainer protects both the internal mechanisms of the valve and the devices downstream of the valve from damage due to contaminants in the water supply and is commonly used in hard water areas.

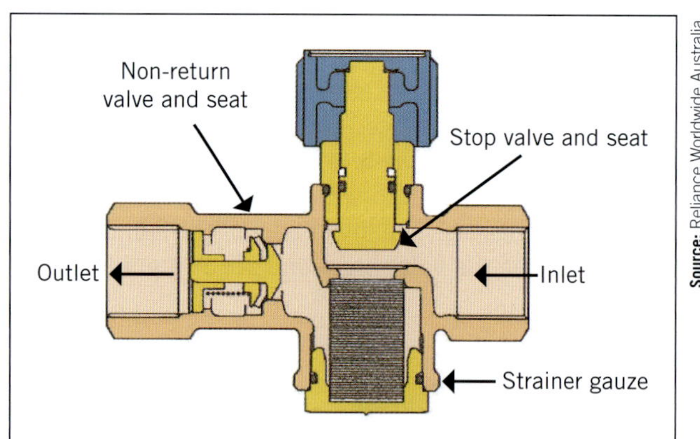

Source: Reliance Worldwide Australia.

FIGURE 5.17 Trio valve

FROM EXPERIENCE

Remember to be aware of the water quality in the area you are working. If the water is high in minerals, usually calcium and magnesium, it is known as hard water. So be sure to install line strainers to protect valves and tapware.

Solenoid valves

A **solenoid valve** (see Figure 5.18) can be described as an electrically controlled or electromechanical valve for use in the control of the conveyed materials by allowing or preventing an electrical current to pass through a **solenoid**. A solenoid is basically a coil of wire that creates a magnetic field when an electrical current is passed through it and is connected to a magnet that controls the valve. When the current is active, the solenoid will open/close the valve, and the opposite will happen when the circuit is broken. Solenoid valves may be used to control automatic flushing systems, infra-red tapware and a range of controls in the plumbing industry. They are commonly used in clothes washing machines and dishwashers to control the water cycles. As mentioned earlier, solenoid valves can cause water hammer because they are a fast-closing valve.

FIGURE 5.18 Solenoid valve

LEARNING TASK 5.1

1. What is the difference between a storage water heater and a continuous water heater?
2. What position must the spindle be in when changing a loose jumper valve (tap washer)?
3. What are the two functions of a duo valve?

COMPLETE WORKSHEET 1

Ancillary valves

Ancillary valves are valves that are installed upstream of equipment to assist in the operation and performance of a particular fixture. They play an important role by protecting downstream valves and equipment.

Line strainers

Line strainers (see Figure 5.19) are installed to protect downstream valves and fittings from impurities and debris that may be contained in the water supply line. Valves are often expensive and are delicate precision devices. The most common cause of malfunction in most controls and devices is foreign matter such as sand, hemp, teflon, dirt or rust getting stuck under the valve seats. The installation of a line strainer is the most cost-effective way to prevent the foreign matter from entering the valve. In many products, line strainers are built into the body of the valve and are available in several designs for specific applications. Line strainers are mandatory when installing backflow prevention devices and thermostatic mixing valves.

Source: Reliance Worldwide Australia.

FIGURE 5.19 Line strainer

Non-return valves

A **non-return valve** is a valve to prevent reverse flow from the downstream section of a pipe to the section of pipe upstream of the valve. Non-return valves are also called 'check' valves, and are installed on mains pressure storage water heaters and thermostatic mixing valves. Non-return valves are available in several basic types, as shown in Figure 5.20.

Non-return valves are also available for specialised applications, such as a high-temperature version suitable for use in solar water heater systems. Solar and uncontrolled heat source water heaters present an increased hazard to downstream components, as water can become superheated and potentially flash to steam. Only high-temperature non-return valves should be used in these applications.

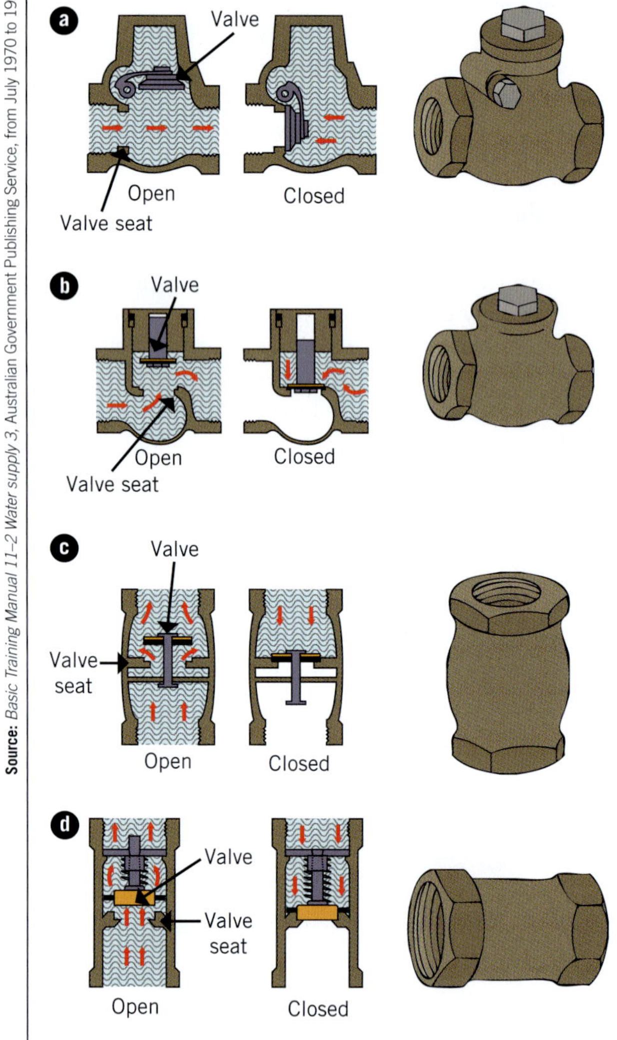

FIGURE 5.20 Types of non-return valve

(a) The swing check – must only be installed in a horizontal position.
(b) The horizontal lift – must only be installed in a horizontal position.
(c) The vertical lift – can only be installed in a vertical position.
(d) The spring loaded (the most common type of non-return valve used today) – may be installed in a horizontal or a vertical position.

Pressure-limiting valves

Pressure-limiting valves are designed to reduce a high inlet pressure to a standard specified outlet pressure. When being purchased for installation in a domestic situation, they are usually manufactured to limit the outlet pressures to 350 kPa, 500 kPa and 600 kPa settings. They are ideal for use in domestic applications such as water heaters and water meter assemblies.

Pressure-limiting valves control inlet supply line pressure to a fixture at a preset maximum pressure and will remain in the open position if the pressure in the supply line to the valve falls below that preset pressure. These valves are suitable in areas with high water pressure and are installed:

- on the inlet side of the hot water system
- on the inlet side of the dishwasher
- at the boundary of the house on the outlet of the water meter.

They can be installed either horizontally or vertically and are available in either a barrel-design or a T-design (see Figure 5.21). Figure 5.22 shows a cross-section of a barrel-design pressure-limiting valve.

FIGURE 5.21 Pressure-limiting valves

(a) T-design pressure-limiting valve
(b) Barrel-design pressure-limiting valve

Pressure-reducing valves

Pressure-reducing valves deliver a preset pressure, which avoids pressure fluctuations throughout the home installation, and are suitable for use in areas of high water pressure. Pressure-reducing valves are designed to be installed at the boundary of the property, in conjunction with the water meter.

Pressure-reducing valves (see Figure 5.23) automatically reduce a higher inlet pressure to a steady lower downstream pressure, regardless of any changes in the upstream flow rate or pressure. They are ideal for domestic installations because they create less resistance to water flow.

Pressure-reducing valves come in either a right-angle or a straight-through configuration and are used in residential installations. Using a pressure-reducing valve can also minimise water wastage.

Domestic pressure-reducing valves can deliver high flow rates with minimal head loss. These are available in both adjustable pressure (150–600 kPa) and pressure-locked at 500 kPa.

Pressure-reducing valves also can be used to reduce pressure upstream of some commercial and industrial devices (see Figure 5.24), such as dosing apparatus, high-pressure cleaners and laboratory equipment. Pressure-reducing valves are favoured in these situations because they deliver a more accurate preset outlet pressure compared to pressure-limiting valves.

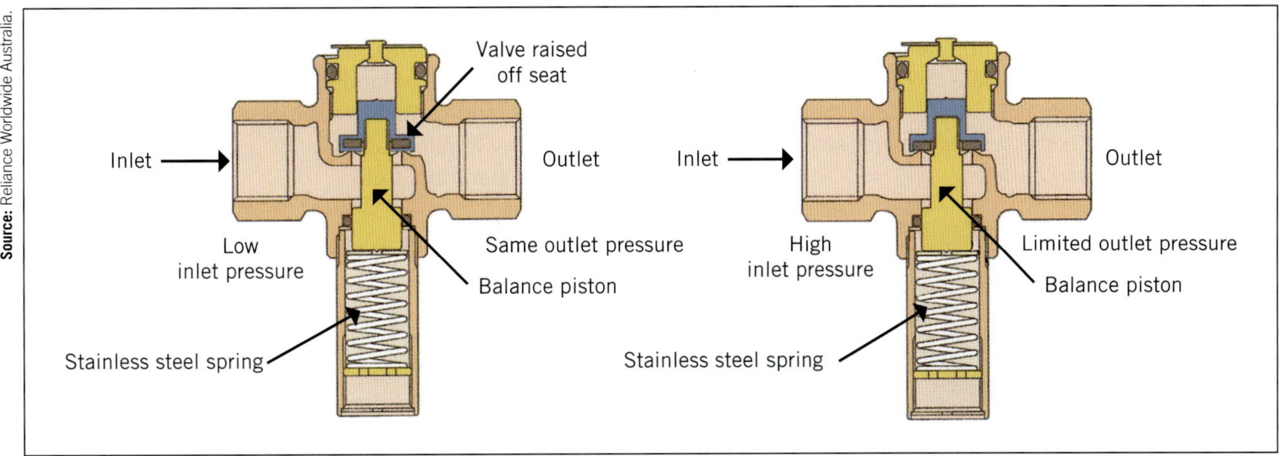

FIGURE 5.22 Cross-section of a T-design pressure-limiting valve, showing pressure-limiting valve: fully open (left) and in operation (right)

FIGURE 5.23 Pressure-reducing valve

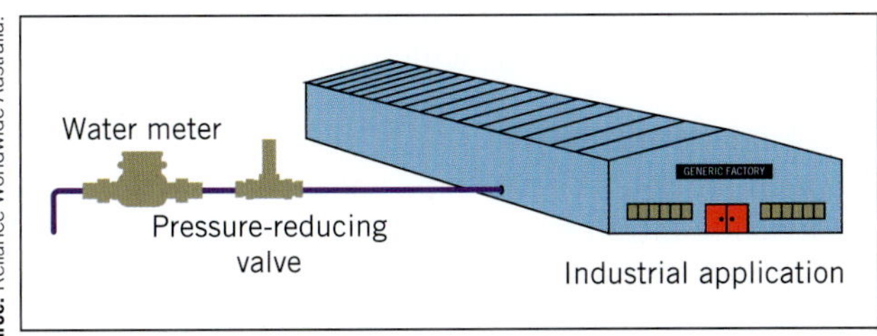

FIGURE 5.24 Installation of a pressure-reducing valve for industrial service

Pressure ratio valves

Pressure ratio valves, or mains proportion valves as they are sometimes referred to, are designed to reduce the outlet pressure of the valve by a set ratio to that of the inlet pressure. For example, a valve set at a 2:1 ratio with an inlet pressure of, say, 100 kPa would deliver an outlet pressure of 50 kPa.

As the outlet pressure varies with the inlet pressure, these valves are not suitable where excessive pressure fluctuations occur, such as in domestic situations or applications where there is a peak demand period. However, they are ideal to break down the pressure from storage tanks supplying water in high-rise buildings.

Spring-loaded taps

Spring-loaded taps, or self-closing taps, are used mostly for drinking fountains and pillar taps in public places where there is a danger of taps being left open with water running. Spring-loaded taps are quick-acting valves, and their sudden closure may produce water hammer in the pipes. Anti-hammer devices may be installed in the service pipes close to the offending fitting, or they can be built into the fitting itself (for example, **dashpots**). These devices act as shock-absorbers to the force of water.

Leakage past the spindle requires tightening of the stuffing box gland or replacement of the greased hemp in older types. In more modern types, the hemp and gland have been replaced by a rubber O-ring fitted into a groove that must be lubricated with a silicone-based grease.

Aerators

Essentially, **aerators** are fitted to tapware outlets to add air to the water stream, resulting in a softer flow to reduce the effects of splashing and, more importantly, reduce the flow of the water. Aerators come in several sizes and styles and can reduce flow rates by up to 50% while maintaining washing effectiveness.

Aerators have three basic components (see Figure 5.25):

1. *Housing* – the housing connects to the thread of the tap and provides the outer covering for the insert.
2. *Insert* (flow restrictor) – the insert regulates the tap's water flow. It is the insert that is chosen and

fitted to suit the various flow-rate requirements. For example, a 5 L per minute aerator should be fitted to hand basins and a 7.6 L per minute aerator is best for kitchen and laundry applications.

3 *Rubber washer* – the sealing washer seals the aerator to the tap outlet.

FIGURE 5.25 Three components of aerators

> **FROM EXPERIENCE**
>
> It is important to remove all aerators and flush all debris before refitting when commissioning a water service.

Sensor-controlled devices

Infra-red sensing tapware is used in many situations for the operation of tapware without having to physically touch the tap. This helps to prevent cross-contamination in sterile areas.

Wall-mounted sensors

Hands-free infra-red tapware uses pulsed infra-red light beams to offer protection from false triggering while being easy for the user to operate (see **Figure 5.26**). The power supply is usually a 24 V AC power supply, which includes surge suppressors. To protect from damage and ensure the tap gives years of reliable service, it is recommended that a strainer is installed before the solenoid valve.

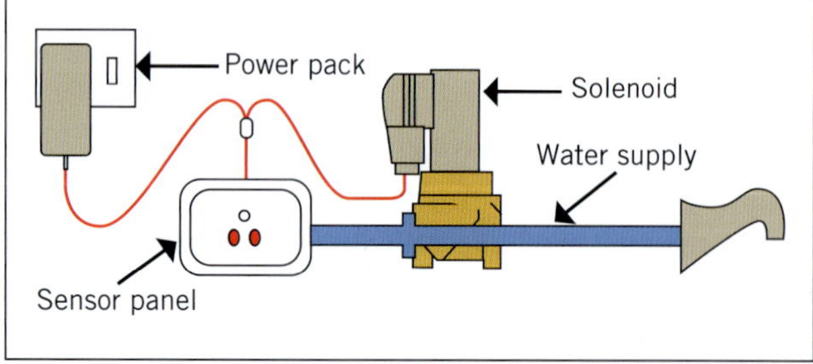

FIGURE 5.26 Typical configuration when installing wall-mounted infra-red-operated tapware

Infra-red sensing tapware can help reduce the risk of cross-contamination caused by physical contact between the operator's hands and a tap handle after another person has touched the same handle. This is of particular importance when a sterile environment must be maintained. Hands-free infra-red tapware is suitable for, and can be beneficial when installed in, the following applications:

- food preparation areas
- factories
- scrub-up areas
- operating theatres
- dental clinics
- laboratories
- sterilising rooms
- abattoirs
- restaurants
- veterinary clinics
- photography labs
- disabled-access bathrooms
- institutions (schools, kindergartens, etc.)
- public amenities.

To operate the tap, the user will move their hand within 5–10 cm of the sensor to turn on the water. A second pass in front of the sensor will turn the water off or, if the water is allowed to run, the sensor will turn itself off after approximately 45 seconds. An LED (light-emitting diode) indicator on the sensor plate lights up when an object is being sensed, so it is easy for the user to place their hand in the correct position to operate the tap.

When choosing a location to install a wall-mounted sensor, ease of operation, passing traffic and obstructions directly in front or within possible range of the sensor must be considered. Up to 500 mm clearance may be necessary from reflective surfaces such as ceramic tiles and stainless steel, directly in front of and parallel to the front face of the sensor. Leads should be installed through a conduit for easy removal. Installation instructions are supplied with every product and the manufacturer should be consulted if any problems with installation arise.

Bench-mounted sensors

A bench-mounted sensor tap is also controlled by an invisible infra-red beam (see **Figure 5.27**). This unit includes simple installation and compact design with the options of battery or mains power. The bench-mounted spout has the sensor at the base of the body, which also encapsulates the electronics and solenoid.

When the sensor detects motion, this triggers the solenoid valve and the water will flow as long as the user's hands remain in the vicinity of the sensor, directly under the spout. When the user removes their hands from the sensor range, the water will turn off after a second. No movement in front of the sensor will stop the flow of water.

FIGURE 5.27 Bench-mounted sensor

The unit is suitable for installation in most single-hole basins, but is not recommended for stainless steel sinks due to beam reflections. The base diameter is 50 mm and the tap includes flexible hose for water connection and fixing plate with surface protection, O-ring seal and hex nut.

Programmed and on-demand sensor flushing

In commercial applications, such as office buildings, stadiums, public buildings and educational institutions, flushing to urinals is often electronically controlled using sensors. This plays an essential part of water conservation, particularly in large buildings. Note that due to significant water wastage, automatic or set-cycle cisterns are no longer permitted.

GREEN TIP

On-demand sensor flushing has dramatically reduced water wastage to conserve water.

Urinals (individual, grouped or slab) are the most likely fixtures to have automatic or programmed flushing systems. It should be noted that urinals have different flushing volumes from toilet pans. AS/NZS 3500.1 states that the quantity of water discharged for sanitary flushing must be not more than 2.5 L and not less than 1.5 L for each single stall or each 600 mm length of continuous urinal wall.

Automatic flushing systems can also be fitted to existing or newly installed cisterns.

AS/NZS 3500.1 PLUMBING AND DRAINAGE: WATER SERVICE

HOW TO

INSTALL A SENSOR TO ACTIVATE URINAL FLUSH

1 Position the ceiling sensor not more than 500 mm from the urinal wall. Sensor coverage from a 2700 mm ceiling is approximately 2400 mm × 3600 mm.
2 For a single stall urinal, position the sensor above the centre of the stall. Position the sensor slots at right angles to the urinal wall.
3 For a double stall urinal, position the sensor midway between the stalls. Position the sensor slots at right angles to the urinal wall.
4 For a triple stall urinal, position the sensor above the centre of all three stalls. Position the sensor slots parallel to the urinal wall.
5 For more than three stalls, use additional sensors. Remember not to connect the battery or power pack until all plumbing connections are completed.

The power must be connected last, as its connection activates the system test mode.

Flushing cistern

A **cistern** is a tank in which water is stored at atmospheric pressure. The tank incorporates a means of both controlling the water level and releasing the water when it is required to flush the fixture, with the water entering the cistern controlled by a float valve; it also has an air gap to prevent cross connection to the water supply. The water leaving the cistern is controlled by a flush valve assembly.

Float valves

Even though there are a variety of designs of **float valves**, their basic operation remains essentially the same. The float, which is usually a hollow plastic ball, is attached to a lever arm that operates a plunger fitted with a rubber washer. The float valve is designed to regulate the flow of water into a **flushing cistern** or tank to a predetermined level. When the required water level has been reached, the float forces the rubber-faced plunger against the seat of the inlet valve to shut off the water supply to the tank. If the water level in the tank drops, the float also drops along with the water level, and the lever pivoting on its fulcrum pin follows the float, lifts the plunger off its seat and so allows water to enter and top up the tank/cistern. As the water level gradually rises, the process is reversed: the float with its arm gradually closes the inlet valve, until the required water level has been reached (see Figures 5.28 and 5.29).

FIGURE 5.28 The Plymouth-type float valve

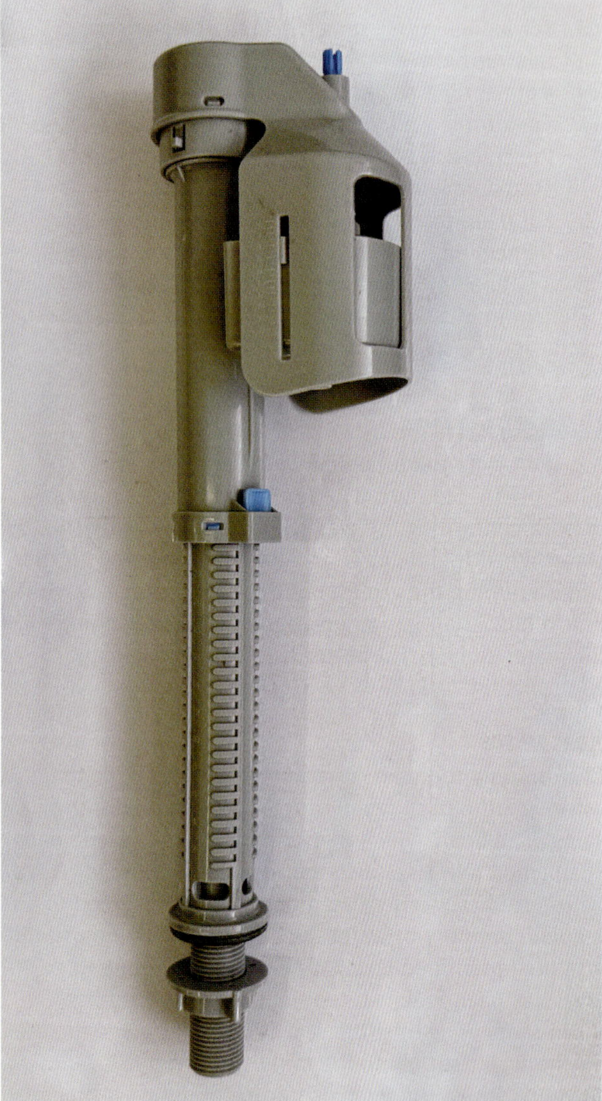

FIGURE 5.29 Modern style float valve

Flush valve

A **flush valve** is a valve that controls the water that flushes a fixture such as a toilet or urinal. The water is supplied from a pressurised water supply system or an elevated storage tank. It is also known as a **flushometer** or **flusherette**.

Once activated, the flush valve allows a predetermined amount of water to pass through it into the fixture and then automatically shuts off when the flush is complete.

Flush valves can be specified instead of cisterns in high-demand situations such as office blocks, sports stadiums and theatres. This is because, unlike cisterns, they can be flushed more often as there is a constant water supply. They do not suffer from the time delay required to fill the cistern.

Flush valves can be activated by either manual push button/lever or sensor. The sensor is usually set to activate the valve when the user has moved away from the fixture.

When the water is supplied by a storage tank, the tank is installed at a higher level to produce the necessary head (a minimum of 3 m – approximately 30 kPa; and a maximum of 30 m – approximately 300 kPa). There is a section in AS/NZS 3500.1 that states the minimum pipe sizing requirements and pressure requirements.

An example of a tank-fed flush valve is shown in Figure 5.30.

Mains pressure flush valves are more common on new installations than tank-fed flush valves these days and require a minimum DN 25 pipe to supply the volume of water required for a scouring flush (see Figure 5.31). Because mains pressure flush valves are supplied directly from the mains pressure water supply, they must have an integral vacuum breaker valve fitted for backflow prevention. See Chapter 9 'Install backflow prevention devices' for more information on the backflow prevention devices.

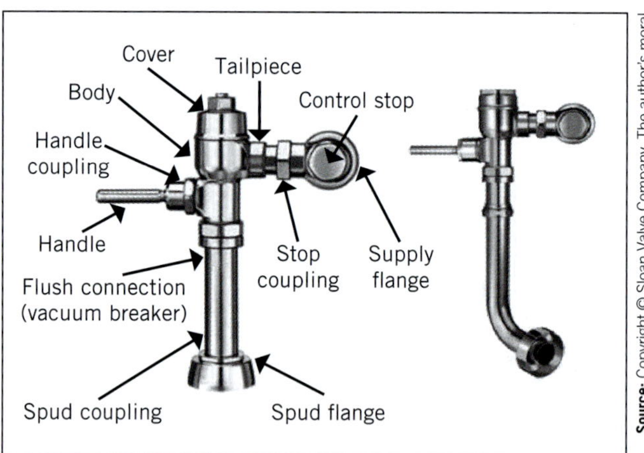

FIGURE 5.30 Tank-fed flush valve

AS/NZS 3500.1 PLUMBING AND DRAINAGE: WATER SERVICES

Backflow prevention devices

Backflow prevention is an extremely important part of any drinking water system design. A **cross connection** between the drinkable water supply and a potential source of contamination can pose serious health risks to the individual user and/or the wider community. There are numerous well-documented cases where cross

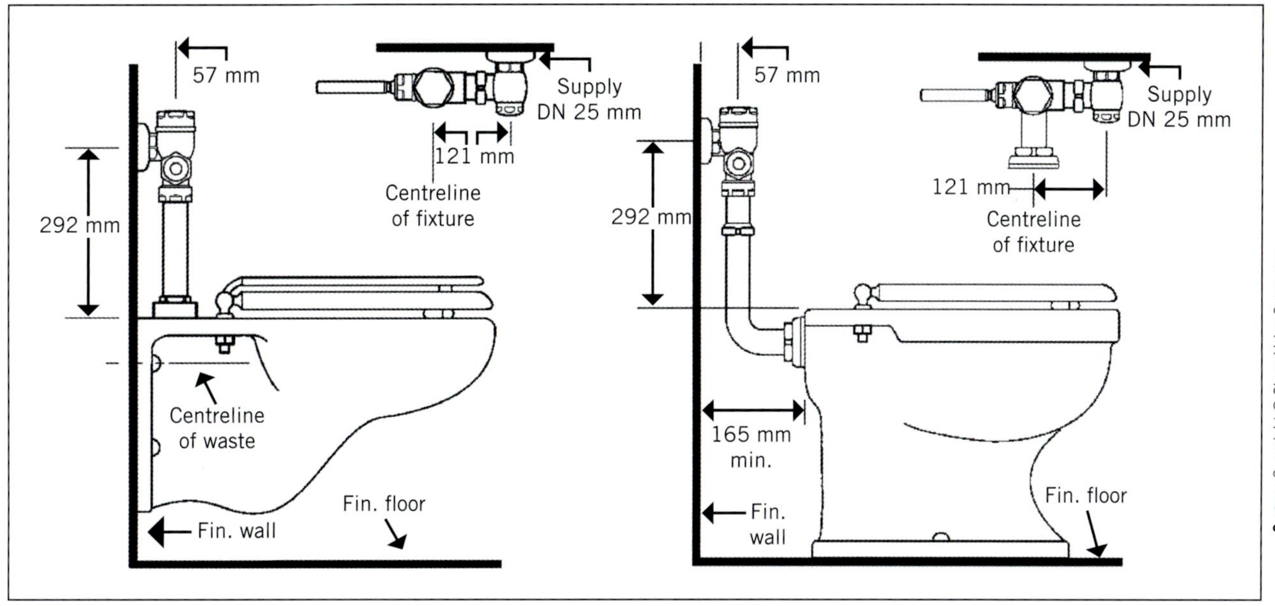

FIGURE 5.31 Flush valve connected to toilet pans: (Left) mains pressure flush valve top inlet pan, (Right) flush valve rear inlet pan

connection have been responsible for the contamination of drinking water supplies, resulting in the spread of disease. The devices and valves used in the prevention of cross connection are explained in depth in Chapter 9 'Install backflow prevention devices'.

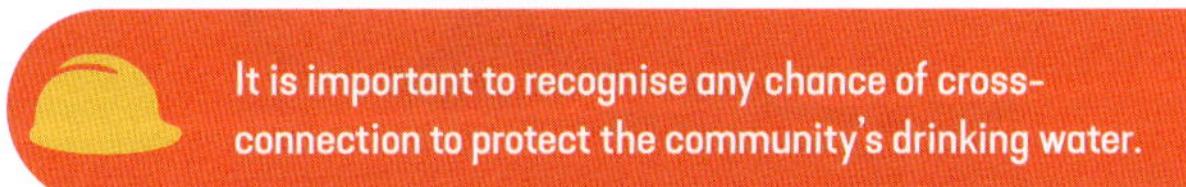

Water hammer and noise transmission

Water hammer is sometimes referred to as **line hammer** and it is due to the sudden stop of a column of water moving at speed through the pipes. This occurs because water is non-compressible. The condition is identified by a noise similar to a hammer blow on the pipes and over time can weaken or even burst them. Water hammer and noise transmission are commonly caused by excessive water pressure/velocity, as well as poorly constructed or supported pipework. Water hammer also occurs from jumper valves vibrating in an open-valve situation, such as a washing machine tap, dishwasher tap, cistern tap or in-line isolating valves. Fast-closing valves – such as solenoid valves on clothes washing machines and dishwashers, single-lever mixer taps and quarter-turn ceramic disc tapware – also can create water hammer.

Water hammer arrestors

Water hammer arrestors or **line hammer arrestors** are devices that are installed on water services to eliminate water hammer. They are chambers in which air is sealed, and operate similarly to a shock absorber, either via a flexible diaphragm or via a piston. This reduces the shock of the water stopping quickly, which reduces the noise transmission. Some arrestors have air recharged through a Schrader valve (tyre valve), but the most common types are completely sealed and are not rechargeable. They can be spherical or cylindrical in shape and may be mounted either vertically or horizontally (see Figure 5.32). They must be installed as close as possible to the offending source of line hammer and preferably be without any intermediate bends.

FIGURE 5.32 Water hammer arrestor

FROM EXPERIENCE

Using problem-solving skills to locate and eliminate water hammer impresses the client.

Mixing valves

Mixing valves are used to provide water of a desired temperature at outlets for domestic, commercial and industrial buildings. They are connected to a hot and a cold water supply source, with the desired water temperature at the outlet being obtained through the intermixing of the two water supply sources by the manipulation of one or two valves.

The breeching piece

The simplest way of providing water mixing is by connecting the outlets of hot and cold water with a **breeching piece (shower)** (see Figure 5.33) to a common outlet. Mixing takes place at the junction of the two branches and the flow to the required temperature is controlled by the degree to which each of the two taps is opened.

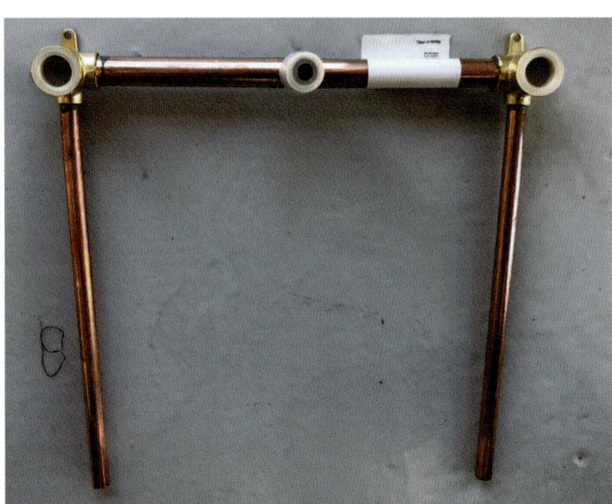

FIGURE 5.33 Typical breeching piece

Single-lever mixing valves

Single-lever mixers, or flickmixers as they are more commonly known (see Figure 5.34), are available for installation on kitchen sinks, laundry troughs, vanity basins, baths and showers (see Figure 5.35). They are available in many styles and qualities. Most single-lever mixing valves use ceramic disc technology to control the water flow. Most of the ceramic disc assemblies are contained in a cartridge that is replaceable.

The movement of a central lever controls the basic operation of a single-lever mixer. When the lever is lifted, the water is allowed to flow to the outlet. Single-lever mixers should be connected to the supply service in such a manner that when the lever is moved towards the left, the water at the outlet should become hotter, and when the lever is moved towards the right, the water should become colder.

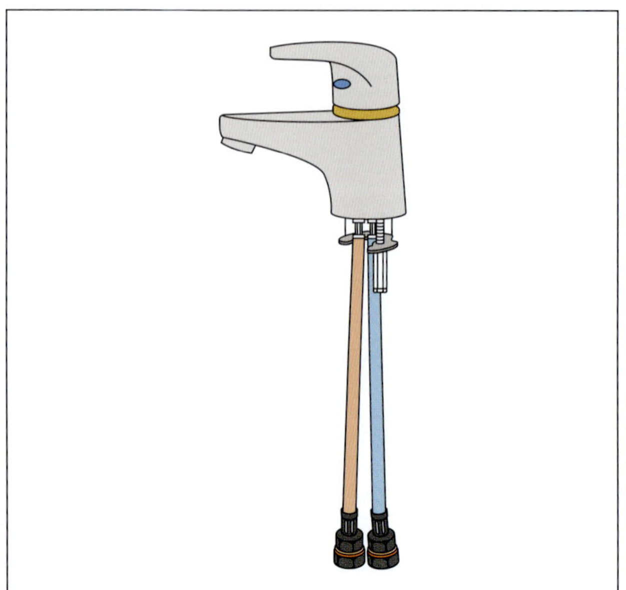

FIGURE 5.34 Single-lever mixing valves

FIGURE 5.35 Basin and sink mixers

When installing single-lever mixing valves, they should be connected with isolating valves. This provides limited backflow prevention and will also allow the mixer to be isolated should the ceramic disc fail. A backflow prevention device such as a check valve should also be fitted, particularly if the mixer has a hose as an outlet, as this can potentially cause a cross connection.

Water temperature

Hot water must be stored at a minimum temperature of 60°C; this is designed to inhibit the growth of *Legionella* bacteria.

To avoid scalding, hot water installations must be designed to deliver heated water at maximum temperatures used for personal hygiene (ablution fixtures), such as baths, showers, basins and bidets. As per AS/NZS 3500.4, they must not exceed:

- 45°C for buildings such as:
  - early childhood centres
  - primary or secondary schools
  - nursing homes or similar facilities looking after young, elderly or sick people or people with a disability (a check of state or territory temperature requirements for the above should be made prior to any installation proceeding)
- 50°C in all other buildings, including residential buildings.

Laundries and kitchen sinks are not a part of these water temperature requirements.

Tempering valves

Tempering valves are used to mix hot water with cold water as it leaves the hot water service to supply ablution fixtures (basin, bath and shower) at a temperature of 50°C (see **Figures 5.36**, **5.37** and **5.38**). AS/NZS 3500.4 requires hot water to be delivered to all domestic fixtures that are used for ablution purposes to be at 50°C to prevent severe scalding. Water at 60°C will cause third-degree burns within five seconds, and at 70°C in one second.

A tempering valve operates where cold water enters the mixing chamber above the piston and hot water enters the mixing chamber below the piston. The thermostatic controlling device (wax element) is positioned in the mixing chamber connected to the piston. The expansion or contraction in the length of the wax element will cause the piston to move either up or down. In the event of an increase in the hot

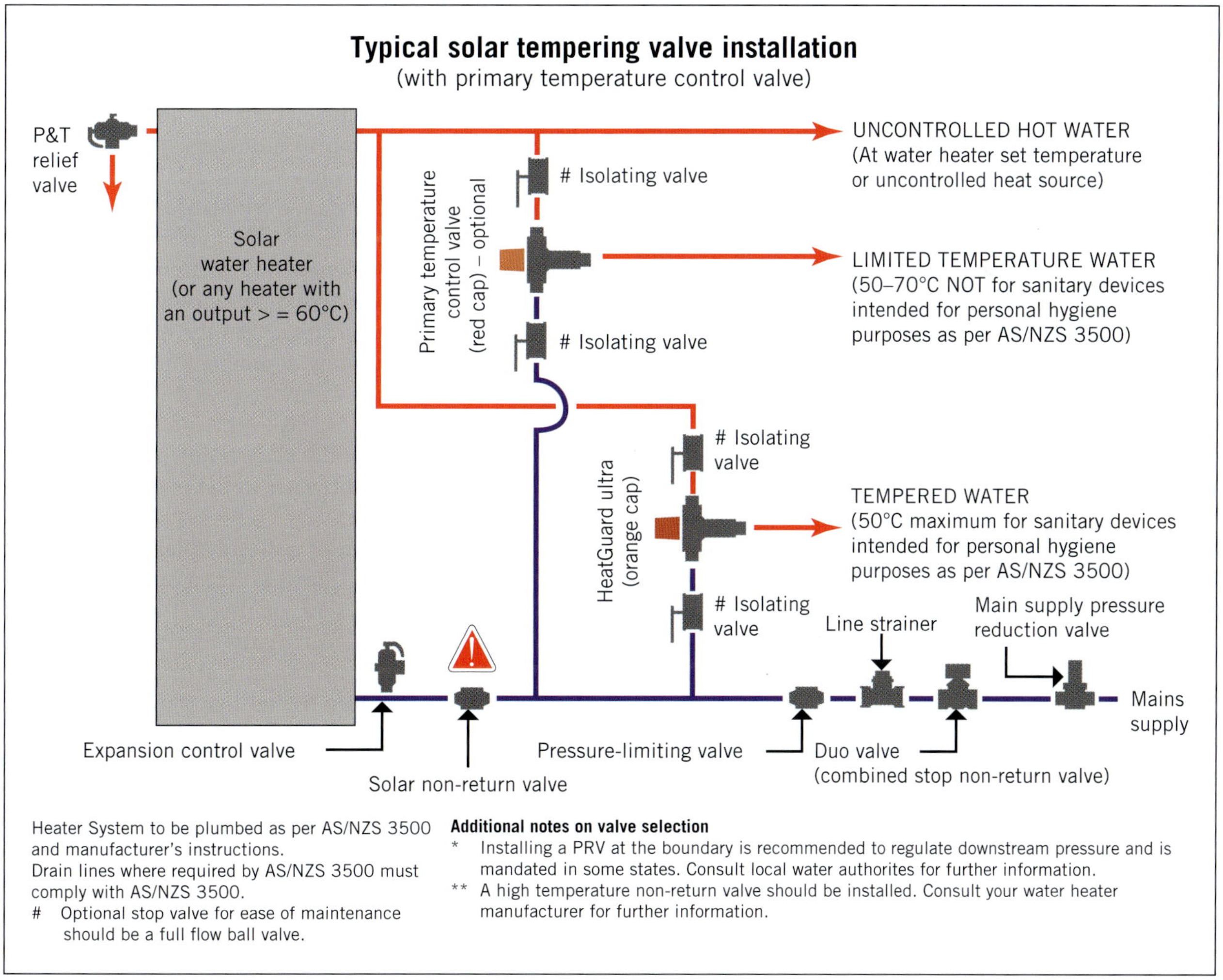

Typical solar tempering valve installation

(with primary temperature control valve)

P&T relief valve

Solar water heater (or any heater with an output > = 60°C)

Primary temperature control valve (red cap) – optional

Isolating valve

UNCONTROLLED HOT WATER (At water heater set temperature or uncontrolled heat source)

LIMITED TEMPERATURE WATER (50–70°C NOT for sanitary devices intended for personal hygiene purposes as per AS/NZS 3500)

Isolating valve

HeatGuard ultra (orange cap)

Isolating valve

TEMPERED WATER (50°C maximum for sanitary devices intended for personal hygiene purposes as per AS/NZS 3500)

Isolating valve

Line strainer

Main supply pressure reduction valve

Mains supply

Expansion control valve

Solar non-return valve

Pressure-limiting valve

Duo valve (combined stop non-return valve)

Heater System to be plumbed as per AS/NZS 3500 and manufacturer's instructions.
Drain lines where required by AS/NZS 3500 must comply with AS/NZS 3500.
Optional stop valve for ease of maintenance should be a full flow ball valve.

Additional notes on valve selection

* Installing a PRV at the boundary is recommended to regulate downstream pressure and is mandated in some states. Consult local water authorites for further information.

** A high temperature non-return valve should be installed. Consult your water heater manufacturer for further information.

Source: Reliance Worldwide Australia.

FIGURE 5.36 Typical solar hot water tempering valve set-up

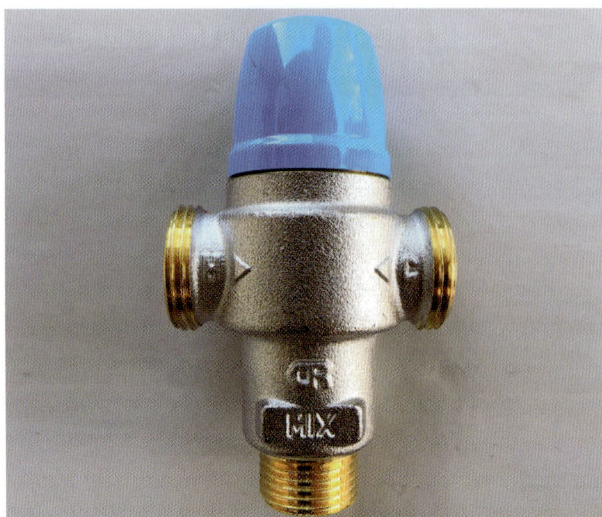

FIGURE 5.37 Blue cap tempering valve

FIGURE 5.38 Orange (high temperature) cap tempering valve

water temperature in the mixing chamber, the element will expand, decreasing the hot water supply. If the water temperature in the mixing chamber decreases, the element will contract, allowing more hot water into the mixing chamber. The sensitivity of the wax element ensures instant movement and therefore a minimum in temperature fluctuation. If there is a failure in cold water supply, the rapid expansion of the wax element will completely shut off the flow of hot water. This is called a *thermal shutdown*.

Tempering valves are non-serviceable and must be replaced as required by the manufacturer, generally every five years, and cleaning of the valve filters may be necessary to prevent the loss of hot water pressure.

FROM EXPERIENCE

Tempering valves used on solar hot water systems must be of the high performance type to handle higher temperatures. They are easily identified by their orange caps (see Figure 5.37).

Thermostatic mixing valve (TMV)

The operation of a **thermostatic mixing valve (TMV)** and a tempering valve is very similar, with the same function to prevent scalding by providing a thermal shutdown if the hot or cold water supply is interrupted. The differences are that in most states a TMV must be serviced annually by a TMV accredited licensed plumber and the temperature settings are more accurate than those of a tempering valve.

As the PCA states,TMVs must be installed in healthcare buildings such as hospitals and nursing homes, as well as childcare facilities and disabled bathrooms. They must be serviced annually and have their log books updated by a TMV accredited licensed plumber.

For optimum performance, it is recommended that the TMV has a balanced cold and hot water inlet supply pressure to within 10% of each other. TMVs must be installed with check valves, strainers and isolating valves in accordance with AS/NZS 3500.4.

'Dead legs' should be kept to a minimum, and the maximum permissible length of pipe from the valve to the furthest fixture must be no more than 10 m or 2 L in volume. This is to prevent the breeding of *Legionella* bacteria forming at temperatures between 20°C and 45°C (which warm water is delivered at) and to avoid temperature loss over a longer distance. The minimum length of the branch line is 1 m. This is to avoid spiking of hot water in the event of a thermal shutdown.

There is a selection of styles available to suit different applications. The TMV can be set up in a recess box (see Figure 5.39) or there is the mini TMV for single point applications (see Figure 5.40). Thermostatic mixing at the point of use is very popular due to reduced installation time. It reduces the risk of *Legionella* and other water-borne bacteria, improving water system health because the warm water is mixed at the outlet. The tap is designed ergonomically for people with disabilities and for surgical and clinical hand washing (see Figure 5.41).

FIGURE 5.39 Thermostatic mixing valve (TMV) in a recess box

FIGURE 5.40 Mini thermostatic mixing valve (TMV)

FIGURE 5.41 Point-of-use thermostatic mixing valve (TMV)

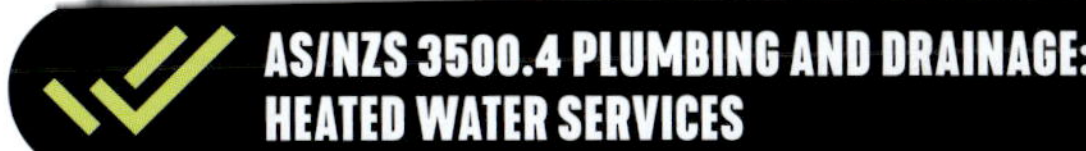

Temperature and pressure relief valve (TPR)

Mains pressure water storage water heaters require a temperature and pressure relief valve (TPR valve) and drain to be installed as part of the installation requirements. The function of the TPR valve is to relieve excessive temperature should the thermostat fail, relieve any excessive pressure from the cold water supply and relieve the expanded water from temperature increase (see Figure 5.42).

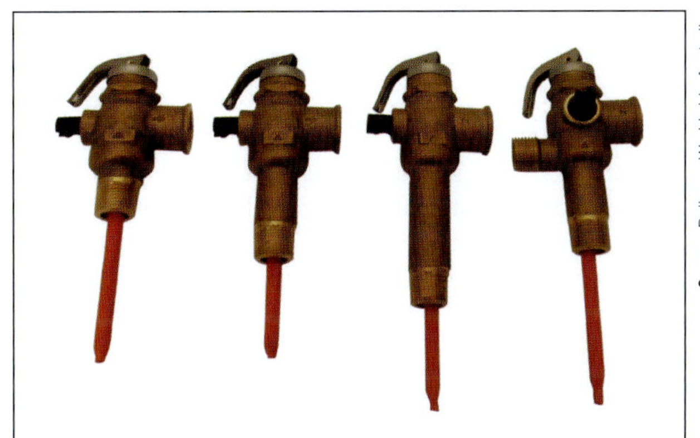

Source: Reliance Worldwide Australia.

FIGURE 5.42 Typical temperature and pressure relief valves

TPR valves incorporate a vacuum breaking (air-inlet) device to protect the cylinder against implosion during the contraction cycle in lower working pressure hot water cylinders. There is a further safety precaution that operates to relieve pressure build-up in the cylinder if the drain line becomes blocked or damaged. It will also relieve a partial vacuum in the drain line. In both cases, the vacuum breaker will blow out to relieve pressure.

When replacing the TPR valve on a water heater, it must be replaced with the exact same model, including maximum temperature and pressure ratings. Under no circumstances should a TPR valve be fitted that has higher maximum pressure or temperature ratings than the original one. They usually have a lifespan of five years.

Expansion control valve (ECV)

As water is heated, it expands and increases in pressure. This expanded water will relieve through an expansion control valve (see **Figure 5.43**), which is installed on the cold water inlet and is a requirement in some Australian states. This is more efficient as the heated water is not wasted and much safer because it is set to relieve the cooler water from the bottom of the tank at a lower pressure than the TPR valve. Because of the water heating cycle, it is normal for mains pressure storage water heaters to relieve the expanded water (2–3 litres) every day through the TPR valve or, preferably, the expansion control valve when installed.

FIGURE 5.43 Expansion control valve

It is for this reason that expansion control valves and temperature and pressure relief valves are installed on water heaters. If this volume was not allowed for, an increase in pressure would occur with a corresponding increase in temperature, which could result in vaporisation or boiling point. It is important to understand that 1 L of water can create approximately 1600 L of steam if that water reaches vaporisation or boiling point within a vessel such as a hot water heater.

FROM EXPERIENCE

It is extremely important that plumbers understand valve requirements, their function and the factors that influence their installation.

Expansion of water

When water in its liquid form is heated, the corresponding expansion must be allowed for to prevent dangerous situations from occurring. The amount of expansion can be calculated easily using the following formula:

$E = M \times T \times C$

where:

E = expansion in litres

M = mass of water

T = temperature variation in °C

C = coefficient of expansion = 0.000375

(The coefficient 0.000375 is how much 1 L of water will expand when raised through a 1°C temperature rise.)

EXAMPLE 5.1

HOW TO CALCULATE EXPANSION

Calculate the amount of expansion that would occur if the water in a 270 L hot water storage unit was heated from 15°C to 68°C.

Formula: $E = M \times T \times C$

where:

E = expansion in litres = ?

M = mass of water = 270

T = temperature variation (68 – 15 = 53)

C = coefficient of expansion = 0.000375

Workings: $E = 270 \times 53 \times 0.000375 = 5.36625$

So, 5.36625 L of expansion would occur over a temperature increase of 15°C. This is the amount of water that will relieve in the heating cycle through the temperature pressure relief valve or, preferably, the expansion control valve if fitted.

LEARNING TASK 5.2

1 What causes water hammer?
2 What is an ablution fixture?
3 What is the minimum temperature that hot water can be stored at?
4 Why is backflow prevention important?

COMPLETE WORKSHEET 2

Prepare for work

To ensure a safely executed and quality outcome, it is essential to prepare properly for the work, as this will help to reduce delays and/or prevent accidents from occurring. A site inspection will ensure any special requirements, such as access, and any existing services that may affect the work. Services search via Before You Dig Australia (https://www.byda.com.au or phone 1100) should be undertaken if excavation work is needed.

It is important that plumbers understand the different types of hot water heaters available so they can provide the client with accurate information to choose the most appropriate water heater for the situation. Valves and water heaters need to be correctly located and accessible during the design and set-out stage of the works.

Quality assurance requirements

More and more companies are adopting quality management systems. The systems contain policies and procedures that assure the quality of the product and/or an installation for the customer. The quality assurance policies are designed to control the processes used in a company to ensure that there is uniformity and reliability in record-keeping, the quality of the materials used and the way the work is carried out. Essentially, it is a standardisation of the processes that will guarantee the customer a quality job that conforms to all the relevant standards and specifications required.

Select the appropriate tools, equipment and PPE

When installing and adjusting water service controls and devices and hot water heaters, some standard plumbing tools and specialised equipment may be needed. Select and check serviceability of the appropriate tools and equipment before using them. Ensure electrical tools and equipment are currently tagged. Check the personal protective equipment (PPE) being used is intact and in good condition.

Create a material list and collect materials

A typical material list for a mains pressure storage water heater installation would be similar to this when ordering to collect from a plumbing supplier:

- 1 × 250 L mains pressure storage water heater (brand)
- 1 × 15 mm duo valve
- 1 × 15 mm pressure limiting valve – 500 kPa
- 1 × 15 mm expansion control valve
- 1 × 15 mm tempering valve
- 1 × 3 m length 20 mm copper tube (type B)
- 1 × 3 m length 15 mm copper tube (type B)
- 1 × 3 m length 20 mm pipe insulation
- 1 × 3 m length 15 mm pipe insulation
- 2 × 20 mm Cu tees (press fit)
- 2 × 20 mm no. 62s (press fit)
- 2 × 20 × 15 mm no. 62s (press fit)
- 3 × 20 × 15 mm no. 3 (press fit)
- 2 × 20 mm no. 1 (press fit)
- 2 × 20 mm hex. nipples
- 2 × 15 mm hex. nipples
- 2 × 20 × 15 mm hex. nipples.

Mains pressure storage water heaters

Mains pressure storage water heaters can be heated by gas, electricity, solar or heat pump. The water connection requirements are generally the same (see Figure 5.44). The functions of the valves required have been explained earlier in this chapter.

Mains pressure storage water heaters work on the **displacement principle**. As hot water is less dense than cold water, it sits on top of the cold water. As a hot tap is opened, cold water enters the cylinder from the bottom, forcing the hot water out at the top of the cylinder. The cold water continues to flow into the water heater until the hot tap is turned off and the cylinder is re-pressurised.

These water heaters need to hold enough hot water for the requirements of all the occupants of the building. The water is stored in an insulated container and is ready for immediate use. As heated water is used, it is replaced with cold water, which is then heated (unless it is outside the off-peak heating times). The heating continues after the cylinder is re-pressurised, until the whole contents reach the prescribed temperature, which is a minimum of 60°C and is controlled by an adjustable thermostat.

Continuous flow water heaters

Continuous flow water heaters (previously known as instantaneous water heaters) may be connected to either electricity or gas. This is illustrated in AS/NZS 3500.4.

When a hot tap is opened, the flow of water activates a valve (for a gas model) or switch (for an electric model), allowing the gas burner or electric

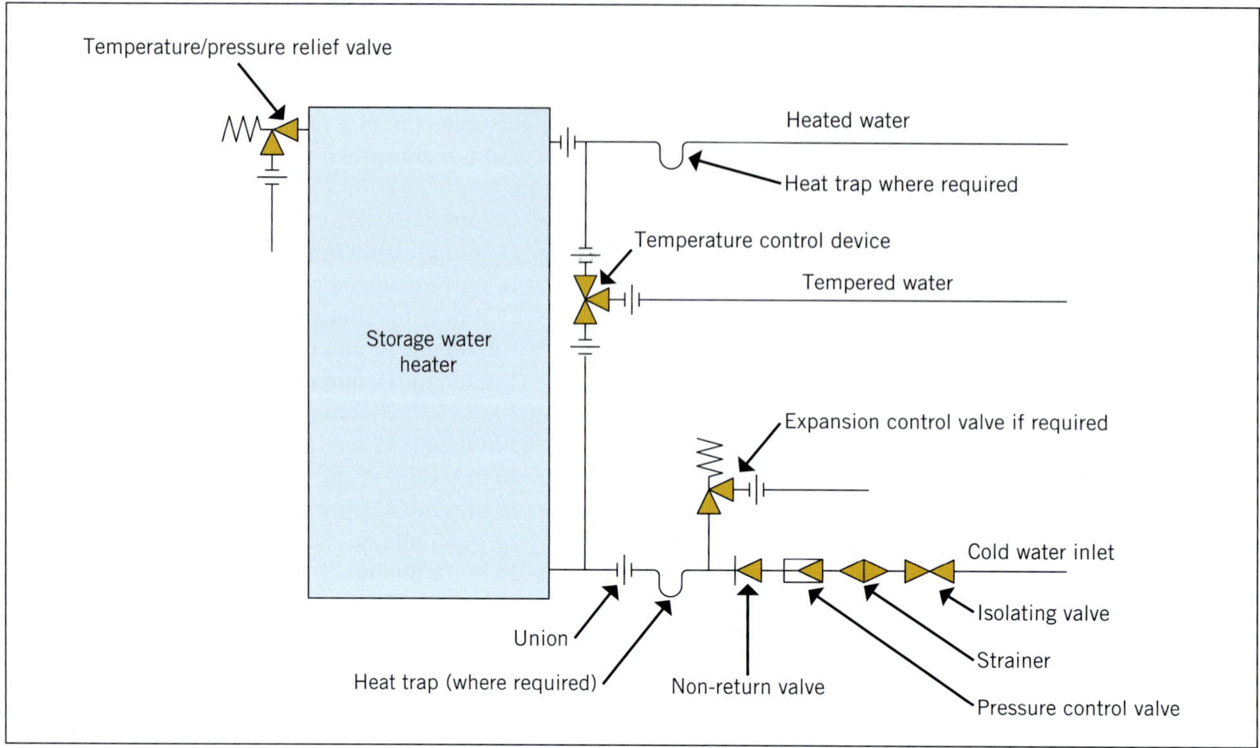

FIGURE 5.44 Mains pressure storage hot water system

element to heat the water as it passes through a coil in the water heater and then shuts down when the hot tap is closed. Continuous flow water heaters can be a single point-of-use water heater for one outlet only or, more commonly, serve multiple outlets.

New models available can be fitted with remote electronic controllers to allow the user to adjust the end-use water temperature. This conserves energy, as water is not 'overheated', and is then cooled to a safer temperature by adding cold water at the mixer or shower set. While continuous flow water heaters (see Figure 5.45) are seen to be rather water efficient, they are higher-level energy consumers with an average gas rate of 188 MJ/h compared with a storage unit's average rate of 38 MJ/h.

FIGURE 5.45 Continuous hot water heater

Solar water heaters

There are two main types of solar water heating systems: close coupled (storage tank and collectors on roof together) and split systems (storage tank at ground level and collectors on roof). These systems can be further described in two other categories:

1 **Direct solar water heating system** – the water is directly heated by the solar collectors and is circulated between the storage tank and the collectors.
2 **Indirect solar water heating system** – a heat exchange fluid (glycol) is heated by the solar collectors and circulates between the heat exchanger within the storage tank and the collectors. Heat is transferred from the heat exchange fluid to the water via conduction in the heat exchanger. The heat exchange fluid is resistant to freezing at temperatures as low as –28°C. Indirect solar water heating systems are used in areas subject to freezing temperatures.

GREEN TIP

The development of technology using the sun to harness renewable energy is increasingly becoming more efficient and is constantly improving over time.

There are three main components to a solar hot water system: the **collectors**, the **water storage tank** and the **booster**. Australia is in the Southern Hemisphere so the collectors should be installed facing north to point more directly at the sun to collect the most heat. Extreme care must be taken with the placement of a roof storage system, as the roof structure may not be designed for the added weight of the storage cylinder and water. The roof structure may need additional supports prior to placement of the storage cylinder and filling with water. Any penetration of a roof covering must be made waterproof in an approved manner.

Where the building is not occupied, all solar collectors not filled with water (or heat exchange fluid) *must* be covered to prevent overheating of the system until the building is occupied and the solar system is being used.

Only metal pipe, usually copper, is to be used between the solar collectors and the storage cylinder or between the water heater outlet and the temperature control valve.

The water temperature gets extremely high between the collectors and the storage cylinder. Therefore, plastic pipe is not suitable.

Solar energy has a greater input into the proportion of water heated when the system is installed closer to the equator. Thus, Darwin can produce between 90% and 95% of all water heating requirements, whereas Hobart will produce only 50% to 55% from solar energy. Sydney will produce between 65% and 75% of all hot water requirements.

A **solar non-return (SNR) valve** is installed to prevent backflow of heated water from the water heater to the mains line. This serves the dual purpose of protecting the mains supply from contamination and the loss of water that has already been heated. It must be installed between the cold water branch to the tempering valve and the cold water heater inlet (see **Figure 5.46**).

A further SNR valve is installed on the hot water line between the solar collectors and the hot water storage tank; this will restrict hot water leaving the tank (mainly at night). Without this additional SNR valve, hot water tends to flow back to the top of the solar collectors where the hot sensor probe will sense the warmer water. In turn, the hot sensor may activate the circulating pump to move water around the entire system (at night) when it is not required, ultimately cooling it down.

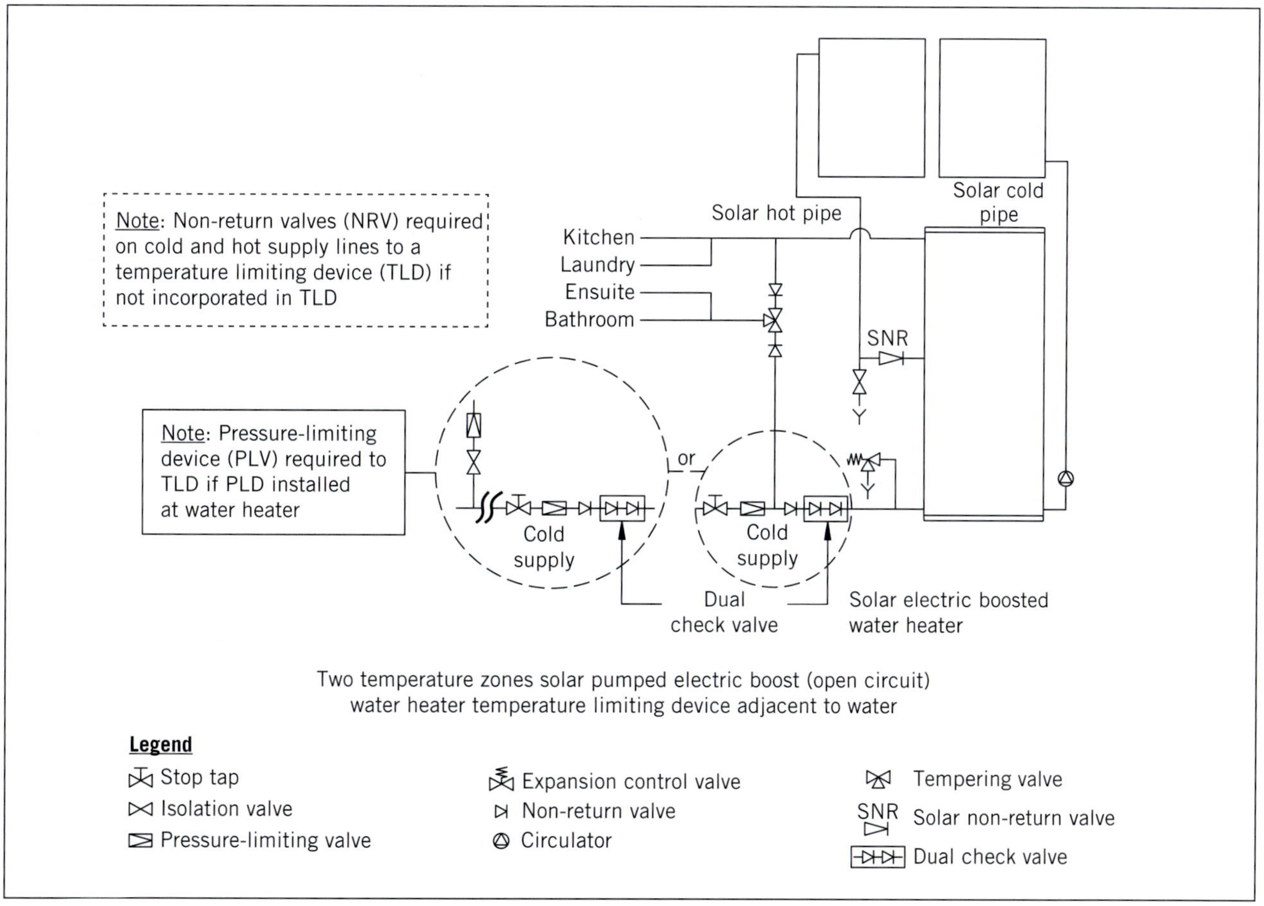

Note: Non-return valves (NRV) required on cold and hot supply lines to a temperature limiting device (TLD) if not incorporated in TLD

Note: Pressure-limiting device (PLV) required to TLD if PLD installed at water heater

Kitchen
Laundry
Ensuite
Bathroom
Solar hot pipe
Solar cold pipe
SNR
or
Cold supply
Cold supply
Dual check valve
Solar electric boosted water heater

Two temperature zones solar pumped electric boost (open circuit) water heater temperature limiting device adjacent to water

Legend

Stop tap
Isolation valve
Pressure-limiting valve
Expansion control valve
Non-return valve
Circulator
Tempering valve
SNR Solar non-return valve
Dual check valve

Source: © Rheem Australia Pty Ltd.

FIGURE 5.46 Solar Lo-line installations showing correct positioning of valve train and cold water take-off to tempering device

Source: © Rheem Australia Pty Ltd.

FIGURE 5.47 Rheem close coupled solar system

Solar and uncontrolled-heat heating systems present an increased hazard to downstream components as water can become superheated and potentially flash to steam.

Only high-temperature non-return valves, such as solar non-return valves, should be used on solar hot water installations.

The positioning of the collectors is critical to the efficiency of the system. Collectors need to take advantage of the maximum solar efficiency in the daytime sunlight and should therefore be positioned so that excessive shading from trees or buildings does not occur during this time period. Solar collectors have a chemical coating that has a purply/black metallic colour to obtain the greatest absorption of the sun's radiation. Collectors need to be installed a minimum of 150 mm below the bottom of the storage container for close coupled storage systems. For maximum solar efficiency, the collectors need to be installed at a similar angle to the area's latitude angle. So, the further from the sun, the higher the angle. The optimum angle for each Australian city may be found in AS/NZS 3500.4.

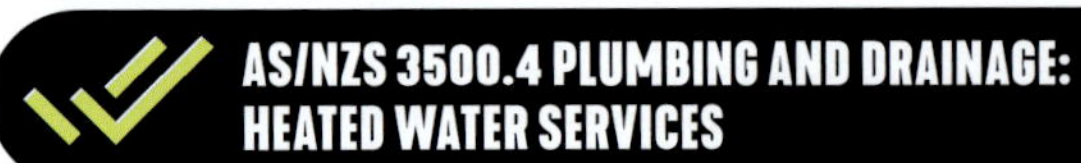

AS/NZS 3500.4 PLUMBING AND DRAINAGE: HEATED WATER SERVICES

Solar systems may have either a gas or an electric booster system installed to supplement the heating requirements.

All solar hot water systems, as with any type of water heater, must be installed to meet the manufacturer's installation instructions. Failure to do so may result in the cancellation of the product warranty.

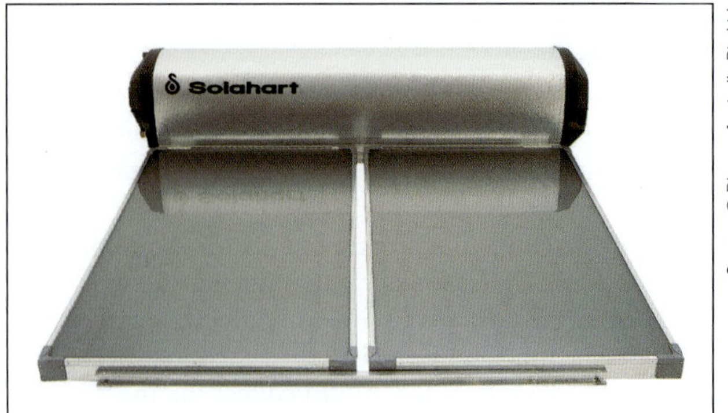

Source: © Rheem Australia Pty Ltd.

FIGURE 5.48 Solahart close coupled solar system

Close coupled solar systems

Close coupled solar systems (see Figures 5.47 and 5.48) use natural **thermosiphon circulation** to circulate the water (direct) or heat exchange fluid (indirect) for heating through the system. Further information on close coupled solar systems may be found online at http://www.rheem.com.au/rheem/products/Residential/Solar/c/solar and https://www.solahart.com.au/landing-pages/solar-hot-water.

Thermosiphon circulation

The heating of the water or heat exchange fluid relies on the three forms of heat transfer: **radiation**, **conduction** and **convection**. This system relies on the fact that hot water rises (convection), which allows water to circulate from the solar collectors to the storage cylinder and back again. This system can be installed either as a close coupled or split system. Whichever type is preferred, they both must have the collectors installed below the storage tank to enable the thermosiphon system to work; a circulating pump is not required.

The system's fluid (either water or heat exchange fluid) in the collectors is heated by radiation from the sun; this in turn heats the fluid in the collectors via conduction. The heated fluid then flows up to the top of the collector and then into the storage cylinder (convection). The fluid in the collector is replaced by cooler (denser) fluid from the base of the storage cylinder; this fluid enters the bottom of the solar collector, allowing circulation to start again.

The system control valves (see Figure 5.49) must be installed in a readily accessible location from ground or floor level. It is good practice to install an isolation valve at the inlet to the cylinder on the roof for servicing; however, it must be a full way valve only, such as a ball valve.

These systems, like a mains pressure storage water heater, have a temperature and pressure relief valve

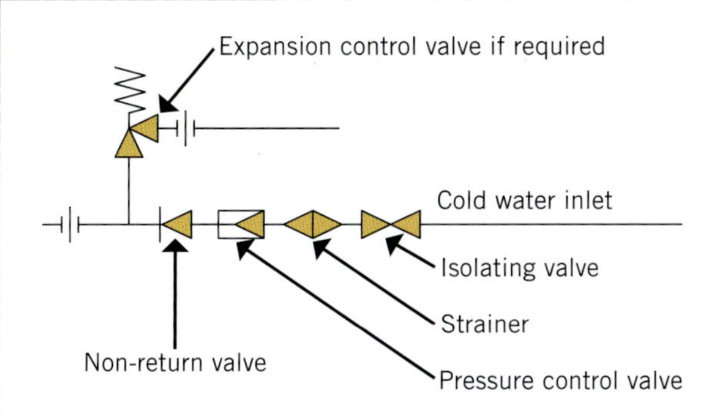

FIGURE 5.49 Valve requirements for a close coupled system to be located at ground or floor level

installed to prevent overheating or over-pressurising the water heater. It must terminate as described later in this chapter and must not discharge onto the roof covering.

Split (forced circulation) solar systems

Split (forced circulation) solar systems use a pump to circulate the system's fluid (water or heat exchange fluid) for heating through the system (see Figure 5.50). The position of the solar collectors in relation to the storage cylinder is usually no more than 9–10 metres apart. A small circulating pump is installed to create a forced circulation of the fluid between the solar collectors and the storage cylinder. This type of system is much more flexible and is more common.

The flow of fluid between the collectors and the storage cylinder is controlled by a temperature sensor. This senses if the temperature in the collectors is hotter

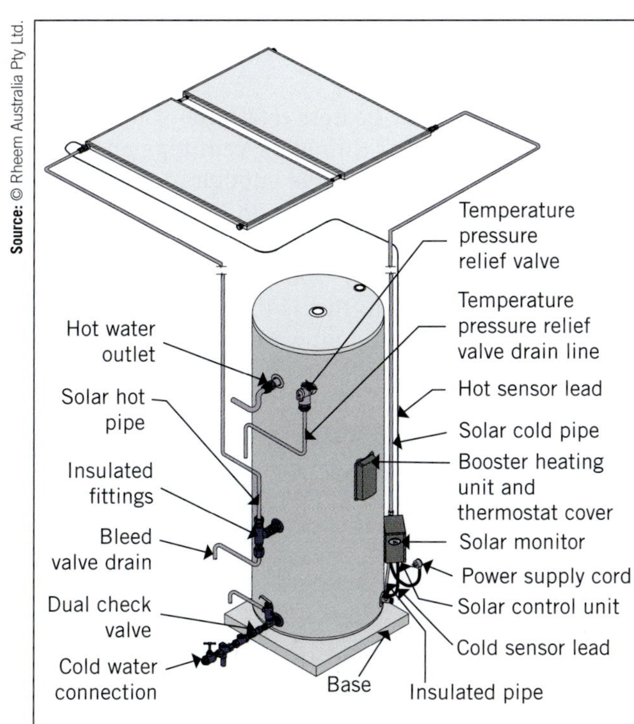

FIGURE 5.50 Rheem forced circulation solar system

(about 4°C) than that in the storage tank, turning on a pump to circulate fluid between the two parts.

Solar indirect systems are a conventional split (mains pressure water heater) indirect (glycol) heat exchange solar system with or without a continuous flow water heater attached (see Figure 5.51). The split indirect solar system is installed as you would intall a conventional indirect solar installation.

Gas-boosted solar indirect system

FIGURE 5.51 Gas-boosted solar indirect system

When a hot tap is opened, the water from the storage tank passes through the continuous flow gas water heater. If the water is below the required outlet temperature, the continuous flow gas water heater cuts in, heating the water to the required temperature.

This type of system has a very high energy efficiency rating as it has the advantage of solar energy use and only boosts the water temperature as required as it is being used. During periods of solar gain, the heat exchange fluid is pumped through the collectors to capture the sun's radiation. When heating is complete, or during periods of frost, the fluid drains back to the tank providing total frost protection to -17°C. This works like a car radiator, protecting the fluid from freezing in winter and overheating in summer. The solar cold water (flow) and hot water (return) (see Figure 5.52) connect at the top of the solar storage tank. Flow and return lines must be insulated metallic pipes. Polymer (plastic pipes) *must not* be used.

Evacuated tube solar collectors

Instead of using flat solar collectors, **evacuated tube solar collectors** can be used to heat water (see Figure 5.53). They consist of two glass tubes fused at top and bottom and installed in series. The space between the two tubes is evacuated to form a vacuum. A copper pipe (called a heat pipe) running through the

Source: © Rheem Australia Pty Ltd.

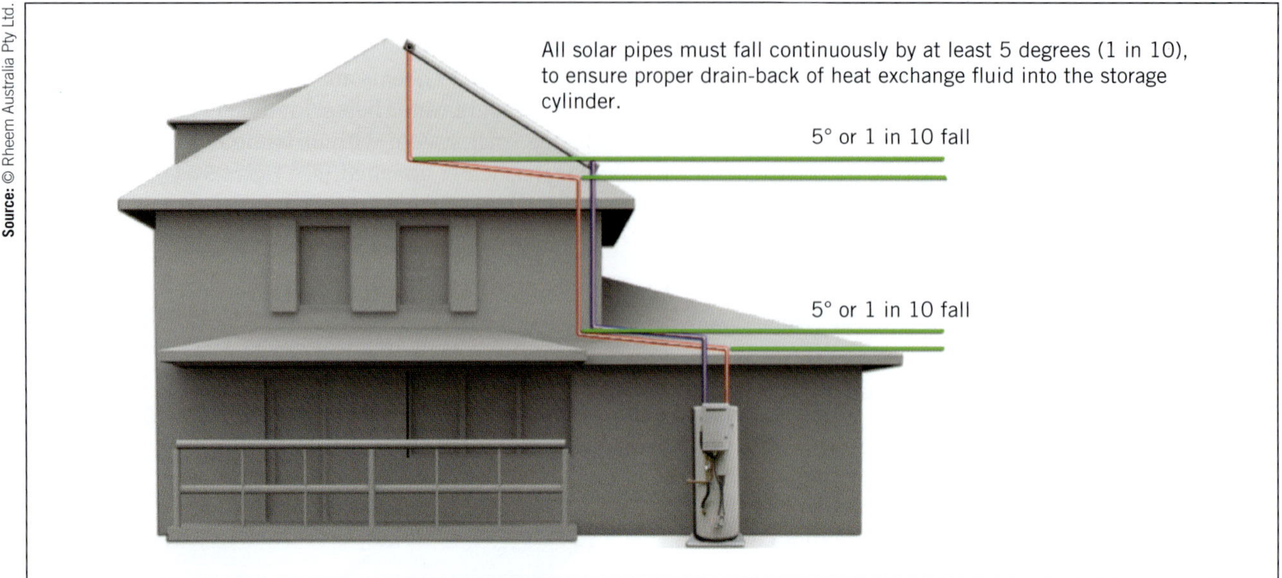

FIGURE 5.52 Indirect split system flow and return line

Beautiful landscape/Shutterstock.com

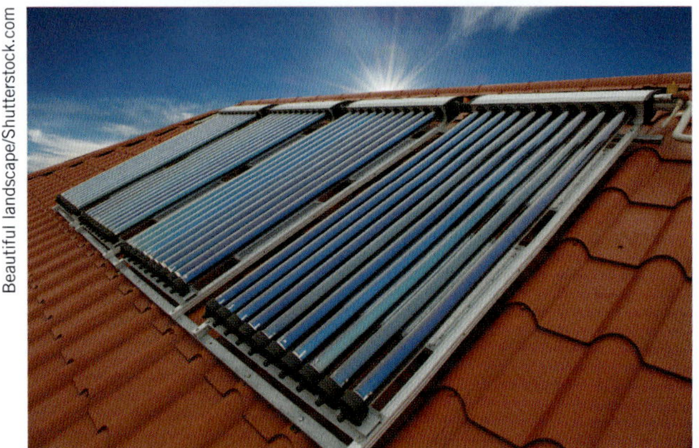

FIGURE 5.53 Evacuated tube solar collectors

centre of the tube meets a common manifold that is then connected to a slow flow circulation pump that pumps water to a storage tank below, thus heating the water during the day.

The evacuation tube systems are more efficient than the flat plate collectors at high water temperatures. Due to the vacuum inside the glass tube, the total efficiency of the glass surface is higher and increased performance is achieved in early morning and late afternoon.

Solar systems installed in problem areas

When installed in a frost-prone area, a solar hot water system must be fitted with a frost protection system or device as recommended by the manufacturer and approved in AS/NZS 2712 Solar and Heat Pump Water Heaters.

When installing solar systems in cyclone-prone areas, the collectors and mounting system must be approved by the manufacturer and the local authority.

Low-pressure (gravity feed) water heater systems

Low-pressure (gravity feed) hot water systems (see **Figure 5.54**) are not very common these days because of their low-pressure operation. They are located in the roof space to create the head pressure for the outlets below and are designed to take advantage of electrical off-peak rates. The cold water is supplied via a feed tank or cistern (make-up tank), in which the water is controlled by a float valve. The feed tank is usually mounted on the side or top of the storage tank (water heater) to supply a small amount of pressure, making the water flow through the water heater.

Gravity feed water heaters retain the water that expands by directing it through a vent pipe and by having the make-up tank large enough to accept the extra water. The cold water feed to these storage tanks requires an isolation valve located on the inlet to the tank and a full way isolation valve on the hot water outlet connection; this is to enable the water heater, its ancillaries and tapware to be serviced. A further control valve must be provided in a position that is readily accessible from the floor or ground level to isolate the cold water feed to the system, and so enable the property owner to turn off the water supply without needing to access the roof area.

GREEN TIP

Recycling expanded water in low-pressure systems can save up to 3000 L of water annually.

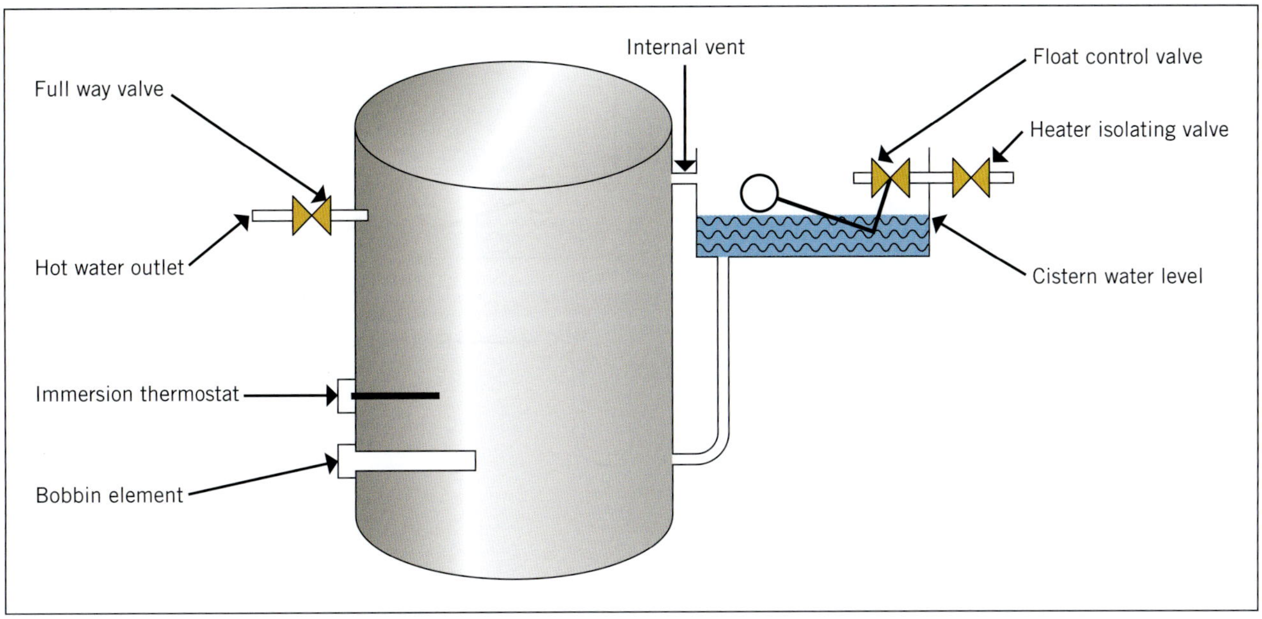

FIGURE 5.54 Low-pressure (gravity feed) hot water system

The advantages of low-pressure (gravity feed) hot water heaters are that they:

- are long-lasting
- have a minimum of operating parts and are trouble-free
- are easy to service
- occupy no floor space.

The disadvantages are:

- only low pressure is received at the outlet
- the installation may require larger diameter piping (sizing)
- replacement is more difficult due to access in the roof space
- the warm water blend on single-lever mixing taps is difficult because of the high pressure difference between the hot and cold water supply.

The water in a **falling level displacement water heater** (see **Figure 5.55**) is heated by an off-peak electrical supply. The heated water may be used as required, but replacement water will not enter the water heater until there is a power supply to operate the solenoid valve.

This type of water heater may not only run out of heated water, but also may run out of water altogether, until it refills when power is available.

The hot water pipework for low-pressure (gravity feed) hot water systems must be sized correctly to give an adequate volume of water to the various hot water outlets. (This is because the supply pressure is created from the storage tank due to the difference in height between the tank inlet and the various outlets.) Generally, the most disadvantaged outlet is the shower outlet. AS/NZS 3500.1 requires a minimum supply pressure to an outlet of 50 kPa.

Formula: P = 9.81 – approximately 10 kPa per 1 m head

Suggested minimum pipe sizes

For a low-pressure water heater less than 85 kPa:

- a DN 20 (15.0 mm internal bore) service from the water heater to the first branch
- a DN 15 (10.0 mm internal bore) branch to a kitchen sink or basin
- a DN 15 (10.0 mm internal bore) branch to a sink and laundry
- a DN 20 (15.0 mm internal bore) branch to a bathroom and one other room
- a DN 18 (12.5 mm internal bore) branch picking up all fixtures within a bathroom.

For a low-pressure water heater 85–170 kPa:

- a DN 18 (12.5 mm internal bore) service from the water heater to the first branch
- a DN 15 (10.0 mm internal bore) branch to a kitchen sink or basin
- a DN 15 (10.0 mm internal bore) branch to a sink and laundry
- a DN 18 (12.5 mm internal bore) branch to a bathroom and one other room
- a DN 15 (10.0 mm internal bore) branch picking up all fixtures within a bathroom.

The above sizes are a suggested minimum only; the service must be sized to meet the minimum flow rate outlet requirements. This may require a larger pipe to be used.

The hot water pipes should be installed with continuous fall from the tank outlet to the fixture outlet to avoid airlocks occurring, as this can block the hot water flow.

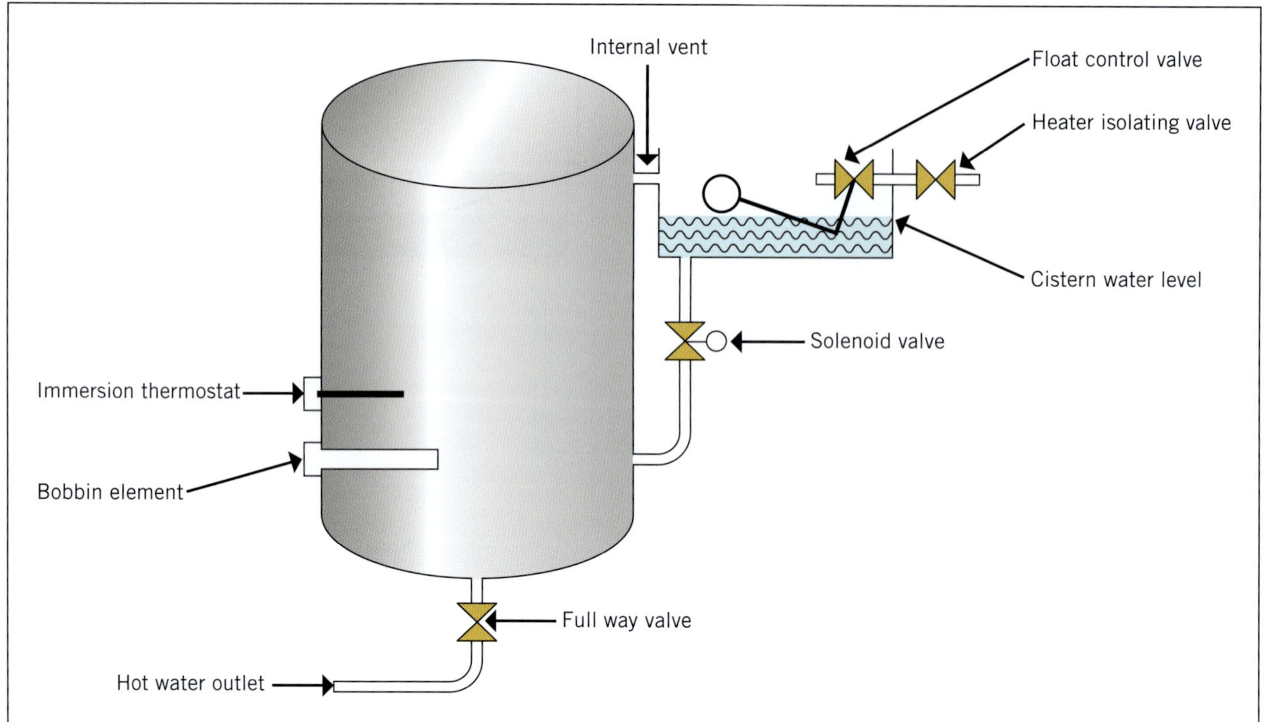

FIGURE 5.55 Falling level displacement water heater

FROM EXPERIENCE

A large pressure difference between the hot and cold water causes major problems in achieving a blend with single-lever mixer taps and tempering valves when installed with gravity feed and falling level displacement systems.

Heat pump air-sourced water heater (close coupled system)

A **heat pump air-sourced water heater** (see Figures 5.56 and 5.57) is a conventional mains pressure water storage tank with a heat pump attached. The heat pump converts the outside air temperature to heating energy in a similar manner to the way that an air conditioner heats or cools a house or car.

Heat pump systems are very popular and a good solution for the replacement water heater market. It allows property owners to take advantage of converting external air temperature into heat energy, without the expense and inconvenience of having solar collectors and pipework installed, to reduce their energy consumption. A typical heat pump system uses up to two-thirds less energy than a traditional electric water heater. The water connection to this type of system is the same as for mains pressure water heaters.

Source: © Rheem Australia Pty Ltd.

FIGURE 5.56 Mains pressure heat pump water heater with minimum installation clearances

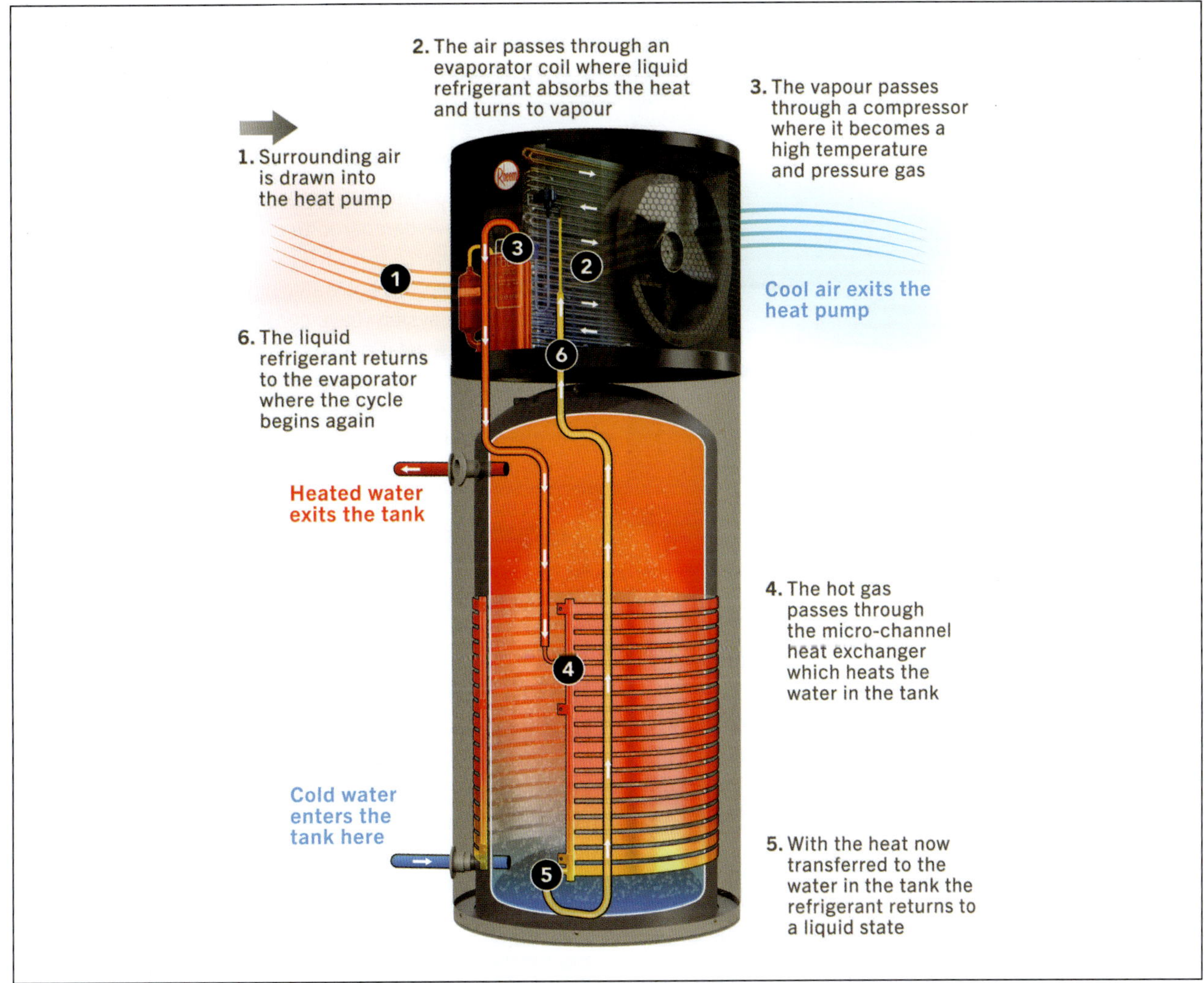

Source: © Rheem Australia Pty Ltd.

FIGURE 5.57 Working diagram of a heat pump air sourced electric boosted water heater

For best results, these systems should ideally be installed in an outdoor location, and be connected to a continuous electrical tariff.

A heat pump air-sourced system must have a 300 mm clearance around the air louvres of the module (see Figure 5.56).

GREEN TIP

A heat pump hot water system is claimed to reduce the carbon footprint of a domestic property by up to 2.7 tonnes of carbon dioxide (CO_2) per year.

Push-through 'free outlet under-sink' water heaters

These units are designed to service a single outlet only, usually a sink (see Figure 5.58). Cold water is connected to the hot tap, which acts as the cold water inlet valve for the water heater, as shown in AS/NZS 3500.4. When the hot tap is turned on, cold water enters the water heater, pushing hot water out the free outlet (sink spout). As these water heaters are not under pressure and have a free outlet allowing for hot water expansion, they do not need a temperature and pressure relief valve, and do not need to be installed in a safe tray.

Source: Dreamstime.com/Valentyn Bilou.

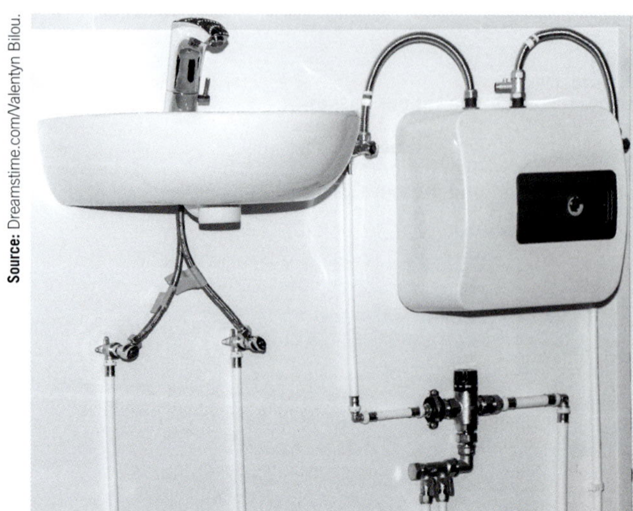

FIGURE 5.58 A modern 'push through' water heater

Water heater construction

Water heaters available today may be constructed from the following materials:

1 *Copper*. While copper has been traditionally used for the construction of low-pressure water heaters, it is not suitable for pressure systems above 85 kPa. Its use is therefore restricted to gravity-fed systems. Copper provides a natural corrosion resistance and it is not unusual for these types of water heater to last for 20 or more years.
2 *Stainless steel*. Stainless steel is being used more often these days by manufacturers; it provides for a lighter water heater and gives a similar life to a quality manufactured, glass-lined steel water heater. However, it is more expensive.
3 *Mild steel vitreous enamel*. Mild steel vitreous enamel water heaters are the most common type of water heater; they are manufactured as a lower-cost alternative to stainless steel. The vitreous enamel coating used on water heaters is similar to that applied to cooking equipment and used on barbecues.

Vitreous enamel water heaters require a *sacrificial anode* to be installed to help protect and prolong the life of the water heater. Anodes are manufactured of either magnesium or aluminium with a mild steel wire core; some water heaters may have two anodes installed. When selecting the anode, particular attention should be given to its compatibility with the local water supply. Water heater manufacturers can advise on the correct anode for their water heater in an area. Anode replacement forms part of the servicing requirements for vitreous enamel water heaters to ensure the longevity of the water heater's life.

Manufacturers may offer two grades of vitreous enamel coating. For example, Rheem Australia offers two types of vitreous enamel water heaters: Rheemglas water heaters, which have a seven-year cylinder warranty, and Optima and Stellar water heaters, which have a 10-year cylinder warranty.

LEARNING TASK 5.3

1 What advantage does a continuous hot water heater have compared to a mains pressure storage water heater?
2 Why should the hot water piping from a gravity-fed hot water system have continuous fall?
3 Where is a solar non-return valve located on a solar hot water system installation?
4 Why is the evacuation tube system more efficient than the flat plate collector?
5 Where does a heat pump get its heat source?

COMPLETE WORKSHEET 3

Install water heating system and service control device

When installing hot water heaters, it is important to check the plans and specifications to ensure that the installation complies with the regulatory requirements. If there are any issues, they need to be negotiated at this time.

FROM EXPERIENCE

Always check the approved plans and BASIX specification for a nominated water heater on new developments.

Sizing the water heater

When selecting the correct type of water heater, the following points need to be considered:

- government regulations
- the type of water heater required – continuous or storage
- energy source – gas, solar or electric
- running costs
- purchase and installation costs
- the type of warranty (five or 10 years)
- water heater recovery and performance (the WELS rating)
- temperature requirements
- pressure and flow rate requirements
- the location of the water heater
- the number of bathrooms
- the number of bedrooms

- the number of people using the system and any possible family expansion
- whether the dishwasher is connected to the water heater
- whether the clothes washing machine uses hot water
- whether there is a spa bath.

Table 5.2 is a general summary of water heater sizing; it is recommended that manufacturers' guides are consulted for more specific details of sizes. Rheem Australia has a web-based sizing system that may help when sizing water heaters for domestic use (see the References at the end of this chapter). Other suppliers also readily supply information about their heating units.

Position water heating system in accordance to relevant Australian Standards

Water heaters should be located as close as possible to the most frequently used outlet; this is generally the kitchen sink. Remember that it is good practice for a hot water service to have a maximum of 2 L of cold water draw-off from a hot tap (this equals about 25 m of DN 15 copper pipe). If this cannot be achieved, then either an additional water heater should be installed or a flow and return pipe layout installed to reduce water wastage ('deadleg').

The water heater must be installed so that:

- its rating plate and instructions are readily visible
- there is unobstructed access to its burner, element, thermostat, controls and any equipment requiring maintenance
- its valves and easing gear are accessible
- it has a minimum of 150 mm clearance for the removal and replacement of the temperature and pressure control valve
- it can be removed and replaced without major structural alteration to the building or major alteration to the pipework.

An additional point to note is that if a water heater is installed in a known earthquake area, it must be restrained against movement.

GREEN TIP

Reducing the cold water draw-off when opening a hot water tap will save a large amount of water wastage over time. Use the options available to achieve this.

Gas water heaters

When installing any gas water heater, its location and method of installation *must* comply with all relevant standards and building codes, specifically AS/NZS 5601.1 Gas Installations, and is dependent on the unit's gas capacity in MJ/h and if the flue terminal is fan assisted or natural draught. (The diagrams in Figure 5.59 are a guide only.)

Concealed water heaters

All water heaters installed in a location such as a roof space or a cupboard must be placed on a **safe tray** (see Figure 5.60). For an example of this, refer to AS/NZS 3500.4. When installing a water heater on a safe tray, provision must be made for draining and prevention of overflow. If the safe tray is installed under a sink, a minimum size of DN 25 **safe waste drain** must be installed, with a minimum of DN 50 (DN 40 in New Zealand only) for all other installations. If a safe waste drain is fabricated from sheet metal it must be lapped in the direction of the flow and all lapped joints must be watertight. Any seam in the sheet metal must be installed on the top.

TABLE 5.2 General summary of water heater sizing

Sizing guide							
Electric continuous tariff		Electric off-peak tariff		Gas		Solar	
Heater delivery capacity (litres)	Number of users	Number of users	Number of bedrooms	Storage (litres)	Continuous flow	Solar storage (litres)	Heat pump storage (litres)
25							
50	1						
80	1–2						
125	2–3						
160	2–4						
250	3–5	1–3	1–2	90	18	270	
315	4–6	2–4	3	130	24	325	310
400	5–9	4–6	4–5	160	26	410	325

Source: © Rheem Australia Pty Ltd.

Source: © Rheem Australia Pty Ltd.

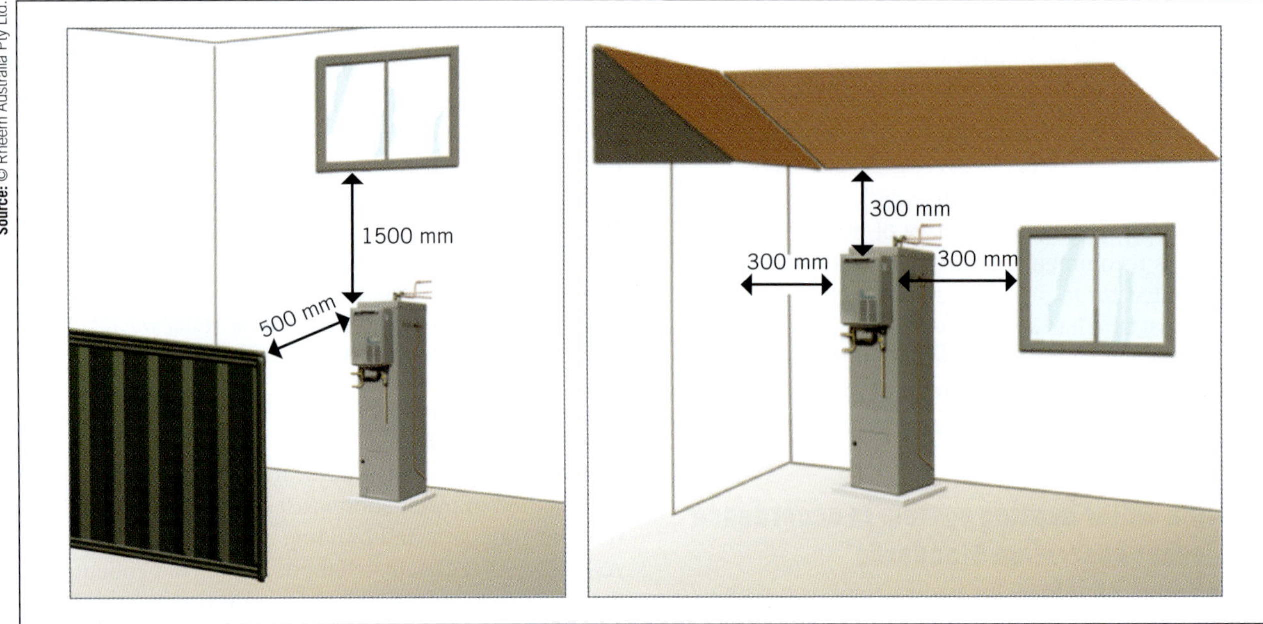

FIGURE 5.59 Installation location for continuous flow gas water heaters

FIGURE 5.60 Electric water heater installed in a safe tray and hanging brackets

FIGURE 5.61 Terminator valve installed on the cold water inlet of a pressure water heater

A safe waste drain must:

- be installed with a constant fall to its point of discharge
- be supported adjacent to the point of connection to the safe tray and at a maximum distance of 1 m on a grade and 2.4 m when installed vertically
- be readily visible if discharging within a building and not cause any damage to the property or injury to people
- discharge to an external point of a building and within the property boundary that is readily visible and clear of any openings into a building such as windows or doors.

When installing a mains pressure water heater in a safe tray, provision for protection against damage from leaking water can be made by installing an approved shut-off device (such as a **terminator valve**) instead of a safe waste drain. The terminator valve (see Figure 5.61) is installed between the cold water inlet isolation valve and the water heater inlet, upstream of any expansion control valve.

When a leak is detected (by water entering the safe tray), the valve mechanically shuts off the water entering the water heater.

An unconcealed water heater located inside any building on or above an impervious floor, and draining to a floor waste gully or an external door, does not require a safe tray.

Water heaters located in a roof space

Any water heater installed in a roof space must be installed on a safe tray and supported by a hardwood platform, as outlined in AS/NZS 3500.4.

Externally installed water heaters

When a water heater is installed in an external location it must be supported by:

- 75 mm thick bonded bricks or concrete (cast in situ) support
- 50 mm precast concrete support slab.

The top of the base must be a minimum of 50 mm above the surrounding finished ground or surface level.

AS/NZS 3500.4 PLUMBING AND DRAINAGE: HEATED WATER SERVICES

Materials selection

Unless otherwise stated in the plans or specifications, the type of materials that may be used is covered by AS/NZS 3500 and includes:

- copper
- polybutylene (PB)
- cross-linked polyethylene (PE-X)
- polypropylene (PP)
- polyethylene (PE)
- unplasticised polyvinyl chloride (PVC-U).

When selecting material you need to consider:

- what the pipework is going to be used for
- the water quality and its temperature
- the compatibility of the material and products
- frost protection
- the water pressure within the water utility's supply system
- any special material installation requirements; for example:
  - polymer (plastic) pipes and fittings may not be installed in direct sunlight
  - polymer pipes may not be used between the cold water inlet isolation valve and the water heater
  - polymer pipes may not be used as part of a temperature and pressure relief valve
  - polymer pipes and fittings may not be used within 1 m of the outlet of a water heater
  - polymer pipes and fittings may be installed on the outlet side of a temperature control device.

Any material or product used as part of the heated water installation must comply with the PCA and AS 5200.000–2006 Technical Specification for Plumbing and Drainage Products. A database of authorised products may be found in the References section at the end of this chapter.

Requirements for water connections

Every water heater must have:

- a union or other similar coupling connection on its inlet and outlet
- the required valves as specified by the manufacturer and described below
- the required temperature and pressure relief valves supplied by or recommended by the manufacturer.

Install and connect pipework and valves to water heating systems

All valves used to control the water supply to and from a water heater must be installed in accordance with both the water heater manufacturer's requirements and the following requirements:

- The water heater isolation valve must be installed in a location that is readily accessible from the ground or floor level.
- The valves on the cold water inlet to a mains pressure storage hot water heater are installed in the following sequence (see Figure 5.62):
  1. isolating valve
  2. line strainer
  3. pressure control valve
  4. non-return valve
  5. expansion control valve where required by the local water utility or water heater manufacturer.
- The isolating valve on the cold water inlet to a continuous hot water heater must be the full flow type, such as a ball valve.

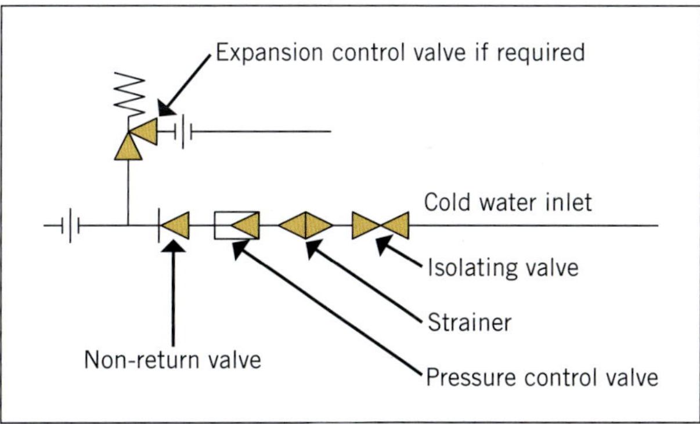

FIGURE 5.62 Valve requirements for a mains pressure storage hot water heater

Temperature control devices

The type of temperature control device is determined by the type of building.

A thermostatic mixing valve must be fitted to ablution fixtures with a maximum temperature of 45°C in hospitals, medical centres, nursing homes, pre-schools, primary schools, secondary schools and tertiary educational facilities. The diagram in Figure 5.63 shows the required valve sequence for a TMV installation.

A tempering valve or a preset temperature water heater is installed to supply hot water at a maximum of 50°C to ablution fixtures in other buildings such as residential and domestic dwellings.

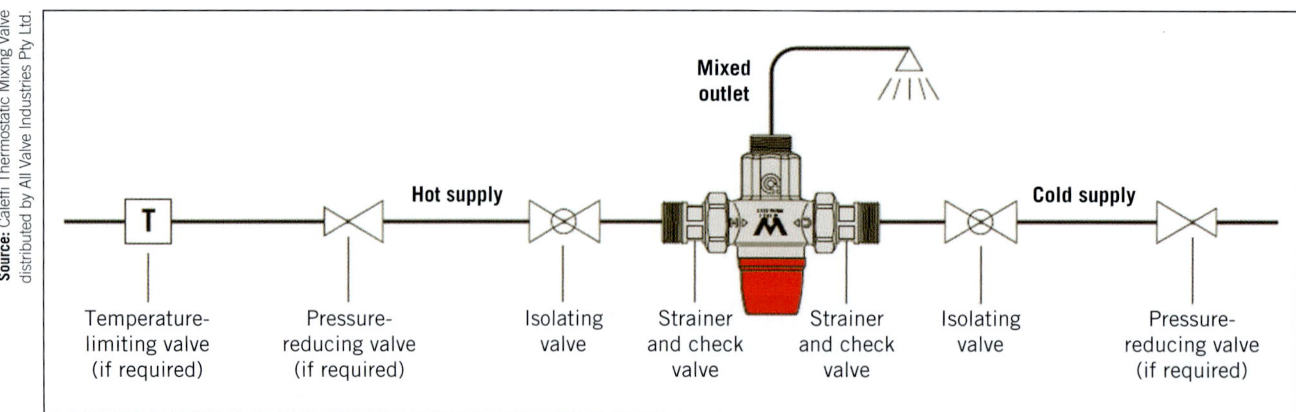

FIGURE 5.63 Thermostatic mixing valve installation

Temperature/pressure relief valve and expansion control valve drains

Temperature and pressure relief and expansion control valve drains are sized no smaller than the valve outlet and will be of copper or other approved materials. They must not exceed the maximum length, as shown in AS/NZS 3500.4.

Where the drain exceeds the maximum length stated in AS/NZS 3500.4, it must discharge via a **tundish** (see **Figure 5.64**).

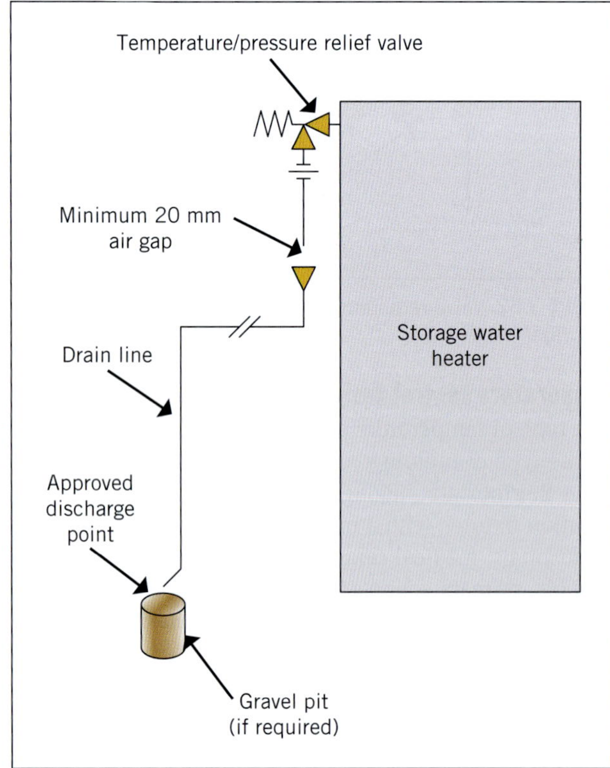

FIGURE 5.64 Temperature and pressure relief drain discharging via a tundish

Temperature and pressure relief valve drains must be installed:

- with no taps, valves or restrictions installed within the drain
- with a continuous fall from the valve to the approved point of free discharge
- so as not to discharge into a safe tray
- to discharge to a point readily visible
- so as not to cause any damage to a building
- so as not to cause injury to persons
- so as not to cause a nuisance by the release of steam or hot water
- with an air gap of twice the diameter of the drain if discharging over a tundish, and be a minimum of 20 mm
- so as not to discharge onto any roof covering.

Where the drain terminates outside a building, it must have its point of free discharge between a minimum of 200 mm and a maximum of 300 mm above an unpaved area, and a minimum of 75 mm and a maximum of 300 mm above an overflow/disconnector gully or a DN 100 gravel pit in paved areas. This is so visual contact can be made when the temperature and pressure relief valve is discharging. The drain line from a tundish must be a minimum of DN 20 and be one size larger than the largest drain line discharging into the tundish. If the water heater is installed externally, the drain line must discharge away from the water heater to protect any person operating the temperature and pressure relief valve and so as not to cause damage to building footings or foundations.

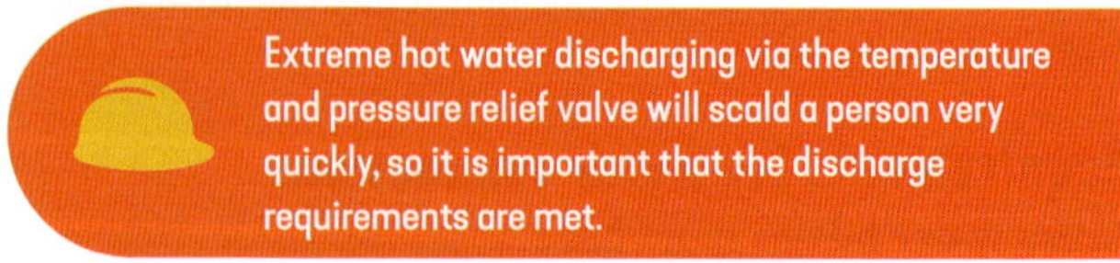

In areas subject to water pipe freezing, the drain line must not exceed 300 mm in length and must be insulated. It must discharge into a tundish with an air gap of between 75 mm and 150 mm. The tundish must discharge as described above.

Under no circumstances is the temperature and pressure relief valve to be blocked off or removed. This would lead to the water heater failing and under certain circumstances would create an extremely dangerous and explosive situation.

FROM EXPERIENCE

Always operate the temperature pressure relief valve when commissioning the hot water heater to ensure it is functioning correctly.

AS/NZS 3500.4 PLUMBING AND DRAINAGE: HEATED WATER SERVICES

Sizing connections to the water heater

The minimum sizes of hot water branches for a storage water heater in excess of 170 kPa (mains pressure) are:

- a DN 18 (12.5 mm internal bore) service from the water heater to the first branch
- a DN 15 (10.0 mm internal bore) branch to a kitchen sink or basin
- a DN 15 (10.0 mm internal bore) branch to a sink and laundry
- a DN 15 (10.0 mm internal bore) branch to a bathroom and one other room
- a DN 15 (10.0 mm internal bore) branch picking up all fixtures within a bathroom.

For other types of heated water installation, refer to AS/NZS 3500.4.

Insulation of hot water pipes

Thermal insulation of hot water pipes must meet the minimum thermal insulation R-value for the area. This information may be found in AS/NZS 3500.4 (see Figure 5.65). Insulation is placed on hot water pipes for several reasons, including to:

- prevent heat loss
- protect pipes 'chased' into solid walls (also to help prevent adverse effects from expansion when sealed in these walls)
- help prevent burns from exposed pipework.

Areas subject to freezing

If a water heater is to be installed in an area subject to regular low temperatures (below 0°C), the service must be protected by preventing the water from freezing, as outlined earlier in this chapter. Installing a water heater in an external location in these conditions is not recommended, unless it has built-in protection against freezing.

FIGURE 5.65 Insulated pipework

LEARNING TASK 5.4

1. What would be the correct size gas continuous water heater for a three-bedroom house?
2. How much clearance is required around a temperature and pressure relief valve so it can be easily replaced?
3. What is the minimum size of a safe waste drain in a cupboard?
4. Why is insulation important on a hot water pipe?

COMPLETE WORKSHEET 4

Large residential and commercial/industrial hot water supply services

If a large volume of hot water is required, water heaters can be manifolded together. It is not uncommon to see two or more water heaters manifolded together, which will dramatically raise the delivery capability of the system. **Manifold systems** are generally installed as flow and return systems. This piping configuration is known as an equaflow system.

Manifolding storage water heaters

The 'equal-flow' installation method means that the demand on each water heater in the bank is the same.

The manifold must be sized to meet the installation requirements.

All types of systems, such as mains pressure storage, solar and continuous flow, may be manifolded together to supply larger volumes of heated water, meeting the needs of the property owner/occupier.

HOW TO

INSTALL MULTIPLE HOT WATER HEATERS

- The heated water manifold must be designed to leave the bank of water heaters from the opposite direction to the cold water entering the bank of water heaters, balancing the flow from each unit (see Figure 5.66).
- The cold water manifold must be designed to enter the bank from the opposite direction to the heated water leaving the bank, balancing the flow to each unit (see Figure 5.66).
- The water heaters must have the same storage capacity and energy input.
- The inlet and outlet connections must be the same size and, if possible, the same length.
- Full way (ball or gate) valves must be fitted to the inlet and outlet of each water heater.
- The hot water return line must be connected to the cold water manifold (see Figure 5.67), or to a return connection boss if one has been supplied. It must enter the banks from the opposite direction to the hot water leaving the bank.

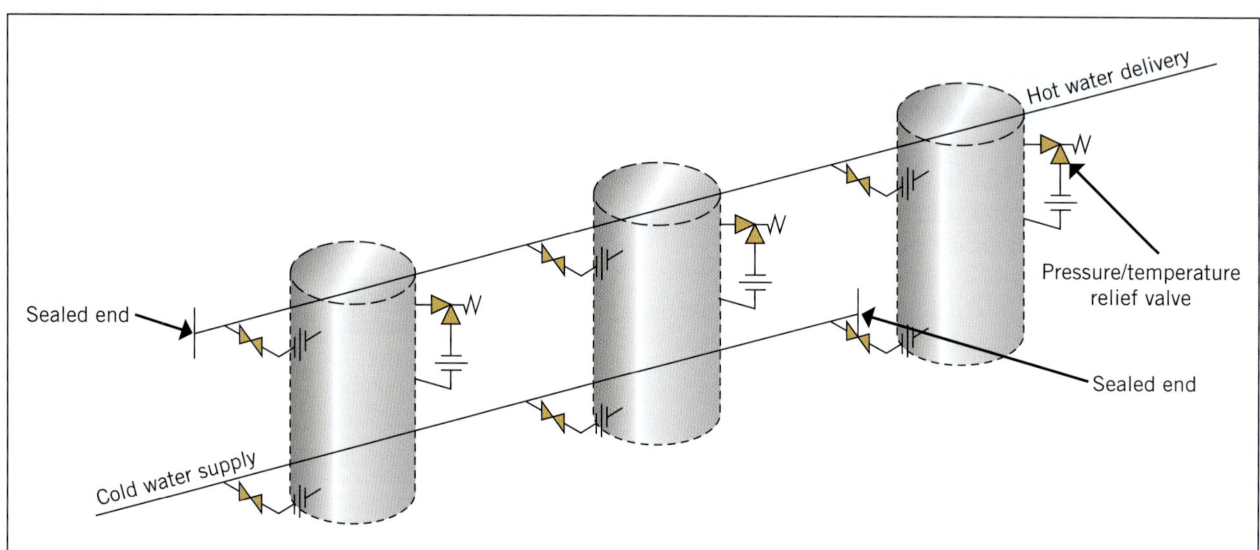

FIGURE 5.66 Manifold (equaflow) system

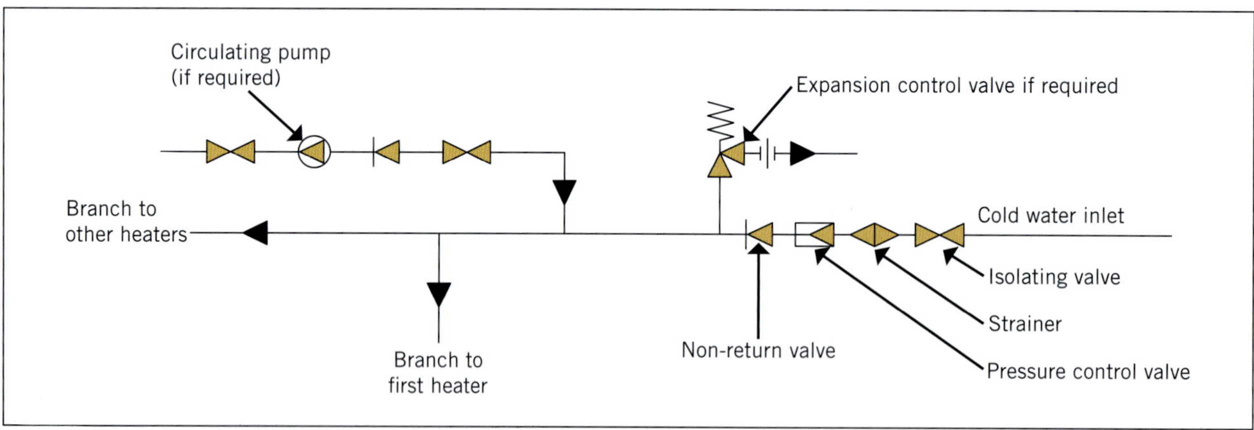

FIGURE 5.67 Manifold (equaflow) system valve requirements

Manifolding continuous flow gas water heaters

Continuous flow water heaters are manifolded in a similar manner to storage water heaters, with the exception that each unit has a **staging valve** (also known as a **pressure responsive control valve**) installed on its cold water inlet (see Figures 5.68 and 5.69). These valves are designed to respond to the flow of hot water through the system by igniting each unit depending on the flow requirements. A check must be made with the water heater manufacturer for any specific design requirements.

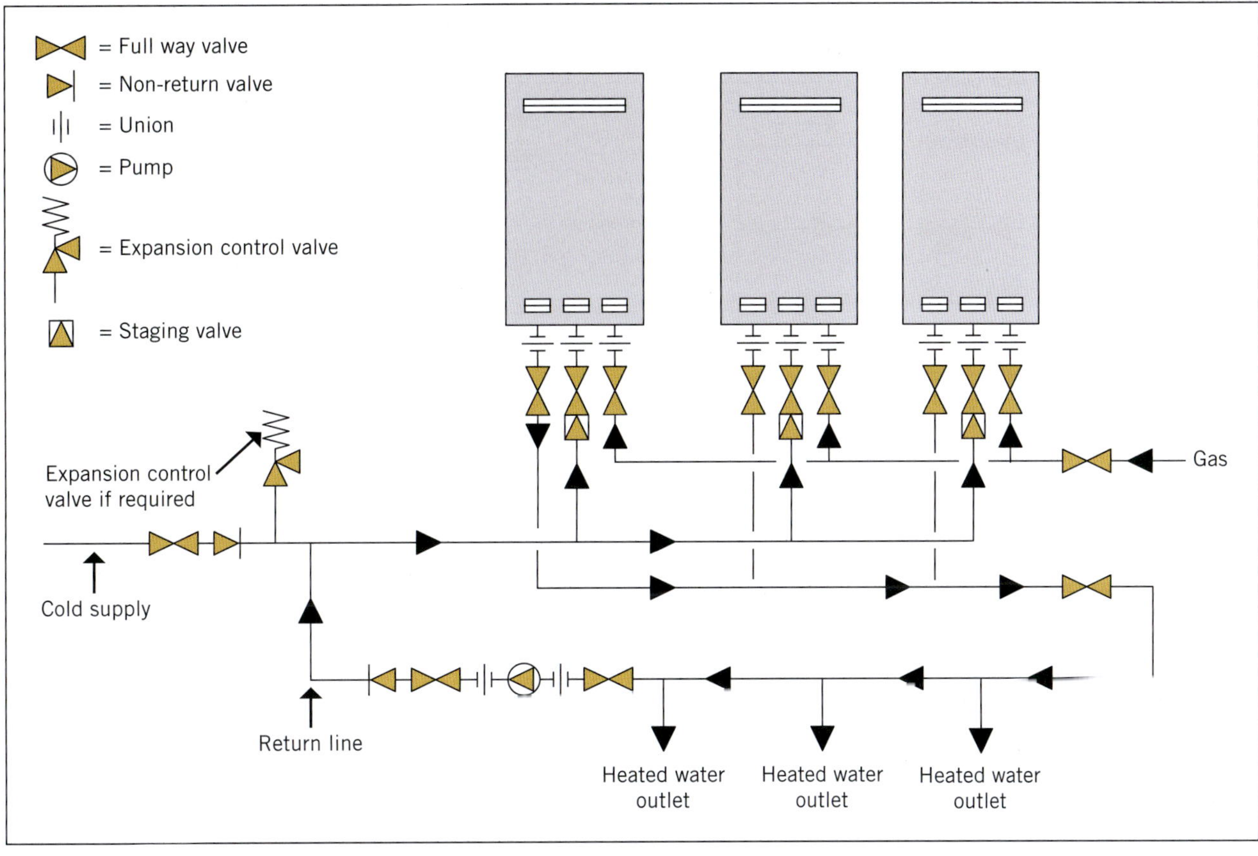

FIGURE 5.68 Gas continuous flow mains pressure manifold (equaflow) system – diagram

FIGURE 5.69 Gas continuous flow mains pressure manifold (equaflow) system

Manifolding of water heaters has several advantages, including:

- allowing individual water heaters on the manifold to be isolated and maintained or replaced while the rest of the system continues to function and provide hot water
- allowing part of the system to be shut down during off-peak periods (such as in caravan parks).

Installing a domestic building ring main for storage water heaters

When installing a building ring main, an additional small mains pressure water heater may need to be installed to maintain the temperature within the piping system, as shown in Figure 5.70. *Do not* return the hot line back through the main storage system for solar, heat pump, electric off-peak or twin (non-simultaneous) water heaters: a separate booster water heater needs to be installed to maintain any heat loss. A 25 L or 50 L 2.4 kW element on a flex and plug storage water heater is more than sufficient.

Most people prefer to use a timer to control the circulating pump, as it suits their lifestyle and needs.

LEARNING TASK 5.5

1 What is the purpose of a flow and return system (ring main)?
2 Explain a manifolded (equaflow) system.

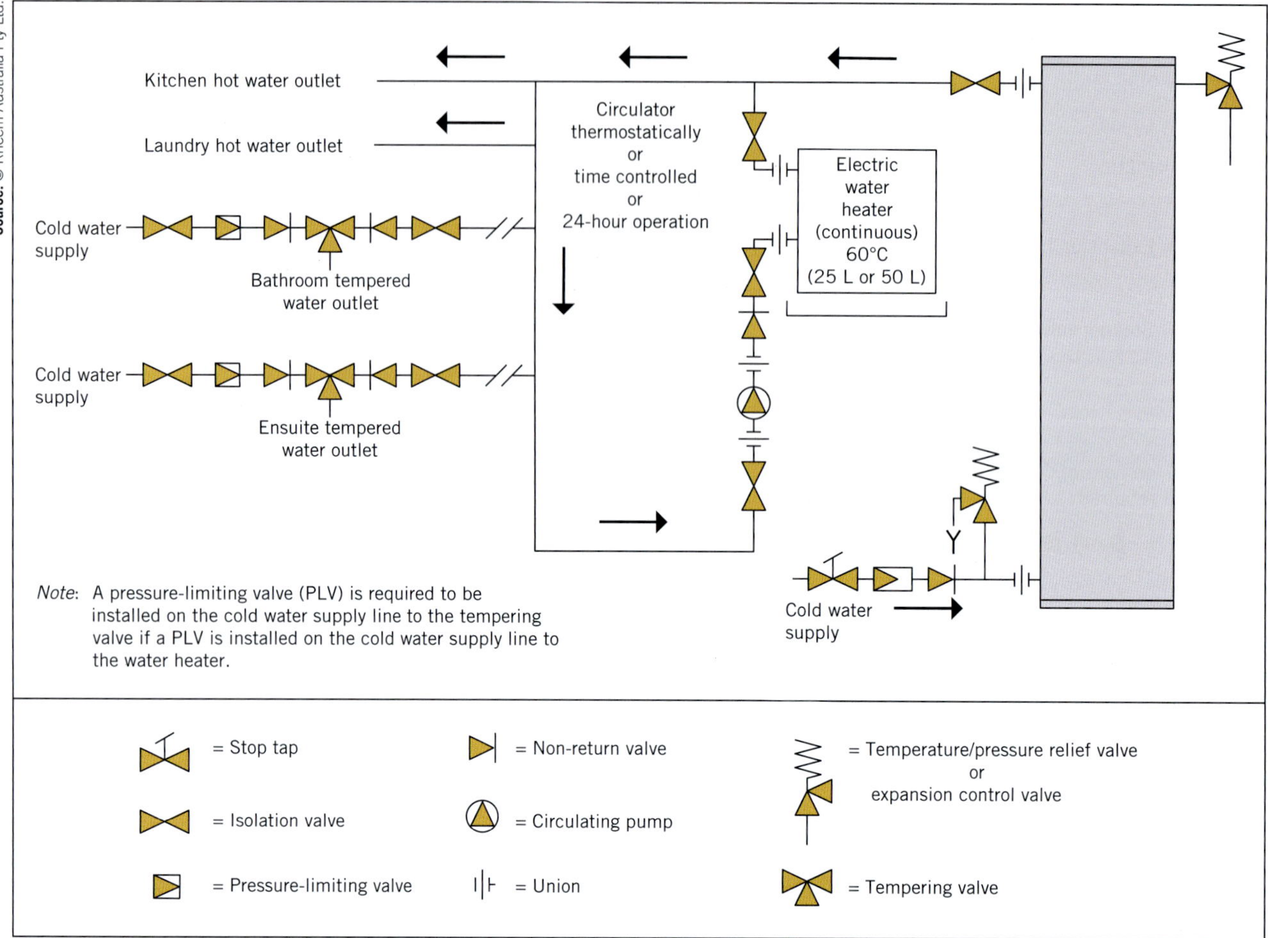

FIGURE 5.70 Domestic building ring mains and circulating loops system for solar, heat pump, electric off-peak or twin (non-simultaneous) element water heaters

Testing the system

The system should be flushed prior to connecting the water heater to remove any foreign matter. The water service should have been tested at the completion of the rough-in stage and left charged with water during construction. Once the water heater is connected, any interconnection between the hot and cold water services must be removed to prevent any damage occurring to the system valves by the backpressure from the cold water on the hot water system. The water heater installation must be visually checked for leaks. *Do not pressure test the water storage cylinder.*

Commissioning the system

Once the water heater has been installed, the system needs to be commissioned. The following is a guide only and the manufacturer's guidelines must be followed.

When installing multiple water heaters in a bank, each water heater must be checked for correct operation independently of the others.

HOW TO

COMMISSION A HOT WATER SYSTEM

1. Fill the system with water and purge it of air by opening all the hot water outlets prior to the heating medium being turned on.
2. Visually check for leaks on water heater connections.
3. Check the following points for correct operation:
   - a temperature and pressure relief valve and expansion control valve (if fitted)
   - b stored water temperature
   - c water delivery temperature
   - d flow and return temperature
   - e water level of a gravity system
   - f that the correct valves are installed, are opened and operate correctly
   - g outlets' minimum flow rates
   - h pump operation
   - i pump vibration
   - j pressures, where a pressure-limiting valve is fitted
   - k noise and water hammer.

Submit documentation

Upon completion, all the relevant documentation required by the regulatory and workplace authorities must be submitted; for example, a certificate of compliance.

The manufacturer's instructions and warranty must be left for the owner/occupier.

Hot water maintenance

Understanding the function and operation of basic components of water heaters will help diagnose and problem solve issues related to maintaining and repairing hot water heaters.

Element selection for electricity

The selection of the element size is critical to the performance of an electric water heater. Elements may be either immersion or ceramic bobbin.

Immersion elements

An immersion element (see Figure 5.71) is installed in direct contact with the water within the water heater. Should the element need replacing, the electricity must be isolated and the water heater drained.

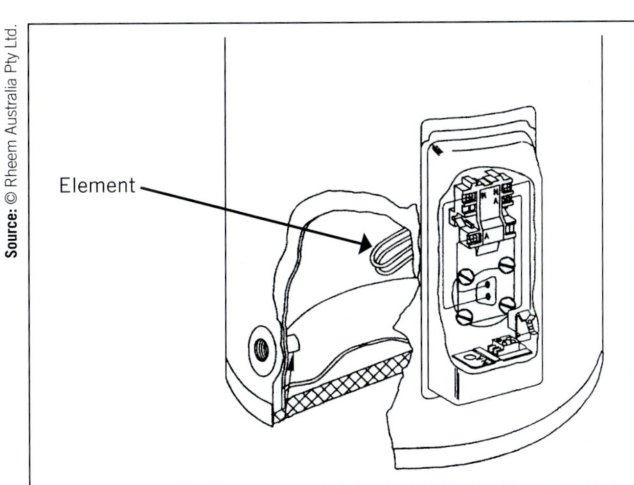

Source: © Rheem Australia Pty Ltd.

FIGURE 5.71 Immersion element

If the water heater is connected to an off-peak power supply, then a 4.8 kW element must be installed. Elements are also available in other sizes, such as 2.4 kW and 3.6 kW. Some electrical authorities may allow a 3.6 kW element to be used; check with the electrical authority for their requirements.

Ceramic bobbin elements

A ceramic bobbin element (see Figure 5.72) has an outer watertight sheath; the (bobbin) element is installed inside the sheath. These elements are not as efficient as an immersion element and may be replaced without emptying the water heater of water. They are mainly used in low-pressure (gravity feed) water heaters.

Lutsenko_Oleksandr/Shutterstock.com

FIGURE 5.72 Ceramic bobbin element

Electric elements are installed at the bottom of the water heater. When the temperature of the water falls below that set on the thermostat, and if power is available (that is, not subject to off-peak limitation), the water is heated. As the water is heated, it becomes less dense and rises, with the colder water falling to the bottom to be heated. This is called convection.

Twin booster (non-simultaneous) elements

Manufacturers can supply water heaters with twin elements (see Figure 5.73). The main element is located at the bottom of the tank, with the second one located in the upper section of the water heater. The top element is connected to a continuous power supply. Its function is to maintain the temperature in the top of the storage water heater at or above 60°C. The top element is designed to heat the top 50 L to

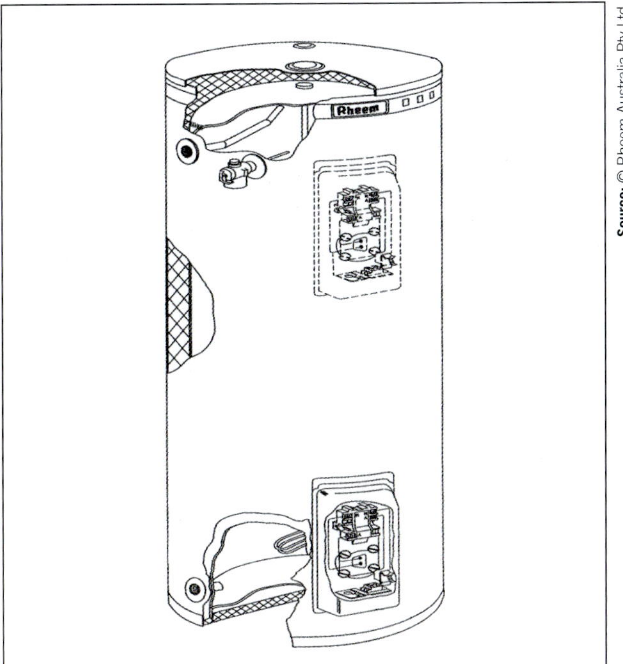

Source: © Rheem Australia Pty Ltd.

FIGURE 5.73 Twin booster elements

90 L only. The bottom element is normally connected to an off-peak tariff. (The elements are designed to heat individually and at no stage will both elements heat simultaneously.)

Thermostats

As discussed at the beginning of this chapter, heated water must be stored above 60°C to prevent the growth of *Legionella* bacteria. All storage water heaters have a thermostat installed to control the water temperature. The thermostat may be of the contact type (see **Figure 5.74**) or the immersion type (see **Figure 5.75**). A **contact thermostat** is attached to the outside wall of the water heater and relies on heat transfer through the wall of the tank. **Immersion thermostats** are used on gas water heaters; the thermostat extends into the heater, giving a quicker response time.

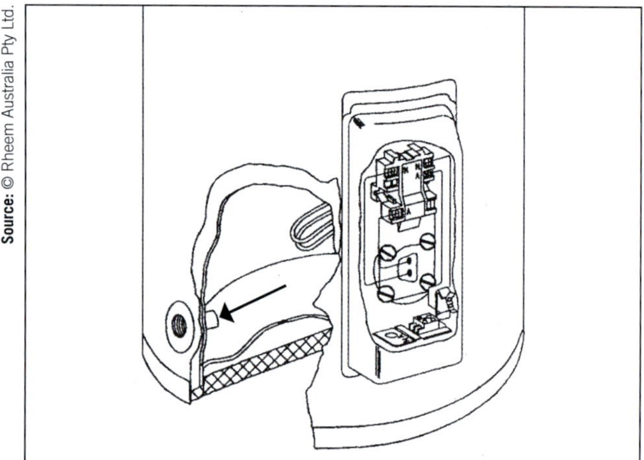

FIGURE 5.74 Contact thermostat on an electric storage hot water heater

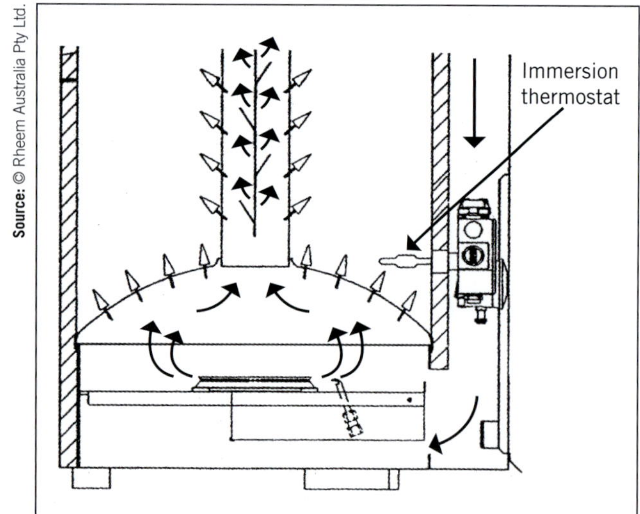

FIGURE 5.75 Immersion thermostat on a gas storage hot water heater

Sacrificial anodes

Sacrificial anodes are easily corroded materials in the form of a long rod installed in hot water storage tanks that are sacrificed to corrosion instead of the steel cylinder (see **Figure 5.76**). Therefore, the steel tank will last much longer.

The anode rod is made from aluminium/zinc or magnesium and slowly dissolves away from galvanic corrosion happening between the water and steel tank. The rod depletes from this corrosion and the steel tank is left alone until there is no more sacrificial metal left.

FIGURE 5.76 Sacrificial anode

Sacrificial anodes generally last around five years in a mains pressure hot water storage unit. So, it is good practice to replace them in that time and extend the life of the hot water heater.

Clean up

The clean-up process is very important, so all rubbish should be disposed of thoughtfully and placed in the correct bins provided. Any leftover material should be stored so it can be used on another job.

Be sure to clean and store all tools and equipment safely so they are ready to use on the next job. Remember to return any equipment on hire in the same condition it was hired out. Any faulty or damaged tools must be reported and tagged so they are not used until repaired.

The old saying 'the job is not finished until the paperwork is done' is always relevant. Submit all relevant documentation and hand over any product warranties to the client on completion.

Show the client the operational procedure and explain the maintenance instructions of the water heater.

LEARNING TASK 5.6

1 Why is it important to remove all the air from a hot water service when commissioning it?
2 When commissioning a hot water system, why is the temperature and pressure relief valve operated?
3 Why is the hot water cylinder not pressure tested?
4 What is the function of a thermostat?
5 What is the purpose of a sacrificial anode?

COMPLETE WORKSHEET 5

SUMMARY

- The three most common water heaters are mains pressure storage heaters, continuous water heaters and solar water heaters.
- Hot water must be stored at a minimum temperature of 60°C to prevent *Legionella* bacteria growth.
- The maximum temperature to an ablution fixture is 50°C.
- It is important to understand the requirements for temperature pressure relief valves.
- There are two types of solar water heating systems: a split system and a close coupled system.
- A heat pump is very energy efficient by transferring the surrounding air into heat.
- An expansion control valve discharges expanded water from the bottom of the tank instead of the top, therefore saving hot water.
- Plastic pipe cannot be used within 1 m of the hot water outlet and between the isolation valve and cold water inlet.
- Hot water heaters must be located close to fixtures to prevent water wastage.
- A heat trap must be installed on the hot water outlet if the hot water heater doesn't have an integral one to prevent heat loss.
- Pipe insulation prevents heat loss and saves energy.
- A flow and return system prevents water wastage as well as providing hot water instantly.
- Be sure to flush pipework and remove all air when commissioning hot water heaters.
- The client must be informed of the hot water heater operation instructions and maintenance requirements upon completion.

REFERENCES

AS/NZS 3500.1 Plumbing and Drainage: Water Services: **http://www.standards.org.au**

AS/NZS 3500.4 Plumbing and Drainage: Heated Water Services: **http://www.standards.org.au**

Energy Safe Victoria: **http://www.esv.vic.gov.au**

Material production standards: **http://www.watermark.standards.org.au**

NCC: Part 3 – Plumbing Code of Australia (PCA): **https://ncc.abcb.gov.au/**

Rheem Australia: **http://www.rheem.com.au**

Standards Australia ('Earthing of earthing installation using the water reticulation system'): **http://www.standards.org.au**

GET IT RIGHT

1 Which photo shows the correct installation?

2 What is the problem in the other photo?

3 Why is that a problem?

WORKSHEET 1

To be completed by teachers

Student competent ☐

Student not yet competent ☐

Student name: ____________________

Enrolment year: ____________________

Class code: ____________________

Unit competency code/title: CPCPWT3022 Install and commission water heating systems and adjust controls and devices

Task: Review 'Identify installation requirements' and answer the following questions.

1 Name two isolating valves that provide full flow.

____

____

2 List four situations where isolating valves must be installed on water services.

____

____

____

____

3 Where must a full flow isolating valve be fitted?

____

4 What is a problem with a gravity-fed hot water heater installation?

____

____

____

5 Explain the term BASIX in relation to water supply.

____

____

____

6 Explain the term WELS in relation to water supply.

____

____

____

7 Name an advantage and a disadvantage of ceramic disc tapware.

8 Name the three functions of a trio valve.

9 Name three situations where a solenoid valve is used.

10 What is the energy source used to operate a solenoid valve?

11 What are the minimum and maximum water pressures permitted in a building?

12 What is the maximum velocity of water within a building and why is that important?

WORKSHEET 2

To be completed by teachers	
Student competent	☐
Student not yet competent	☐

Student name: ____________________

Enrolment year: ____________________

Class code: ____________________

Unit competency code/title: CPCPWT3022 Install and commission water heating systems and adjust controls and devices

Task: Review the section '5.1 Identify installation requirements' and answer the following questions.

1 What is the function of a line strainer? Name two situations where they are used.

____

____

2 What is the function of a non-return valve? Name two situations where they are used.

____

____

3 What is another name for a non-return valve?

____

4 What is the difference between a pressure-limiting valve and a pressure-reducing valve?

____

____

____

5 In what situation would a pressure ratio valve be used?

____

____

6 Where are spring type taps commonly used and why?

____

____

7 What is the function of an aerator?

____

____

____

8 Name two main components that are part of infra-red sensing tapware.

9 What type of power supply is common with infra-red sensors?

10 List five applications where hands-free infra-red sensitive tapware would be beneficial.

11 Name two types of flushing devices.

12 What is the minimum head required to operate a flush valve?

13 A ceiling sensor should be located within how many millimetres of the urinal wall?

14 What is the minimum size of a water service supplying a mains pressure flush valve?

15 How can water hammer be eliminated?

16 Tempering valves are installed to mix hot and cold water together to produce warm water at a certain temperature. What is this temperature?

17 Why is the warm water limited to this temperature?

18 How often should tempering valves be replaced?

19 What fixtures must be supplied with water from a tempering valve?

20 Name three facilities where thermostatic mixing valves must be installed.

21 What is the maximum permissible distance from a thermostatic mixing valve to the furthest outlet?

22 Explain what a thermal shutdown is.

23 What is the function of temperature pressure relief valves?

24 What is the function of an expansion control valve?

25 What is the formula for the expansion of water?

WORKSHEET 3

To be completed by teachers

Student competent ☐

Student not yet competent ☐

Student name: ______________________

Enrolment year: ______________________

Class code: ______________________

Unit competency code/title: CPCPWT3022 Install and commission water heating systems and adjust controls and devices

Task: Review 'Prepare for work' and answer the following questions.

1 Excessive velocity and flow rate can be reduced by what two methods?

____

____

2 Explain the differences between a continuous flow water heater and a storage water heater.

____

____

____

____

3 What is the minimum temperature water must be stored in a hot water tank and why?

____

____

4 There are two main types of solar water heating systems. What are they?

____

____

____

____

5 Name the three main components in a solar hot water system.

____

____

____

6 Describe a 'close coupled solar system'.

7 Describe a 'split (forced circulation) solar system'.

8 Which direction must the solar collectors face when installed in Australia and why?

9 Why can't plastic pipe be used between the solar collectors and the storage cylinder?

10 What are two advantages of a low pressure (gravity feed) hot water heater?

11 What are two disadvantages of a low pressure (gravity feed) hot water heater?

12 Explain the basic operation of a heat pump air-sourced water heater.

13 Why are heat pump storage water heaters preferable to use than the typical electric heated storage water heater?

14 What is the function of a sacrificial anode?

WORKSHEET 4

To be completed by teachers	
Student competent	☐
Student not yet competent	☐

Student name: ____________________

Enrolment year: ____________________

Class code: ____________________

Unit competency code/title: CPCPWT3022 Install and commission water heating systems and adjust controls and devices

Task: Review 'Install water heating system and service control device' and answer the following questions.

1 List four points that should be considered when installing a water heater.

____

____

____

____

2 List two ways in which a water heater must be supported when it is installed in an external location.

____

____

3 When must a safe tray be installed under a water heater?

____

____

____

4 What is the minimum size of a safe waste drain if installed in a roof space?

____

5 Explain the purpose of a terminator valve on the cold water inlet of a pressure water heater.

____

____

____

6 What are the limitations of polymer pipe with hot water heater connections?

7 Name the valves installed on the cold water inlet to a mains pressure storage hot water heater.

8 Name three ancillary valves installed on the hot and cold inlets to a thermostatic mixing valve.

9 What is the purpose of installing flow and return piping on a hot water service?

10 State three conditions relating to how temperature and pressure relief valve drains must be installed.

11 Using the water heater sizing table, what capacity (size) heat pump storage water heater is recommended for a three-bedroom house?

12 What is the minimum size of the hot water pipe from the water heater outlet to the first branch?

WORKSHEET 5

To be completed by teachers

Student competent ☐

Student not yet competent ☐

Student name: ______________________

Enrolment year: ______________________

Class code: ______________________

Unit competency code/title: CPCPWT3022 Install and commission water heating systems and adjust controls and devices

Task: Review 'Install water heating system and service control device' and answer the following questions.

1 What is the main principle of the manifold (equaflow) system?

____

____

____

2 What type of isolating valves must be fitted to the inlets and outlets of water heaters in a manifolded system?

____

____

____

3 What is the purpose of a staging valve installed on manifolded continuous gas water heaters?

____

____

4 Once the water heater is connected, any interconnection between the hot and cold water services must be removed. Why?

____

____

____

5 Should the storage tank be pressure tested when commissioning a hot water system? Explain your answer.

____

____

____

WORKSHEETS 5

6 List five points that should be checked when commissioning a water heater.

7 What is the function of a thermostat in a water heater?

8 Neatly draw the symbols for the following valves and fittings: Full way valve; Non-return valve; Union; Pump; Expansion control valve

FABRICATE AND INSTALL FIRE HYDRANT AND HOSE REEL SYSTEMS

6

Chapter overview

The installation of fire hydrants and hose reels plays a major role in preventing the spread of fire and protecting people and property throughout the community. The information in this chapter will help you gain the skills and knowledge to competently fabricate and install fire hydrants and hose reel systems using different materials and jointing systems.

Firefighting existed before the **hydrant**, and the idea of getting the wet stuff onto the red stuff is very old. The inventor of the first device that we'd recognise today as a fire hydrant can't be told, because the hydrant was developed over a period of many years by many people.

We do know that the first pillar or post street fire hydrant was developed in America by Frederick Graff in 1801. Prior to that, 'cisterns' or underground tanks were used to supply water for firefighting, and these are still used today.

The first practical fire hose was invented in Holland by Jan van der Heyden in 1673. The hose was made from 15 m lengths of leather or sail cloth sewn together in a single seam.

Fire hydrants, fire hose reels, fire sprinklers, fire extinguishers and smoke alarms are essential for effective fire protection to save lives and protect properties. Fire hydrants are installed within properties for use by the fire brigade. Fire hydrant systems should only be used for firefighting purposes.

Learning objectives

Areas addressed in this chapter include:

- identify installation requirements
- prepare for work
- fabricate, install and test system
- clean up.

Identify installation requirements

This section is aimed at understanding different types of fire service systems and installations, and the requirements necessary to meet the authorities' standards. It also covers work health and safety (WHS) requirements and accessing codes and standards, as well as reading plans and specifications.

Design drawings, job specifications, codes and standards

Hydraulic plans and design drawings for fire hydrant and hose reel system installations are designed, drawn and endorsed by hydraulic consultants. They specify the location of hydrants and hose reels as well as the pipe and fitting material to be used. The pipe sizes are also determined from the available water main supply pressure and flow rate. These plans are established at the design stage and issued at the tender/quote stage.

The Plumbing Code of Australia (PCA), which is volume three in the National Construction Code (NCC), is the overarching guideline and refers to the Australian Standards for the minimum standard requirements. Refer to Chapter 4 'Carry out interactive workplace communication' in *Basic Plumbing Skills* for further information on how to access relevant jurisdictional regulations and other standards. The NCC can be accessed online for no charge. The Australian Standards can be purchased online as electronic copies or hard copies.

The Australian Standards that state the minimum requirements for the fabrication and installation of fire hydrants and fire hose reels are:

- AS 2419.1 Fire Hydrant Installations
- AS 2441 Installation of Fire Hose Reels
- AS 1221 Fire Hose Reels (Design, Construction and Performance)
- AS 2118 Automatic Fire Sprinkler Systems
- AS 4118 Fire Sprinkler Components
- AS/NZS 3500.1 Plumbing and Drainage: Water Services
- AS/NZS 3500.0 Glossary
- National Construction Code (NCC).

Work health and safety (WHS) and environmental requirements

Legislation requires that WHS requirements be observed and adhered to. At the minimum, a job safety analysis (JSA) must be completed on all jobs – identifying the hazards, assessing the risks and applying control measures to minimise the risks.

Working at heights, working in confined spaces, hot works and excavation are considered high-risk activities involved in the fabrication and installation of fire hydrants and hose reel systems. Therefore, a safe work method statement (SWMS) must be completed with a more detailed risk assessment, as well as the JSA, before starting these activities. For example, the correct techniques for safe trench excavation must be carried out to avoid any chance of trench collapse.

Environmental requirements involve taking the appropriate measures to reduce excessive noise and dust when drilling, cutting and sawing different materials. Also, care must be taken when excavating that the spoil and fill is stockpiled safely so as not to cause silt and sedimentation damage to the surrounding area.

Manual lifting and handling techniques must always be followed according to WHS requirements. A correct size-up of the load to be lifted is important to determine whether a mechanical lifting device, a crane or a two-person lift is required. Try to avoid excessive manual handling and double handling of materials wherever possible.

When working at heights or in confined spaces, the correct procedures must be in place. Workers must be appropriately trained and accredited to perform these duties. Heavy fines apply to those who do not follow the correct procedures.

Types of systems and requirements

There are several different systems that incorporate fire hydrants and fire hose reels. As specified in AS/NZS 2419.1, a fire hydrant system must be a wet pipe system. This means that the service is fully charged with water at all times at the required pressure and is allowed to be boosted by an approved pumping system if required.

Hydrant system

The hydrant system is a separate dedicated service connected to the water main. This service supplies water only to fire hydrants via a booster assembly and is generally a minimum of 100 mm in diameter. High-rise buildings usually have a storage tank on the highest level dedicated to supply hydrants (see Figure 6.1). The purpose of the 'booster assembly' is to allow firefighters to connect to the hydrant system and boost the pressure from the pump on the fire truck. The water supply to the fire hose reels usually connects to the domestic water supply metered system.

Combined fire hydrant and fire hose reel system

On certain sites, the fire hose reel service is combined with the fire hydrant service (see Figures 6.2 and 6.3). This is appropriate where the fire hose reels are located near the fire hydrants to keep pipework to a minimum. The service type is designed by the

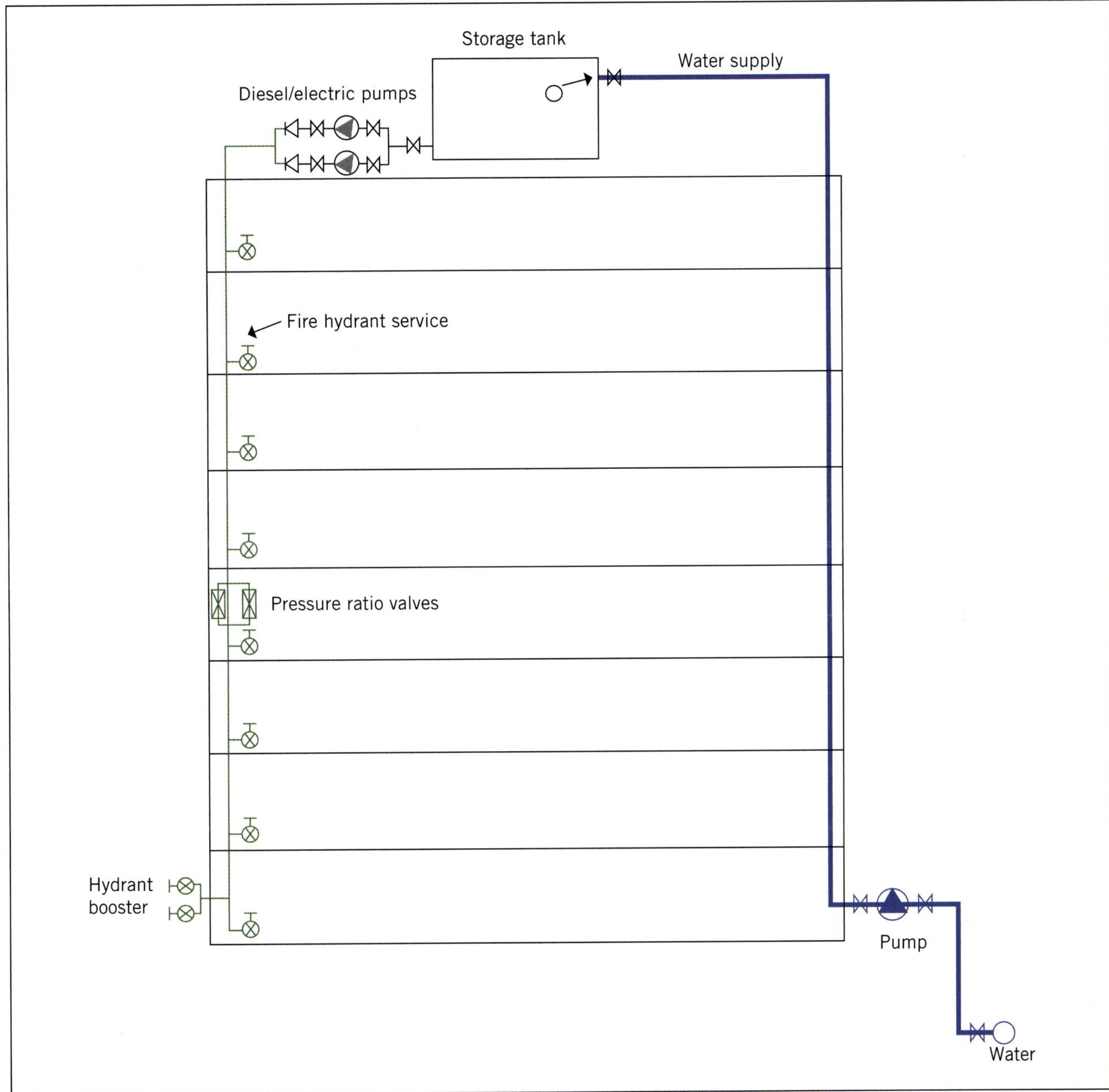

FIGURE 6.1 High-rise hydrant installation

hydraulic consultant and sized accordingly to supply the minimum pressure and flow rates. A combined fire hydrant and fire hose reel system is designed to deliver water at the minimum required pressure and flow rate for four hours.

Combined sprinkler and hydrant system

This system combines the fire sprinkler service with the fire hydrant service and is designed to protect a wide range of building types. Flow in the system is controlled by valves, switches, actuating devices, alarms and pumps. The fire sprinkler system is designed to deliver water at a minimum flow rate and pressure for 30 minutes. By then, the fire brigade should have connected to the system and increased the water supply and pressure through the booster assembly.

Fire hydrant system components

A fire hydrant is a valve with a fire hose connection suited to connect to the hose coupling used by the firefighting agency responsible for the area.

A **feed hydrant** supplies water via a hose to the inlet of the pump on a fire truck. Another hose carries water at an increased pressure from the outlet of the fire truck pump, which can either spray water directly onto the fire or connect to the *booster inlet*, which charges the fire hydrant service on the property.

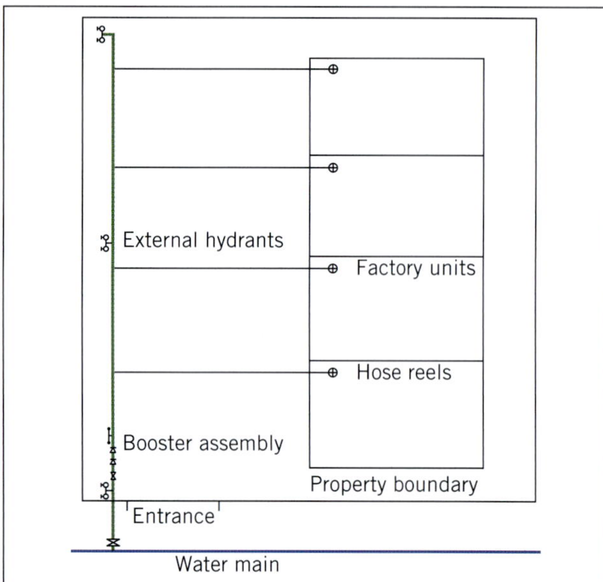

FIGURE 6.2 Low-rise fire hydrant and fire hose reel installation

FIGURE 6.3 Combined fire hydrant and fire hose reel service

An **attack hydrant** is the hydrant to which a firefighter connects a hose to extinguish the fire (see **Figure 6.4**).

A fire hydrant system can be supplied to either a potable or recycled water main with a tee and valve. Also, large buildings require additional water supply stored in tanks dedicated for firefighting.

FIGURE 6.4 External attack fire hydrant

Where the service enters the property, a fire brigade booster and suction connection is set up to supply the fire brigade with water and increase the system pressure with the fire brigade pump. AS 2419.1 states that certain building classes require a ring supply main, so water is fed from both directions (see **Figure 6.5**). Having the water supplied in both directions helps to prevent the water supply from being starved at any hydrant operating.

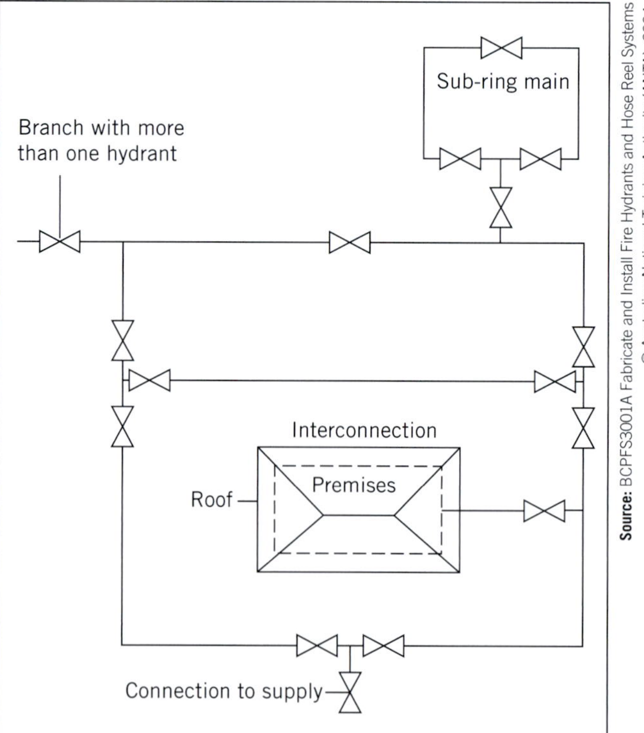

FIGURE 6.5 Fire hydrant ring main

Source: BCPFS3001A Fabricate and Install Fire Hydrants and Hose Reel Systems © Australian National Training Authority (ANTA), 2004.

AS 2419.1 FIRE HYDRANT INSTALLATIONS

An isolating valve is installed inside the property boundary downstream of the water authority valve. It is locked or strapped in the open position so it cannot be tampered with and is normally a component of the booster assembly. Its purpose is to isolate the fire hydrant system from the water supply. Additional isolation valves are located so that hydrant system zones can be isolated in 25% sections and so half of the available hydrants can protect each fire compartment.

Pressure gauges in fire hydrant systems, as per AS 2419.1, must be installed with a gauge cock (isolating valve) to permit removal for testing and replacing. The dial face must have a minimum diameter of 100 mm. The gauge must read 25% higher than the hydrostatic test pressure (see **Figure 6.6**).

FIGURE 6.6 Pressure gauge

Fire hydrant pressure gauges shall be installed:

- on the inlet and outlet of any pump
- next to the booster inlet connection
- on the delivery side of any jockey pump
- at each pressure switch
- at any test point.

Internal or external *cabinets* may be required to house fire hydrants and booster connections or for hose reels. The layout and dimensions of the cabinets must comply with AS 2419.1 (see Fire hydrant installations, Part 1: System design, installation and commissioning). Cabinets protect equipment from the weather, unauthorised interference and vandalism.

The cabinet must have a lever-style lock compatible with local fire brigade requirements and be clearly labelled FIRE HYDRANT or FIRE HYDRANT – HOSE REEL or HYDRANT – BOOSTER as appropriate. The words must in capital letters and a minimum of 50 mm in height (see **Figure 6.7**).

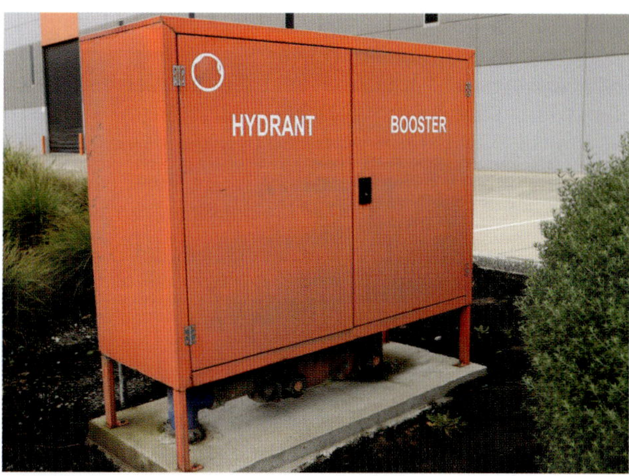

Dorothy Chiron/Shutterstock.com

FIGURE 6.7 Hydrant booster cabinet

Metal cabinets remote from the building should be mounted on legs to provide not less than 50 mm space between the bottom edge of the cabinet and the finished surface level (see **Figure 6.8**).

Source: © Galvin Engineering.

FIGURE 6.8 Tank water supply with booster inlet

All other cabinets, enclosures or recesses should have a sloping floor from the rear of the cabinet to the front so any water on the floor will run off.

AS 2419.1 FIRE HYDRANT INSTALLATIONS

Classes of buildings

There are different fire hydrant systems for different classes of buildings. **Table 6.1** lists how the NCC classes buildings.

AS 2419.1 shows the number of hydrants required to flow simultaneously depending on the building classification, number of floors and floor area. (see AS 2419 Fire Hydrant Installations, Part 1: System Design, Installation and Commissioning).

Fire hydrant coverage should be determined by distances that are measured along the most direct laid-on-ground or floor route to a car park, open yard or marina, and within the protected building. The main measurements are as follows:

- External hydrants are hydrants located outside the building and should be positioned to cover the whole site.

TABLE 6.1 Classes of buildings

Classes of building		
Class 1	**Class 1a**	A single dwelling being a detached house, or one or more attached dwellings, each being a building, separated by a fire-resisting wall, including a row house, terrace house, town house or villa unit.
	Class 1b	A boarding house, guest house, hostel or the like with a total area of all floors not exceeding 300 m^2, and where not more than 12 reside, and is not located above or below another dwelling or another Class of building other than a private garage.
Class 2	A building containing 2 or more sole-occupancy units each being a separate dwelling.	
Class 3	A residential building, other than a Class 1 or 2 building, which is a common place of long term or transient living for a number of unrelated persons. Example: boarding-house, hostel, backpackers accommodation or residential part of a hotel, motel, school or detention centre.	
Class 4	A dwelling in a building that is Class 5, 6, 7, 8 or 9 if it is the only dwelling in the building.	
Class 5	An office building used for professional or commercial purposes, excluding buildings of Class 6, 7, 8 or 9.	
Class 6	A shop or other building for the sale of goods by retail or the supply of services direct to the public. Example: café, restaurant, kiosk, hairdressers, showroom or service station.	
Class 7	**Class 7a**	A building which is a car park.
	Class 7b	A building which is for storage or display of goods or produce for sale by wholesale.
Class 8	A laboratory, or a building in which a handicraft or process for the production, assembling, altering, repairing, packing, finishing or cleaning of goods or produce is carried on for trade, sale or gain.	
Class 9	A building of a public nature.	
	Class 9a	A health care building, including those parts of the building set aside as a laboratory.
	Class 9b	An assembly building, including a trade workshop, laboratory or the like, in a primary or secondary school, but excluding any other parts of the building that are of another class.
	Class 9c	An aged care building.
Class 10	A non-habitable building or structure.	
	Class 10a	A private garage, carport, shed or the like.
	Class 10b	A structure being a fence, mast, antenna, retaining or free standing wall, swimming pool or the like.
	Class 10c	A private bushfire shelter.

Source: National Construction Code (2005), Volume One, pp. 35–6.

- A firefighters' hose is 30 m in length. External coverage is based on using 60 m of hose (two lengths) and a 10 m hose stream (70 m in total) when connected to a dual hydrant. When using a single hydrant, coverage is based on a 30 m hose and 10 m hose stream (40 m in total).
- Where internal hydrants are used to protect parts of the building, all points must be covered by a 10 m hose stream from a 30 m hose (40 m in total), connected across the floor to the hydrant.
- In multistorey buildings, one storey down and one storey up can be covered, provided:
  - there is at least one hydrant supplying each level
  - no more than 30 m of hose is laid in a stairway
  - a minimum of 1 m of hose projects into the area being covered by the hose stream.
- AS 2419.1 explains how it is possible to have a hydrant on a stair landing that is different from the finished floor level it is supplying (see Fire hydrant installations, Part 1: System design, installation and commissioning).
- For roof areas that have occupant access, the area of access must be covered by a 10 m stream from a 30 m hose.
- Open yards must be covered by a 60 m hose and a 10 m water stream.
- Hydrants should be in accessible locations at least 10 m from any building and with hard-standing areas for pump appliance vehicles.
- Feed hydrants should be a maximum 20 m from the hardstand area.

A **hardstand area** is an area for the fire brigade pumping appliance to park on and safely carry out firefighting operations. It must be a reasonably level all-weather surface that is strong enough to take the weight of a fire truck with total height clearance. It must be located no more than 20 m from an external feed fire hydrant.

An **open yard** is a designated area greater than 500 m^2 used for storing or processing combustible material.

Piping material

The piping material used for fire hydrant and fire hose reel services can vary depending on the application, and will depend if the piping is below or above ground. Piping materials used are:

- copper pipe
- polyethylene
- unplasticised polyvinyl chloride (PVC-U)
- galvanised steel
- stainless steel.

Other factors to consider when selecting a piping system are:

- compatibility of materials, fittings and fixings
- the corrosive nature of the ground and environment
- whether the area is frost prone
- access for service and repair.

External pipework

As per AS 2419.1, wherever possible external pipework must be located below ground. Any external pipework above ground that's subject to freezing must be suitably protected.

External above-ground pipework that is attached to an external wall or roof must be protected from fire damage by the supporting structure having either a fire-resistant level (FRL) of 60 minutes or an automatic fire sprinkler system as per AS/NZS 2118.1.

Internal pipework

Any internal above-ground pipework must be protected by the supporting building structure and pipe supports having a FRL of 60 minutes or an automatic fire sprinkler system as per AS/NZS 2118.1. The pipework also can be protected in a fire-isolated stairwell or a fire-resistant shaft.

AS 2419.1 FIRE HYDRANT INSTALLATIONS

Below-ground piping

The most common piping materials used below ground for fire hydrant and hose reel systems are:

- copper
- polyethylene
- PVC-U
- cement-lined cast iron.

Where the ground is subject to ground movement, including mine subsidence, provision must be made to support the pipework.

Galvanised steel pipe and fittings can only be used for hydrant risers or short connection pieces no more than 1.5 m long. Any section of the pipe below the ground must be double-wrapped in petroleum tape.

Galvanised steel pipe cannot be used to connect the inlet service from the water agency main to the property backflow prevention device.

AS 2419.1 FIRE HYDRANT INSTALLATIONS

LEARNING TASK 6.1

1. Who designs the fire service layout?
2. Describe a feed hydrant.
3. Describe an attack hydrant.
4. What is the purpose of a booster valve assembly?
5. Name two pipe materials used below ground for fire hydrant services.

COMPLETE WORKSHEET 1

Prepare for work

Planning and preparing the workplace provides for a smooth-flowing, efficient and cost-effective job. Planning and sequencing tasks with the required tools and equipment will help to ensure quality assurance requirements are met.

Having the job site well prepared helps the work progress more safely and smoothly. A site visit before starting work will help determine access for deliveries, plant and equipment. A storage area can be established for materials, tools and equipment. Ensure site inductions have been carried out. Having the job set and marked out at this time will help identify any problems associated with the installation.

Create a materials list and collect materials

It pays to take the time to work out an accurate order and to have the materials onsite in a timely manner. Be sure to refer to the plans and specifications to get the correct material required. It is important that the specified material complies with the relevant Australian Standards.

Information from specifications and plans can determine the type of pipe and fittings to be used, as well as the brackets, clips and fixing types for the fabrication and installation of fire hydrants and hose reel systems.

Depending on company procedures, the order may be made direct or through a purchasing officer. Several suppliers may be involved to supply the variety of materials, valves and fittings required.

Planning and ordering early will ensure all the materials required are delivered onsite and ready to start the installation when required.

Deliveries must be checked against the order on arrival and missing items documented, along with any items on back-order. This should be part of the company policy assurance process.

Having this process in place shows professionalism and will also help to prevent delays and conflict with suppliers.

Quality assurance requirements

Most companies have quality assurance procedures in place to promote a good standard of workmanship and efficient operations. This also helps to prevent mistakes from happening.

The requirements of this policy would include:

- controls and procedures in the workplace (WHS)
- use and maintenance of power tools and equipment
- interpreting plans and specifications
- quality of materials selected
- handling procedures
- recording information
- clean-up of job site.

Tasks are planned and sequenced

Installing fire hydrant and hose reel systems can involve trench excavation, working at heights and working in confined spaces, so it is vital the correct procedures are taken. This involves planning and sequencing the order of tasks. Preliminary tasks include the following:

- Ensure all fees and permits have been dealt with (e.g. a road opening permit, footpath opening permit and water main drillings).
- Carry out a Before You Dig Australia (BYDA) enquiry.
- Be certain a thorough risk assessment has been done with a completed SWMS.

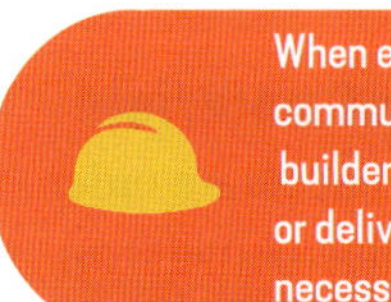

When excavating trenches onsite, be sure to communicate this with the site supervisor or builder well in advance to avoid any clash of trades or deliveries onsite. Trench support also may be necessary, depending on the soil type and depth.

It is essential when planning tasks to inform the site supervisor of the nature and timing of a particular task to avoid clashes with other trades and any access issues. This will help everyone onsite work together and prevent unnecessary arguments. Reputation and respect are earnt for those who work this way, resulting in future work prospects.

Tools and equipment

There are specialised tools and equipment necessary for different jointing systems used in the fabrication and installation of fire hydrant and hose reel systems.

Crimp and press-fit tools used on copper and plastic pipe and fittings need calibrating regularly to ensure the tool is pressing or crimping to the correct tolerances, so the joint is sound. Depending on the tool, recalibration should take place after 50 000 cycles. Some tools have a warning light that alerts when recalibration is due.

Electro-fusion welding machines used for jointing polyethylene pipe also need to be regularly calibrated to ensure they are heating to the required tolerances.

All electrical tools and leads must be tagged for safety regularly to the local authorities' rules. The general requirement for building and construction is to inspect and tag electrical equipment every three months because of the harsh environment the electrical equipment is exposed to.

As with all plumbing hand and power tools, personal protective equipment (PPE) and other associated equipment, it is essential that the correct tool is selected for the task and checked to ensure it is working in accordance with the manufacturer's instructions and used correctly.

Any damage that is found during inspection should be reported to the supervisor immediately and the tool/equipment needs to be tagged out and removed from circulation until the issue has been resolved.

Select and check serviceability of personal protective equipment

The risk of injury can be reduced by the appropriate use of PPE. The types of PPE include the following:

- *Safety boots* have a steel capped toe to protect your feet from falling objects, and a hard-wearing sole that sharp objects cannot pierce.
- *Safety glasses or goggles* will protect your eyes from injury when drilling, cutting or sawing with hand or power tools.
- *Ear plugs or earmuffs* will protect your hearing when using power tools and working in noisy work sites.
- *Dust masks or respirators* will protect your respiratory system from breathing in dust and fine particles when drilling, cutting or sawing and working in a dusty environment.
- *Gloves* will protect your hands from cuts, abrasions and burns when working with different materials, tools and equipment.
- *Overalls* will cover your skin and protect you from exposure and different conditions.
- *High-visibility clothing* allows workers onsite to easily identify where others are and helps to avoid accidents.

FROM EXPERIENCE

Setting out the job at the site inspection identifies problems early so they can be dealt with more easily. For example, other services could be in way of the fire hydrant service. By seeing this at the planning stage, a better solution can be negotiated.

LEARNING TASK 6.2

1 Which documents contain the information to calculate a material list from?
2 How often do press-fit tools need recalibration?
3 How often does electrical equipment need to be tagged and tested in the building and construction industry?

COMPLETE WORKSHEET 2

Fabricate, install and test system

Fabricating, installing and testing fire services must be done with sharp attention to plans and specifications. If not, the results can be disastrous. Working with large pipework equates to large volumes of water, and major leaks can cause major damage.

Set out relevant job plans, specifications and Australian Standards

Fire systems must be designed by a qualified person, usually a hydraulic consultant. This is to ensure the system is suitable for firefighting purposes for any specific site. The fire system design must comply with the NCC and the relevant Australian Standards as a minimum requirement.

Water supply agencies have control over connections to their supply, with backflow prevention being a major consideration (see Figure 6.9). In some instances, onsite storage tanks must be installed as a static water supply for firefighting purposes due to the size of the building. The water authority has control over the materials and products used in fire hydrant and fire hose reel installations.

Therefore, it is important that the plans, specifications and job instructions are set out as designed for the fire service to comply with and perform to the specific site installation.

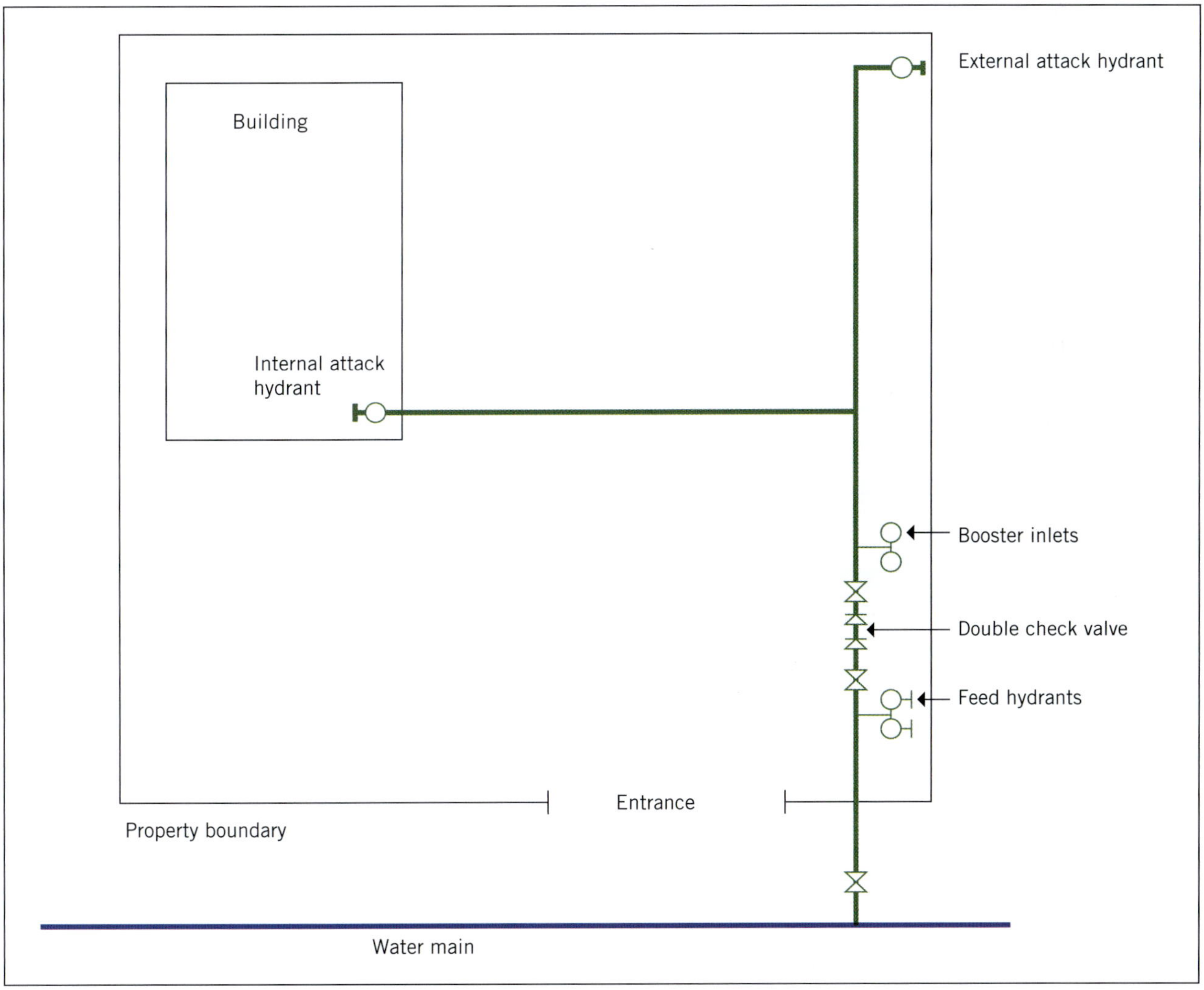

FIGURE 6.9 Fire hydrant service connection to water main with external and internal attack hydrants

When setting out the job, it is important to check plans for dimensions and levels. For underground pipework, check excavation depth and plan trench support, if required. Accurately mark out deck penetrations for vertical pipework to avoid costly core hole drilling and pipe offsets. Set out and mark horizontal pipework effectively to minimise pipe offsets and avoid clashing with other services.

Connect fire brigade suction and booster arrangement

A booster connection point is an assembly of valves on a fire hydrant or fire sprinkler service to allow the fire brigade to connect and pump water into the system at a boosted pressure (see **Figure 6.10**). This helps the fire brigade to work more effectively in controlling a fire, saving lives and protecting property (see **Figure 6.11**).

A fire booster assembly must be installed where:

- the fire hydrant service has internal fire hydrants installed
- two or more external (including street) hydrants are needed to provide coverage
- the fire hydrant service has external hydrants located more than 20 m from the fire brigade pumping appliance hardstand
- a pump set is installed
- a storage tank is installed
- more than one external fire hydrant serves a building with a fire compartment floor area greater than 2000 m^2
- the fire hydrant service has a ring main installed.

FIGURE 6.10 Booster assembly with block plan

Booster assemblies are available in different patterns depending on the number of valves and the space required (see **Figures 6.12**, **6.13** and **6.14**).

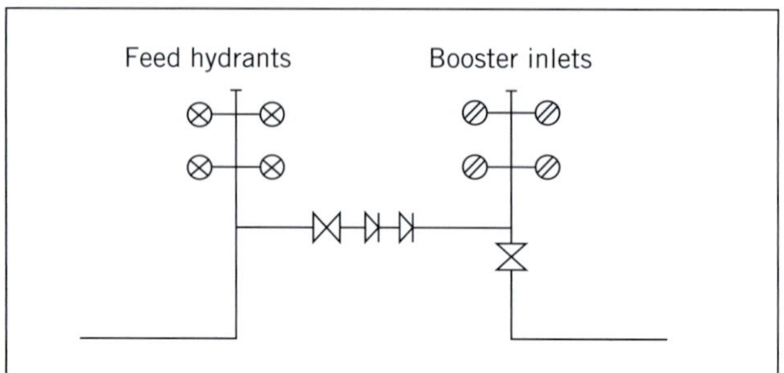

FIGURE 6.12 Typical 'H' pattern booster assembly

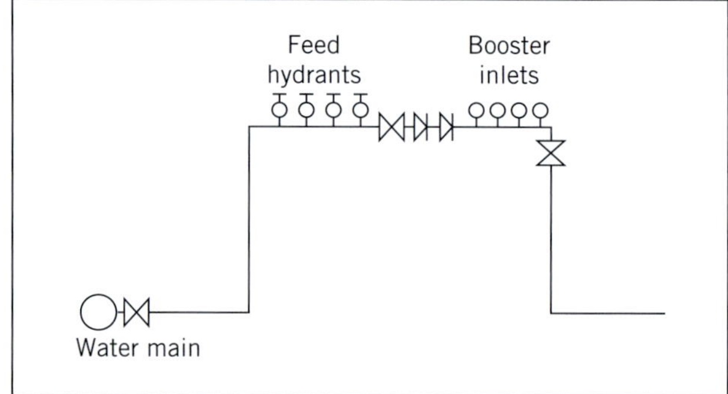

FIGURE 6.13 Typical 'in-line' booster assembly

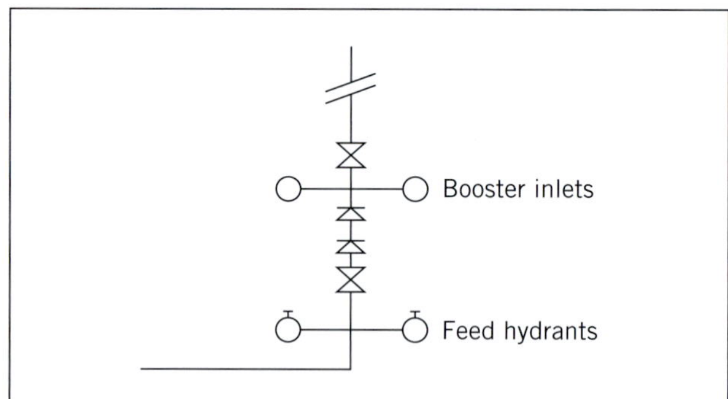

FIGURE 6.14 Typical 'I' pattern booster assembly

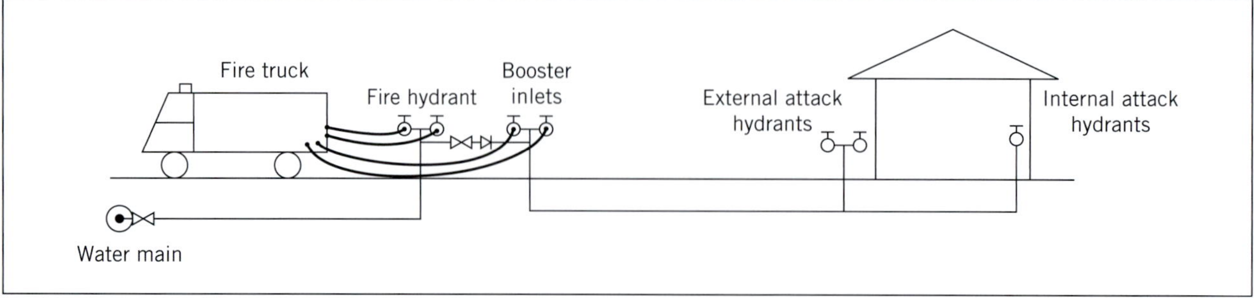

FIGURE 6.11 Diagram of a booster assembly connected to a fire brigade appliance

A **block plan** is a permanent plan located at the **booster assembly** that is weather resistant and cannot fade. It is essentially a site plan with a schematic layout of the fire service showing the location of all the fire hydrants, isolation valves, booster connections, pumps, storage tanks, available pressure and flow rate, along with the highest, most disadvantaged outlet and the location of the main electrical switch room and substation, if applicable (see **Figure 6.15** for an example).

The block plan enables firefighters on arrival at the scene to quickly assess the site layout and the location of all the firefighting, water supply and electrical equipment.

Connect fire hydrant valves, hose reels and ancillary equipment

When installing fire hydrant valves, the following requirements must be followed as per AS2419.1 Fire Hydrant Installations:

- The minimum size pipe supplying a single fire hydrant is DN 80 and a dual hydrant is DN 100.
- The minimum size hydrant valve inlet is DN 65.
- An approved fire brigade quick-connect adaptor must be fitted to the hydrant valve outlet.
- The fire hydrant is installed at a height between 750 mm and 1200 mm.

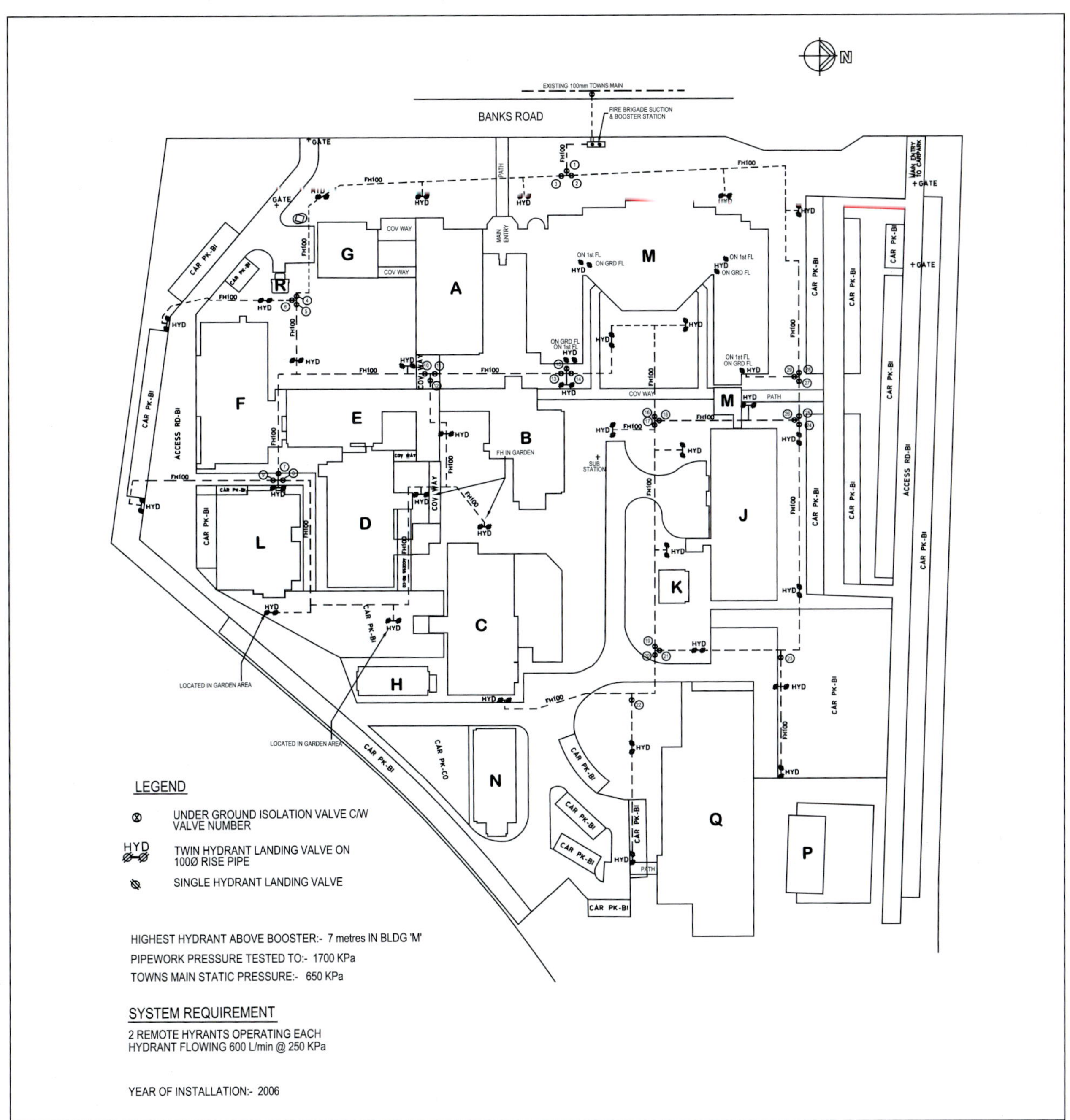

FIGURE 6.15 Block plan for Miller TAFE college

- The angle of the fire hydrant valve is 35° from horizontal so the hose does not kink as it drops from the fire hydrant.
- There must be 100 mm of clearance around the valve handle for easy access, and 300 mm clearance around the valve hose connection (see **Figure 6.16**).

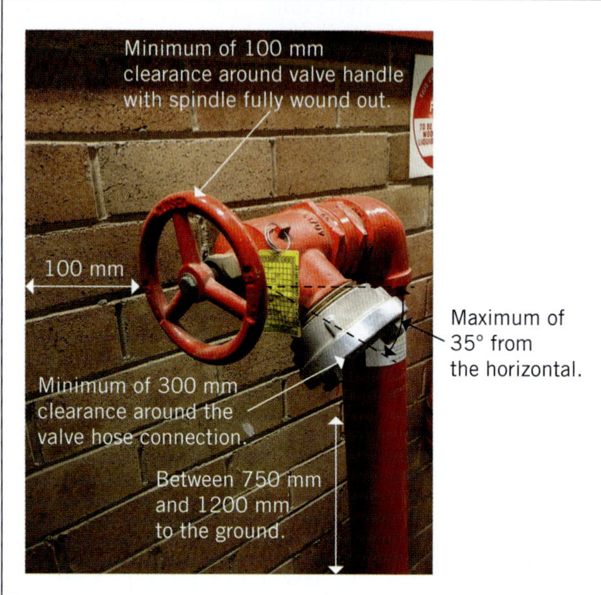

FIGURE 6.16 Fire hydrant clearances

Install fire hose reels

A fire hose reel is used for an initial attack on a Class A fire such as paper, wood, textiles, most plastics and rubber. Unlike a fire hydrant and booster assembly, a hose reel is meant to be used by any member of the public responding quickly to fight the fire at an early stage.

Hoses are made of a non-kinking rubber that is 36 m in length and 19 mm in diameter. They are permanently connected to a water supply and have an isolation valve, a hose guide, a hose and an adjustable shut-off nozzle.

The NCC nominates when and where fire hose reels are required. This depends on the class of building and its uses, the number of floors and the total floor area.

Fire hose reels (see **Figure 6.17**) must be mounted at a height between 1400 mm and 2400 mm from the finished floor level to the centre of the hose reel. The recommended height is 1500 mm.

The stop valve must be mounted between 900 mm and 1100 mm from the finished floor level.

FIGURE 6.17 A typical fire hose reel

The stop valve cannot be mounted more than 2000 mm away from the hose reel and must have a clearance of at least 100 mm around the valve handle so it is easy to access for operation (see **Figure 6.18**).

The minimum size pipe supplying a fire hose reel is DN 25, with a minimum pressure of 220 kPa and a minimum flow rate of 0.33 L per second.

The hose reel assembly must comply with AS 1221 for design, construction and performance. For installation, the hose reel assembly must comply with AS 2441.

Fit all pipework and supports

Plans and specifications will determine what piping material is to be used. This also depends on whether the pipe is installed above or below ground and the size of the fire service.

There is a range of methods used to join pipes, fittings and valves. This depends on the type of piping material used, the pipe diameter and the location.

Copper tube

The installation of copper tube can be above or below ground. It must be either type A or B grade for the appropriate wall thickness. A common method of jointing copper tube and fittings is by press fit or silver soldering. Manipulated joints such as an expanded joint or branch formed joints are silver soldered (see **Figure 6.19**). Copper and brass fittings can be either press fit or silver soldered. The silver content in the rod

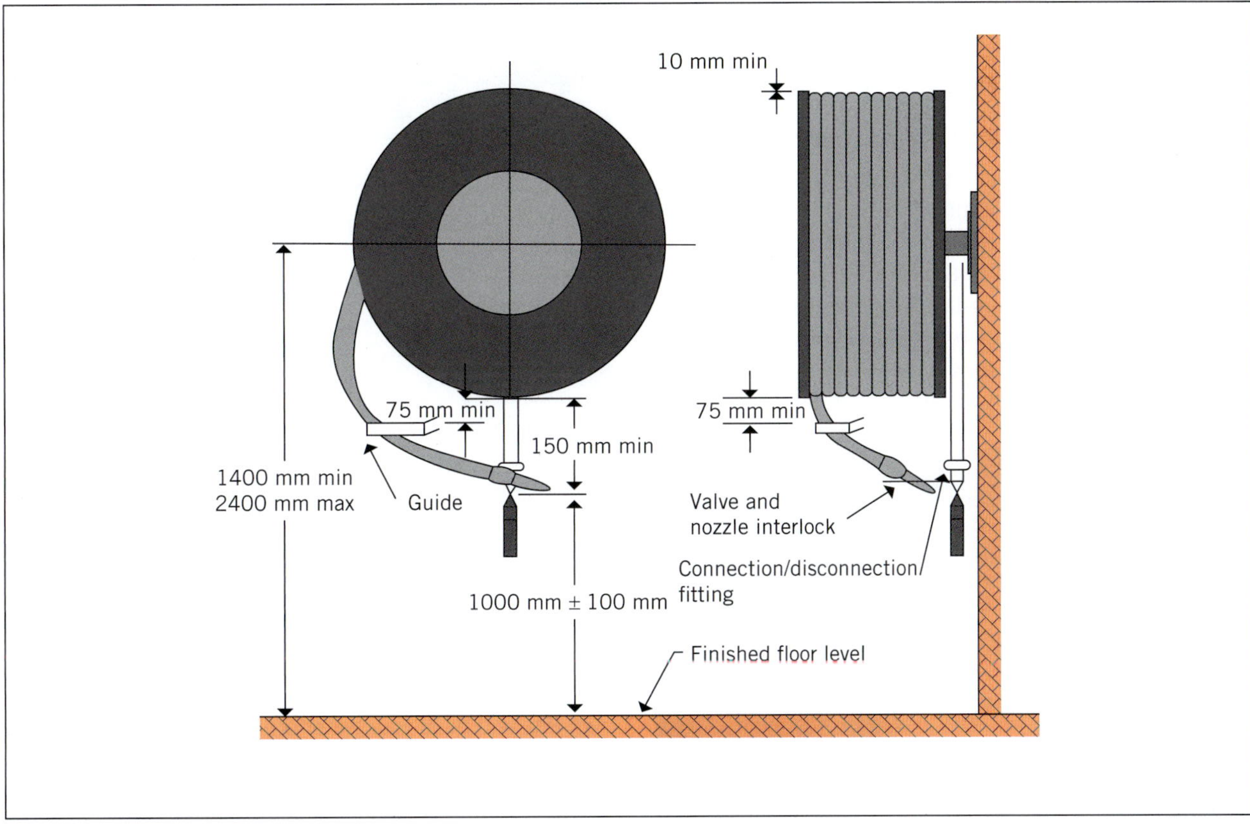

FIGURE 6.18 Fire hose reel measurements

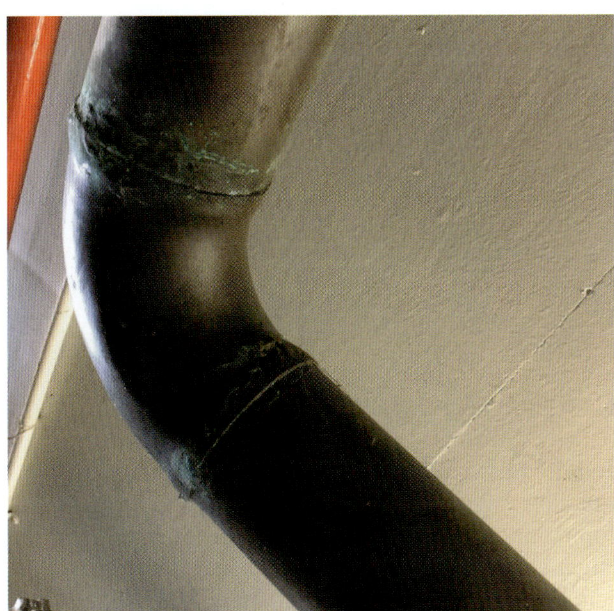

FIGURE 6.19 Silver-soldered joint

must be a minimum of 1.8% as per AS/NZS 3500.1. Valve assemblies larger than 50 mm are usually joined with flanges.

Soft solder cannot be used on fire services.

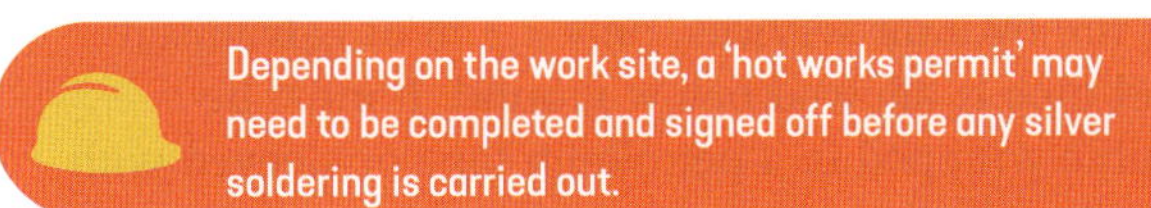

Depending on the work site, a 'hot works permit' may need to be completed and signed off before any silver soldering is carried out.

The most common method used to join copper tube today is press-fit or crimp fittings (see **Figure 6.20**). A special tool crimps the fitting onto the pipe. Although the fittings are quite expensive, much time is saved on the installation and no heat needs to be applied to the pipe or fittings, so certain hazards can be eliminated.

Source: Kembla.

FIGURE 6.20 Press-fit joint

FROM EXPERIENCE

Where copper and steel may come into contact, insulation must be installed to prevent electrolysis (corrosive reaction) from occurring.

PVC-U pipe

Thick-walled PVC-U pipe is commonly used below ground for fire hydrant services for pipe diameters of 100 mm or larger. It is also commonly used for water mains (see **Figure 6.21**). The material is cost-effective in purchase price and installation time. An elastomeric rubber ring jointing method is used for this material (see Chapter 1 for the jointing method). This is a similar method to that used in joining cement-lined cast iron and cement-lined ductile iron pipes. As mentioned earlier, concrete thrust blocks must be installed to prevent any movement from the rubber ring joints and must comply with AS/NZS 3500.1.

FIGURE 6.21 Elastomeric ring joint

Polyethylene pipe

Polyethylene pipe is used underground for fire hydrant and fire hose reel services. This material is joined usually using the electro-fusion jointing method (see **Figure 6.22**). It can also be joined using the heated tool socket method or the butt-welding method. This is explained in more detail in Chapter 1. AS 2419 explains that any polymer (plastic) pipe must have detectable (metal wire) laid in the trench so it can be traced and located.

FROM EXPERIENCE

A common problem with electro-fusion jointing is the joint leaks because it is not prepared properly by shaving and cleaning the pipe with an alcohol wipe prior to electro-fusing the joint.

serato/Shutterstock.com

FIGURE 6.22 Electro-fusion jointing method

HOW TO

SET OUT AND INSTALL BELOW-GROUND PIPEWORK

1. Check your Before You Dig Australia (BYDA) search is complete.
2. Make sure all road/footpath opening permits are paid.
3. Pay and book your water drilling permit.
4. Have necessary barricades and signage in place.
5. Locate the water main in the street in line with the building connection point.
6. Mark out a straight line using a string line and marking paint.
7. Excavate the trench to the required depth and width.
8. Install trench support if required.
9. Install pipework and bedding as per plans and specifications.
10. Pour concrete thrust blocks as required.
11. Test the installation as per AS 2419.1 prior to backfilling.
12. Backfill as soon as possible.

Steel pipe – above ground

Usually, galvanised steel pipe is used for fire hydrant and fire hose reel systems downstream of the main isolating valve. The galvanised steel pipework is commonly joined using the roll grooved method for larger services and screwed for smaller diameter services. Medium black steel grade pipe is used for fire sprinkler services. Black steel pipe is usually prefabricated off-site with welded joints and screwed together onsite. Flanged or roll grooved jointing is used on larger services. The roll grooved jointing technique (see **Figure 6.23**) is set out in Chapter 1.

Steel pipe – below ground

For fire hydrant services, only 1.5 m of galvanised steel pipe is allowed below ground. It must be double-wrapped in a petroleum tape or sleeved with

FIGURE 6.23 Roll grooved joint

a polyethylene coating. It can only be used downstream of the main isolating valve. Below-ground steel pipe is either flanged (see Figure 6.24) or rubber ring jointed.

Screwed joints are made on smaller-diameter steel pipe usually up to 50 mm in diameter.

FIGURE 6.24 Flanged joint

HOW TO

HOW TO MAKE A SCREWED JOINT

1. Thread the end of the pipe with a threading machine or a manual stock and die.
2. Wrap the thread with hemp and soap. Alternatively, thread sealant or teflon tape can be used.
3. Screw the pipe firmly into the fitting, being careful not to cross the thread.
4. Tighten with a Stillson's wrench.

Pipe supports

Fire system pipes, valves and ancillary equipment must be adequately supported. The pipe supports and fasteners should be capable of supporting twice the weight of the pipework filled with water, with an additional load of 115 kg at each point of support. There is a wide range of brands available to suit attaching pipework to structural steel, concrete, masonry (brickwork) and timber.

Other important considerations are as follows:

- Is the structure strong enough to support a pipe full of water?
- Does the bracket and fixing type meet the requirements of AS 2419?
- Is the required bracket spacing adequate as per AS 2419?
- Is extra protection such as galvanising, plastic sheathing or bituminous paint needed because the fire system is exposed to a corrosive/aggressive environment?

Fixings and fasteners

Generally, the fixings and fasteners used on fire services are made from steel (see Figure 6.25). They are bolted or screwed to the supporting structure and have a lifespan of the building. Wooden or plastic fixing inserts are not suitable for fire services.

Nails or explosive powered fasteners are not approved for fixing brackets to fire services. Preferred fixings are:

- expansion bolts
- screw fitting inserts either drilled or cast into concrete
- bolted fixings
- bolted clamps or clips.

Thrust blocks and anchors

Below-ground water/fire services using an elastomeric joining ring (rubber ring joint), as used on PVC-U pipe, must have concrete thrust blocks installed to prevent any movement (see Figure 6.26). Thrust blocks must bear against a firm side of the trench excavation. They are installed at:

- bends and tees
- end caps
- valves
- reducers
- grades in excess of 1:5.

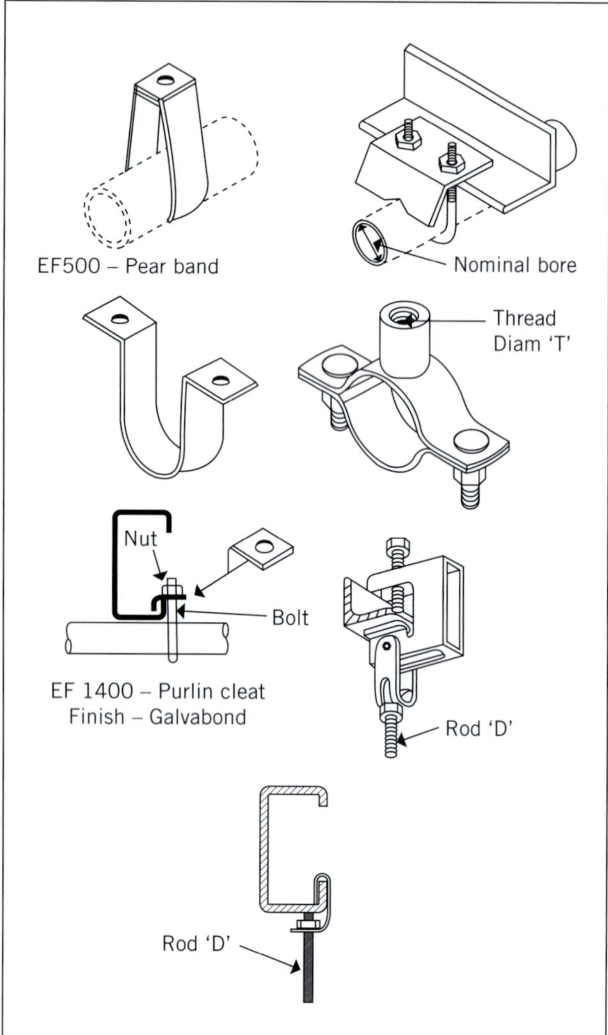

FIGURE 6.25 Examples of brackets and clamps

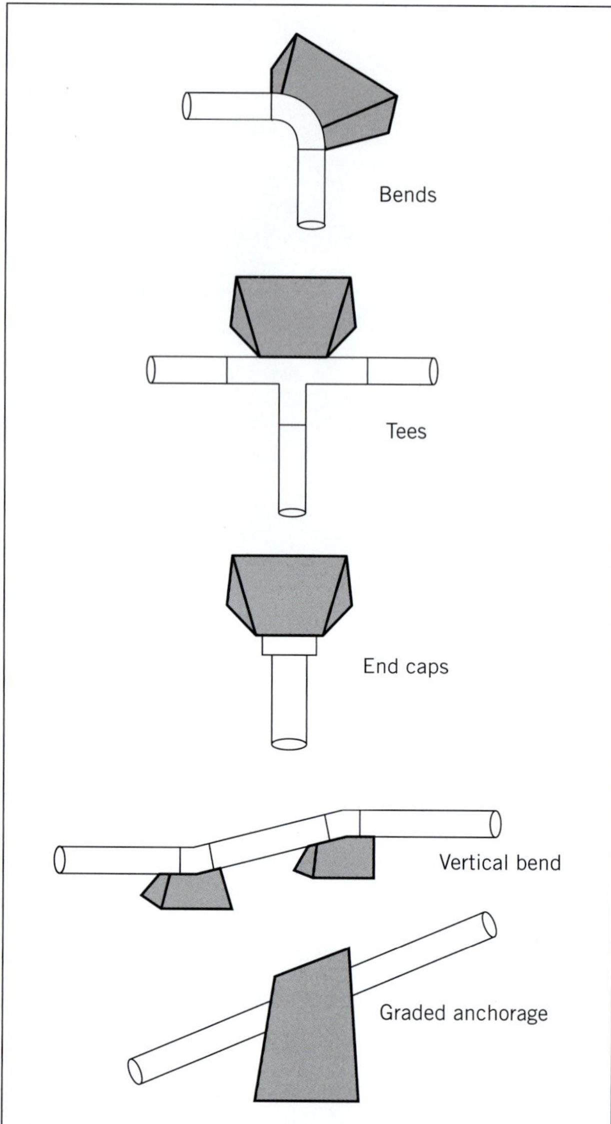

FIGURE 6.26 Typical thrust block positions

Water supply

Water supply for fire services can come from several sources:

- town main water
- recycled water – Class A
- rainwater (see **Figure 6.27**)
- sea, river, lake or dam water
- bore water (treated and untreated).

The water supply, regardless of the source, must be capable of supplying the fire hydrant system with the minimum flow rate and minimum pressure for a minimum of four hours.

A backflow prevention device must be fitted to any fire hydrant service connected to a town water supply to protect the community's drinking water (see **Figure 6.28**). Some water authorities will require the installation of a double-check detector assembly to monitor low water usages; for example, a fire hose reel being used to wash plant or equipment.

Source: Shaun Kristen.

FIGURE 6.27 Large-capacity storage tank used for a fire service

FIGURE 6.28 Double-check detector assembly

GREEN TIP

Plumbers have an important responsibility in protecting our community's drinking water by installing backflow preventers.

Fire hydrant services connect to a water main by a tee and valve. They are a dedicated service for firefighting, and therefore not metered (see Figure 6.29).

FIGURE 6.29 Fire hydrant service connected to water main with a tee and valve

The minimum flow rate for a feed fire hydrant is 10 L per second with a residual pressure of 150 kPa in NSW and 200 kPa in all other states and territories, as per AS 2419.1.

The minimum flow rate for an attack hydrant is 10 L per second with a residual pressure of 250 kPa in NSW and 350 kPa in all other states and territories, as per AS 2419.1.

The minimum required pressure may need to be achieved using a pump if necessary.

Water that is stored for firefighting purposes in tanks must have a specific volume solely dedicated to firefighting. The water storage tanks must conform to AS/NZS 3500.1.

Where water is to be drawn from a source below a pump, such as a dam or a river, the maximum vertical lift shall be 3 m.

The water must not be corrosive to the system. If sea water is used, then the fire service must be flushed with fresh water on completion of use.

Fire hose reel services usually connect to the metered service (see Figure 6.30). The system connects to the domestic supply and should have isolating valves independent of the hose reel system so the supply is not interrupted.

FIGURE 6.30 Fire hose reel service connected to water main with a main tapping

The minimum pressure required for a fire hose reel is between 200 kPa and 230 kPa with a flow rate of 0.33 L per second for a 19 mm hose.

Testing and commissioning

Testing and commissioning the system is the final process of the job prior to clean-up and the final handover. It is a vital process ensuring the system is sound and working to the correct pressures and flow rates.

Testing

Once installed, the fire hydrant and fire hose reel service must be pressure tested. AS 2419.1 states the fire service must be hydraulically (water) tested to a pressure of 1700 kPa or 1.5 times the design pressure, whichever is the greater, for a minimum of two hours.

It is good practice to carry out a preliminary air test to check for any major leaks such as open ends or open valves. Doing this could eliminate possible flooding. This test is not compulsory but highly recommended.

HOW TO

CARRY OUT AN AIR TEST

1 Do a visual check to ensure the installation is complete.
2 If testing sections, ensure the section is isolated.
3 Connect the air compressor to the test point (remove the hose cock on the booster assembly and connect the pressure gauge and isolating valve for the connection point).
4 Raise the air pressure to 300 kPa.
5 Hold for 30 minutes. Air pressure should not drop more than 20–30 kPa over this time.
6 If this test passes, proceed to the hydraulic test.
7 If the air test drops, locate and repair the leak. Then repeat the air test.
8 Record the test data on company records (quality assurance procedure).

HOW TO

CARRY OUT A HYDRAULIC TEST

1 Do a visual check to ensure the installation is complete.
2 If testing sections, ensure the section is isolated.
3 Connect the pump to the test point or use the permanently installed pump with a gauge and isolating valve.
4 Open the control valve and fill the system with water.
5 Bleed air from the system from the furthest outlet.
6 Start the pump and increase pressure at 200 kPa intervals. Hold for 30 minutes at each interval for the system to adjust to the pressure and to allow screwed hemp joints to swell.
7 When pressure is at 1700 kPa, or as specified, hold for two hours.
8 If there are any leaks, rectify the problem and re-test.
9 The owner or representative is to witness the start and finish time of the test and sign off on its successful completion.
10 Record test data on company records (quality assurance procedure).

Commissioning

The fire hydrant and fire hose reel system must be complete, tested and inspected by the hydraulic designer, who will ensure the system meets all the design requirements and issue an inspection report. A copy of this report would be kept by the client and the contractor.

When commissioning the fire hydrant service, all hydrants must be opened to release any trapped air, and flow rates and pressure must be checked at each hydrant. If pumps are installed, they need to be run to ensure they are working properly.

When commissioning the fire hose reel, check the following points:

- Be sure there is no leakage from the hose reel when charged and pressurised with the nozzle closed.
- Check that the hose unwinds easily from the reel in its intended direction.
- Check the required flow rate and pressure.
- Make sure the hose is rewound evenly onto the reel with the hose under pressure.
- Engage the nozzle to the interlock and close the isolating valve.
- Open the nozzle to depressurise the hose and then close the nozzle.

Complete documentation according to relevant regulatory requirements

Company policies must be followed regarding any final documentation. This would include the final design approval certificate, certificate of compliance and any as-built drawings. Also, any written instructions for the operation and maintenance of equipment and fixtures installed must be handed over to the client, along with any warranties.

LEARNING TASK 6.3

1 What is the maximum length of galvanised steel pipe permitted in the ground?
2 What type of pipework must have thrust blocks installed?
3 What is the most common method of joining large-diameter galvanised steel pipe?
4 How is the town's water supply protected from contaminated water?

Clean up

An essential part of the job is to clean up and pack the gear away. Clean-up should be carried out at the end of each day as clutter creates chaos! Having a clean work area at the start of the day creates a positive environment to work in.

Clear work area, remove waste and reuse material

A clean workplace reduces the chance of accidents occurring, such as trip hazards and falling objects. Another reason is that the build-up of waste and debris on a job site creates a problem for storage and access. A clean and tidy work area helps the job run more smoothly, which helps to increase profit.

A thorough sweeping of the work area helps remove dirt, dust and rubbish. At the end of the job, there is usually a large amount of cardboard and paper packaging. This should be placed in the recycling waste so it can be reused and reduce landfill.

Offcuts and leftover material should be appropriately stored and used on another job. This not only prevents waste but also enables savings in cost and increases in profit.

Clean and store tools and equipment

All tools and equipment must be packed up and stored safely every day. This will help to reduce the loss or misplacement of tools. All tools and equipment must be well maintained by regularly cleaning and lubricating them. Any damaged tools must be reported and tagged out. Blades and cutting tools must be kept sharp or replaced if necessary to be effective in use. Any plant and equipment on hire needs to be either taken off hire or returned to the hire company as soon as possible to reduce company costs.

COMPLETE WORKSHEET 3

SUMMARY

- Effective planning and preparing for fire hydrant and fire hose reel installations is important for a cost- and time-effective job.
- There are several types of fire hydrant and fire hose reel systems to be aware of.
- Several types of piping material are permitted to be used, depending whether the fire service is below or above ground and what is stated in the specifications.
- Be sure to place accurate orders for materials to avoid inconvenience and time wasting, and check the order is correct upon delivery.
- When setting out the job, it is important to check plans for dimensions and levels.
- For underground pipework, check excavation depth and plan trench support, if required.
- Check fire hydrant valves and fire hose reels meet the installation requirements.
- Install pipework in the most direct route possible with minimal bends.
- There is a range of methods used to join pipes, fittings and valves. This depends on the type of piping material used, the pipe diameter and the location.
- Be sure to use adequate fixings and brackets to support pipework. Wooden or plastic inserts are not allowed. Nails and explosive fixings cannot be used.
- There are several water sources available for firefighting.
- The minimum testing requirements must be adhered to as per AS 2419.
- Clean up regularly and try to reduce waste.

REFERENCES

AS 1221 Fire Hose Reels (Design, Construction and Performance)

AS 2118 Automatic Fire Sprinkler Systems

AS 2419.1 Fire Hydrant Installations

AS 2441 Installation of Fire Hose Reels

AS 4118 Fire Sprinkler Components

AS/NZS 3500.0 National Plumbing and Drainage, Part 0: Glossary of terms

AS/NZS 3500.1 Plumbing and Drainage, Part 1: Water Services

NCC: Part 3 – Plumbing Code of Australia (PCA): **https://ncc.abcb.gov.au/**

GET IT RIGHT

1 When installing a fire hydrant service, what type of connection to the water main is used?

2 Where is the valve located and why?

WORKSHEET 1

To be completed by teachers	
Student competent	☐
Student not yet competent	☐

Student name: ______________________

Enrolment year: ______________________

Class code: ______________________

Unit competency code/title: CPCPFS3031 Fabricate and install fire hydrant and hose reel systems

Task: Review 'Identify installation requirements' and answer the following questions.

1 What are two important considerations for the correct pipe sizing of a fire hydrant and fire hose reel service?

____

____

2 What are the three main points in a risk assessment?

____

____

____

3 What does a 'wet pipe' system mean?

____

____

4 How long is a combined fire hydrant and hose reel system designed to supply the minimum flow rate and pressure?

____

5 Describe a feed hydrant.

____

____

6 Describe an attack hydrant.

____

____

7 What is the advantage of a ring main?

____

____

____

8 What is the purpose of a booster assembly?

9 What is a hardstand area?

10 Name three types of piping material used below ground on fire hydrant services.

WORKSHEET 2

To be completed by teachers	
Student competent	☐
Student not yet competent	☐

Student name: ______________________

Enrolment year: ______________________

Class code: ______________________

Unit competency code/title: CPCPFS3031 Fabricate and install fire hydrant and hose reel systems

Task: Review 'Prepare for work' and answer the following questions.

1 How are the pipe material, fittings and fixing systems determined to use on a fire hydrant and hose reel system?

____

____

____

2 Name four requirements included in a company quality assurance policy.

____

____

____

____

3 Who must be contacted before excavating in a public place?

____

4 How often do electrical tools and leads need to be tested and tagged on construction sites?

____

____

5 Name four appropriate PPE items for installing fire hydrant and fire hose reel systems.

____

____

____

____

6 Why is it important to communicate with the site supervisor when trenches are being dug and deliveries being made?

____

____

____

WORKSHEET 3

To be completed by teachers	
Student competent	☐
Student not yet competent	☐

Student name: ______________________

Enrolment year: ______________________

Class code: ______________________

Unit competency code/title: CPCPFS3031 Fabricate and install fire hydrant and hose reel systems

Task: Review 'Fabricate, install and test system' and 'Clean up' and answer the following questions.

1 What is a booster inlet valve?

____

2 Why is the isolating valve locked in the open position?

____

____

3 Name four items of information that can be found on a block plan.

____

____

____

____

4 What is the minimum size pipe supplying a single hydrant?

____

5 At what height should a fire hydrant be installed?

____

6 What is the angle at which an attack hydrant outlet is installed?

____

7 Why is the hydrant angle important?

____

____

8 What is the minimum size pipe supplying a fire hose reel?

____

9 What is minimum required flow rate and pressure for a fire hose reel?

____

10 At what height should a fire hose reel be installed?

11 At what height should a fire hose reel isolation valve be installed?

12 Which type of piping systems require thrust blocks?

13 Name two jointing methods for copper tube.

14 What is the most common jointing method for polyethylene pipe?

15 What is the most common jointing method for large diameter galvanised pipe?

16 Is the following statement true or false? 'A maximum of 2 metres of galvanised steel pipe can be below ground for a fire hydrant service.'

17 Name three things to consider for pipe supports.

18 Can plastic or wooden inserts be used for fixings for fire hydrant services?

19 What is the purpose of thrust blocks?

20 Name three water sources for fire hydrant and fire hose reel services.

WORKSHEETS 6

21 Is the following statement true or false? 'The water supply, regardless of the source, must be capable of supplying the fire hydrant system with the minimum flow rate and minimum pressure for a minimum of three hours.'

____

22 What is the maximum vertical lift that a pump can draw water from a water source below the pump as per AS 2419.1?

____

23 What type of backflow prevention device is used for containment on a fire hydrant service?

____

24 What does AS 2419.1 state as the minimum test pressure and duration requirements for a fire hydrant service?

____

____

25 Why is it good practice to carry out an air test before pressurising the system with water?

____

____

26 Why is it important to open all fire hydrants when commissioning the system?

____

27 Name four procedures that are part of the commissioning process for fire hose reels.

____

____

____

____

28 Why is it important to reduce waste?

____

____

INSTALL WATER PUMPSETS AND INSTALL AND COMMISSION DOMESTIC IRRIGATION PUMPS

7

Chapter overview

This chapter addresses the general and basic installation of pumps as part of the supply of water to a property or building and to supply irrigation for domestic systems.

The information and principles mentioned in this chapter directly relate to pumpsets for water supply and domestic irrigation systems, but will also apply to other areas of plumbing, such as wastewater removal. Only a basic knowledge of some pump categories and types is provided here, as well as the common technical issues related to them.

A **water pump** can be simply defined as a mechanical device that is generally driven by a motor and is used for raising water from a lower level to a higher level. In water supply, the pump is also used to raise pressure in a pipework system or circulate water in a pipework system. Pumps are classed according to their principle of action, manner of construction or method of operation. Nearly every pump type is defined by its application, such as circulating or submersible pumps, or by its operation, such as centrifugal or piston pumps. This chapter deals with the two major groups of pumps: **rotodynamic pumps** and **positive displacement pumps**. Their names describe the method of moving a fluid (in this chapter, it is water).

Learning objectives

Areas addressed in this chapter include:

- identify installation requirements
- prepare for work
- Install, test and commission water pumps
- clean up.

Identify installation requirements

A pump can be used for many reasons on a water supply system. The three main purposes of a pump are to:

- increase pressure in the pipeline that is attached to its outlet
- move water from a lower elevation to a higher elevation
- help circulate water in a continuous loop around a water supply system.

For large industrial, commercial or residential systems, the design engineer normally plans and specifies the size and type of pump system, whereas on smaller systems this may be left for the plumber to do. The plumber's task is to order and install a pump that is appropriate for the job required, and to connect the pipework and ancillary equipment to and from the **pump assembly**.

While plumbers are generally involved in the installation of pumps, pump maintenance and repairs are normally left to those who specialise in these areas. However, as the tradesperson most concerned with water supply, plumbers should have a good basic understanding of pumps – their function, how they operate, and when and why they are used in water supply.

The two major groups of pumps are *rotodynamic pumps* and *positive displacement pumps*. Their names describe the method of moving a fluid (the fluid in this chapter is water). Read below to understand how they work.

- Rotodynamic pumps are based on bladed impellers that rotate within the fluid to accelerate the fluid and cause a consequent increase in the energy of the fluid.
- Positive displacement pumps cause the fluid to move by trapping a fixed volume of fluid, then forcing (displacing) that trapped volume into the discharge pipe. These are further discussed in the section 'Types of pumps'.

Access, read and determine installation requirements from codes, standards and specifications

When installing water pumps, the plumber must:

- know how to read plans and specifications to determine the type and location of the pump assembly
- liaise with the client to ensure that the pump installation will meet their needs
- liaise with the pump manufacturer and/or the supplier to ensure the correct pump unit is selected to maintain the range of flow rates, operating head and delivery distance required
- liaise with the local water utility and complete the necessary documentation during the installation process
- check the pump for compliance with standards, docket and order form, and acceptable condition
- safely install the pump assembly, associated pipework and ancillary items in accordance with the relevant codes and standards, the manufacturer's instructions and the job specification
- liaise with other trades to ensure not only that the pump installation is completed satisfactorily, but also that the pump and ancillary equipment are maintained in the future.

It should be noted here that in most large pump installations there is normally more than one pump installed. This is to allow for such things as:

- helping to prevent an individual pump from overworking (that is, to share the load on the pump assembly)
- allowing the system to continue operating while maintenance takes place on part of the pump assembly.

In some areas there may be alternative power sources for the pump assembly. For example, fire services have both diesel and electric power, in case there is a fire and the electricity supply is interrupted or shut down. The pump is still able to provide water to help control the fire by switching to diesel fuel.

In cities and towns, the local water supply utility normally has an obligation to provide water at a minimum of 150 kPa (approximately 15 m of head) at any part of its water supply. Obviously, the pressure within the main varies depending on different sources of water supply and different types of land features within towns and cities. The pressure supplied is normally much higher than the guaranteed minimum.

GREEN TIP

Water utilities normally maintain a conservative approach to their water supply. This is mainly to allow for future growth in particular areas, even if the current available water pressure might appear to be more than adequate.

For this reason, when supplying water to properties that require large volumes of water (such as fire services, some industrial areas or where there are multistorey buildings), the water utility may insist that the property owner takes responsibility by using either pumps or a combination of both pumps and a storage tank, particularly if the main supply has insufficient volume or pressure to feed all parts of the property during peak periods.

AS/NZS 3500.1 states the most disadvantaged point of a drinking water supply system should have a *minimum pressure of 50 kPa* (approximately 5 m of head) and that fixtures connected to the supply require minimum flow rates. So, if there is not enough pressure, a pump will be required. It is the plumber's job to do the installation according to current rules and regulations. This can also be achieved by sourcing the manufacturer's specifications. For further information on how to access manufacturer's instructions and other relevant information, refer to Chapter 4 'Carry out interactive workplace communication' in *Basic Plumbing Skills*.

AS/NZS 3500.1 PLUMBING AND DRAINAGE: WATER SERVICES

Work health and safety (WHS)

Be sure to identify all the hazards of the work area and the tasks involved before starting the job. Then write a job safety analysis (JSA) stating the control measures that will reduce the risks.

Before commencing work, it is important the area is prepared. This includes checking the area is clear and free of anything that may cause a hazard, such as materials left in a walkway causing a trip hazard. Some risks/hazards associated with a pump installation follow.

A safe work method statement (SWMS) must also be completed for any high-risk activity; for example, working at heights.

Manual handling/lifting

Pump units are normally quite heavy and difficult to manoeuvre. Care should be taken when locating the pump in the correct position. If necessary, use lifting equipment to locate the pump, associated valves and equipment.

Working at heights

Although pumps are not usually located in high positions, the associated pipework may be. The plumber must therefore ensure that suitable safe equipment is used while working at heights. If items such as mobile platforms and scissor lifts are used, ensure that the person operating this equipment has the appropriate current qualification and has received adequate training in its use. Care should be taken to protect the safety of co-workers and the public by using barricades and signage.

Fumes

If there is a potential for fumes to be present during the installation process, ensure that suitable respiratory equipment is worn and adequate ventilation is provided and used. Fumes could be generated during welding processes or by other equipment operating in the surrounding work area. Attention should also be paid to correct ventilation of the area if the pump is operated on diesel fuel.

Moving parts

Pumps have moving parts and may therefore need to have suitable guards installed. Mostly, these protective guards are included as part of the pump assembly; however, they may have to be made to suit particular site conditions. Pump guards should be designed and installed to prevent injury to all persons, including those who are working near the pump. It is also a good idea to install warning signs near the pump assembly.

Electrical work

All electrical work should be carried out by a licensed electrician. Before working on the pipework on the pump assembly, the plumber should test the pipework with a neon tester to check for stray current. It is also advisable to bridge any portion of the pipework (or ancillary equipment) that is to be removed with an approved 'bridging conductor' (bonding strap) before starting any maintenance work. If you have any doubts at all, have the system checked by a licensed electrician.

Be sure electricity and any stray current is isolated first! Do not work on pump assemblies or pipework if stray current is present.

Personal protective equipment in the plumbing industry

Personal protective equipment (PPE) considerations for plumbers include boots, gloves and wraparound eye protection as a minimum, along with ear protection when using power tools. Steelcapped boots should be worn, unless working at heights, when Dunlop Volleys will provide better grip. In any work situation, plumbers can reduce their chances of being injured by using PPE. PPE that plumbers may use when installing pump systems includes:

- dust masks
- earmuffs
- fitted gloves
- hard hats
- long-sleeve overalls
- safety boots
- safety glasses or goggles.

FROM EXPERIENCE

Having the knowledge to choose the correct pump for a particular job is an important plumbing skill.

Situations in which pumps are required

Here are some contexts where water pumps are generally required as part of the water supply.

Cities and towns

- In industrial areas, for:
  - circulating water through machinery for cooling purposes
  - circulating hot and cold water

- food processing
  - air-conditioning and other mechanical services
  - irrigation.
- In commercial buildings, for:
  - raising water to storage tanks
  - circulating hot and cold water
  - air-conditioning
  - other mechanical services.
- In residential buildings, for:
  - raising water to storage tanks
  - circulating hot and cold water
  - air-conditioning
  - irrigation
  - pressurising water from rainwater tanks.
- In schools, churches and various other community establishments, for:
  - air-conditioning
  - circulating hot and cold water
  - irrigation.
- In hospitals, for:
  - raising water to storage tanks
  - circulating hot and cold water
  - air-conditioning
  - other mechanical services.
- For fire services:
  - hydrants
  - hose reels
  - sprinklers for any or all of the above building types.
- In other areas, for:
  - irrigation (both domestic and urban)
  - moving water into and out of household appliances (washing machines and dishwashers)
  - circulating water in domestic spas
  - water services – plumbers use a test pump to raise the pressure within a water service when testing the service for soundness.

Rural areas

- Pumping to dwellings and other buildings from:
  - dams
  - storage tanks/rainwater tanks
  - wells
  - streams.
- For irrigation, by pumping from:
  - rivers
  - creeks
  - dams
  - bores
  - other sources.
- For supplying water for stock:
  - by pumping from rivers, creeks, dams, bores and other sources
  - via windmills and other devices.

LEARNING TASK 7.1

1. Name a pump that is of the rotodynamic type.
2. Name a pump that is of the positive displacement type.
3. What are the three main purposes of a pump?

COMPLETE WORKSHEET 1

Pump requirements and basic pumping terminology

To choose the proper pump for a particular system, the plumber must know what size pump is required and/or how many pumps are required. The plumber also must be aware of how much water needs to be moved, how far and how fast it needs to be moved, and how much resistance may be encountered while it is being moved.

Pump manufacturers sum up these factors into two basic measurements and rate their pumps accordingly. The measurements are *flow rate* and *delivery head*, which gives the total *dynamic head*.

Head (pressure)

The International System of Units (SI) units of measurement that are used to express pressure (**head**) throughout this unit include:

- bar
- kilopascals (kPa)
- millimetres (mm head)
- megapascals (MPa)
- metres (m head)
- metres head of water (H)
- pascals (Pa).

Flow rate relates to the amount of water that can be moved in a set period of time and is generally measured in the number of litres per second (L/s), litres per minute (L/m) or cubic metres per hour (m^3/hr).

Delivery head is the vertical height in metres to where the pump delivers water from the pump outlet, which would be the highest outlet from the pump. It can be further classified depending on whether the resistance is encountered on the suction side of a pump (*suction head*) or the delivery or discharge side (*discharge head*); whether it is caused by the standing (or *stationary*) weight of the water (*static head*) or by the movement of water through the system (*dynamic head*); or whether the resistance is brought about by friction as the water passes through the pipes, fittings and valves installed on the system (*friction head*).

The plumber needs to know the delivery head against which the pump has to operate and also whether there is a positive suction head or a negative suction head available, as this is part of the information that has to be provided to the pump supplier when selecting the appropriate pump for a specific application. A positive suction head is where the available water source for the pump is above the pump inlet; see **Figure 7.1** (a). A negative suction head (or suction lift) is where the pump draws the water from a source lower than the pump inlet; see **Figure 7.1** (b). Note that the physics of a pump is that it does not 'suck' but creates a partial vacuum, allowing atmospheric pressure to push water up into the inlet pipe in a negative suction situation.

When discussing pumps and pumping, the term 'head' is used for the pressure generated by the pump or the pressure the pump needs to overcome. The way we calculate pressure, therefore, is to measure the vertical height (head) of a column of water and for each metre multiply by 9.81 (a 1 m vertical column of water exerts a pressure of 9.81 kPa at its base).

The common formula that is used to calculate the pressure in kPa is:

$P = H \times 9.81$

where

P = pressure (kPa)

H = head (height) in metres

9.81 = conversion factor

One simple way to do this is by approximation (in other words, saying that 9.81 is approximately equal to 10). For example, if a building is 30 m high, then at the bottom of the building we can assume that the pressure is approximately 300 kPa; for 40 m, approximately 400 kPa; for 50 m, approximately 500 kPa and so on.

EXAMPLE 7.1

CALCULATE THE WATER PRESSURE AT THE BASE OF A 35 M-HIGH BUILDING

Formula: $P = H \times 9.81$

where: P = pressure (kPa) (to be determined)

H = head (height) in metres (35)

9.81 = conversion factor (9.81)

Workings: $P = 35 \times 9.81$

Answer: P = 343.35 kPa

Examples of technical data related to pump installations can be seen in **Figure 7.2**.

The plumber needs to be aware of suction lift limits when preparing for and installing a pump. In theory, atmospheric pressure (approximately 100 kPa or 1 bar) can support a water column of some 10 m. The suction lift of a pump is limited by the height to which the atmosphere will force water against the partial vacuum formed by the action of the pump. A diagrammatic sketch indicating various terms related to suction lift and suction head can be seen in **Figure 7.3**.

For most pumps it is preferable to have a positive suction head rather than a negative suction head or suction lift. Simply stated, it is better, where possible, to position the pump below the water source. However, in a situation where there is a negative suction head, it is advisable to ensure that the water supply level on the suction side of the pump is kept to a minimum

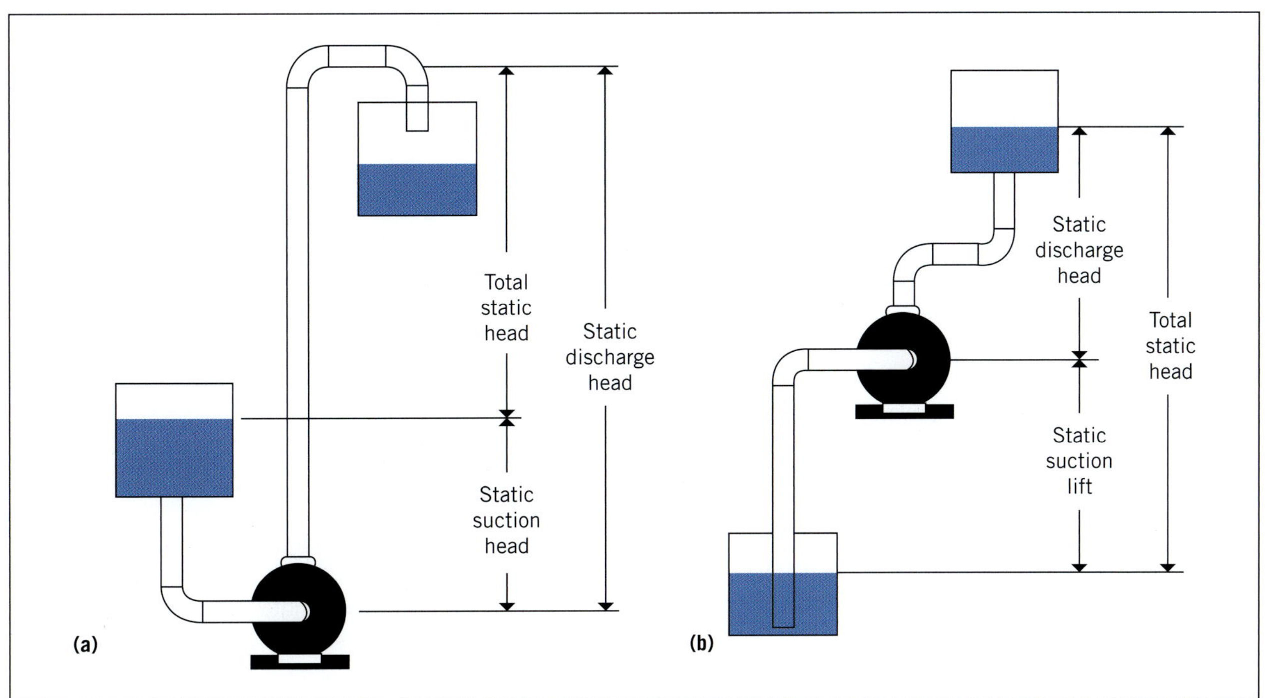

FIGURE 7.1 (a) Positive suction head; (b) Negative suction head/lift

Source: © Grundfos Pumps Pty Ltd.

Temperature conversions

Degrees Centigrade to degrees Fahrenheit = $\frac{°C \times 9}{5} + 32$

Degrees Fahrenheit to degrees Centigrade = $(°F - 32) \times \frac{5}{9}$

Kilowatts to horsepower

kW to HP = kW × 1.341
HP to kW = HP × 0.746

Circle formulas

Area = πr^2
Circumference = $2 \pi r$
Volume of cylinder = $\pi D^2H \times .25$

Volume of sphere = $0.166 \times \pi D^3$ where:
$\pi = 3.14$
r = radius
D = diameter
H = height

Velocity of flow past SP motor

$$V = \frac{Q \times 353}{Db^2 - Dm^2}$$

Where:
V = velocity metres/sec
Q = flow m³/hr
Db = borehole diameter (mm)
Dm = motor diameter (mm)

Torque

Ft lbs to Nm = Ft lbs × 1.355
Nm to Ft lbs = Nm × 0.73757

Weight

Pounds to kilograms = lbs ÷ 2.2
Kilograms to pounds = kg × 2.2
Gallons to pounds = gallons × 10
Litres to kilograms = litres × 1

Affinity laws

(how changes in pump speed affect pump flow, head and input power)

$Q = Q_1 \times N/N_1$
$H = H_1 \times (N/N_1)2$
$P = P_1 \times (N/N_1)3$

where:
N = new speed
N_1 = old speed
Q = flow @ N
Q_1 = flow @ N_1
H = head @ N
H_1 = head @ N_1
P = input power @ N
P_1 = input power @ N_1

UNITS OF PRESSURE

Convert from ↓ / Convert to →	Kilopascal	Metrehead	Bar	lbs per sq. inch	Feet of water	Atmosphere
	Multiply by					
Kilopascal kPa	1	0.102	.01	0.145	0.335	–
Metrehead m	9.804	1	0.098	1.42	3.28	0.098
Bar Bar	100	10.20	1	14.5	33.45	1
lbs/sq. inch psi	6.895	0.704	0.069	1	2.307	0.069
Feet of water ft	2.98	0.3048	0.03	0.4335	1	0.03
Atmosphere At	100	10.33	1.0	14.7	33.9	1

NOZZLE FLOW RATES (litres/sec)

Pressure		Diameter of nozzle (inches*) mm								
		5/64″	5/32″	1/4″	5/16″	25/64″	15/32″	5/8″	25/32″	1″
kPa	psi	2	4	6	8	10	12	16	20	25
200	29	0.06	0.25	0.56	1.01	1.57	2.26	4.02	6.29	9.82
300	43	0.08	0.31	0.69	1.23	1.93	2.77	4.93	7.70	12.03
400	58	0.09	0.36	0.80	1.42	2.22	3.20	5.69	8.89	13.89
500	73	0.10	0.40	0.89	1.59	2.48	3.58	6.36	9.94	15.53
600	87	0.11	0.43	0.98	1.74	2.72	3.92	6.97	10.89	17.02
700	102	0.12	0.47	1.06	1.88	2.94	4.23	7.53	11.76	18.38
800	116	0.12	0.50	1.13	2.01	3.14	4.53	8.05	12.58	19.65
900	131	0.13	0.53	1.20	2.13	3.33	4.80	8.54	13.34	20.84
1000	145	0.14	0.56	1.26	2.25	3.52	5.06	9.00	14.06	21.97

*Approximate imperial equivalent

UNITS OF FLOW RATE

Convert from ↓ / Convert to →	Imperial gallons per min	US gallons per min	Cubic metres per hour	Litres per second	Litres per minute
	Multiply				
Imperial gallons per min	1	1.2	.273	.076	4.546
US gallons per min	.833	1	.227	.063	3.787
Cubic metres per hour	3.666	4.4	1	.278	16.66
Litres per second	13.19	15.85	3.6	1	60
Litres per minute	.219	.264	.06	.016	1

UNITS OF VOLUME

Convert from ↓ / Convert to →	Litre	Kilolitre	Cubic metres	Imperial gallon	US gallon
	Multiply by				
Litre lt	1	0.001	0.001	.220	.264
Kilolitre klt	1000	1	1	220	264
Cubic metre m³	1000	1	1	220	264
Imperial gallon gal	4.546	0.00454	0.00454	1	1.201
US gallon (US) gal	3.785	0.0038	0.0038	0.833	1

UNITS OF LENGTH

Convert from ↓ / Convert to →	Millimetre	Centimetre	Metre	Kilometre	Inch	Foot	Mile
	Multiply by						
Millimetre mm	1	0.1	0.001	–	0.0394	0.0033	–
Centimetre cm	10	1	0.01	–	0.394	0.0328	–
Metre m	1000	100	1	0.001	39.37	3.281	.000621
Kilometre km	–	–	1000	1	–	3281	0.621
Inch in	25.4	2.54	0.0254	–	1	0.083	–
Feet ft	304.8	30.48	0.305	–	12	1	–
Mile mile	–	–	1610	1.61	–	5280	1

Note: The information provided on these pages is for guidance only. Grundfos Pumps Pty Ltd accepts no responsibility for the misuse or misapplication of this information.

FIGURE 7.2 Typical sheet showing relevant engineering data relating to pumps

(normally no more than 6 m below the centre line of the pump, and in extreme cases no more than about 8 m). This is because of friction losses caused by the pipe, fittings, valves and so on, and can vary depending on site conditions and the manufacturer's recommendations. (In all cases, it is recommended to refer to the pump manufacturer's installation instructions before connecting the pumpset so as not to void any warranty. If there is any doubt at all, contact the pump manufacturer.)

LEARNING TASK 7.2

1 What are the two main factors required when choosing a pump?
2 What force pushes water into the pump inlet?
3 How is flow rate measured?

COMPLETE WORKSHEET 2

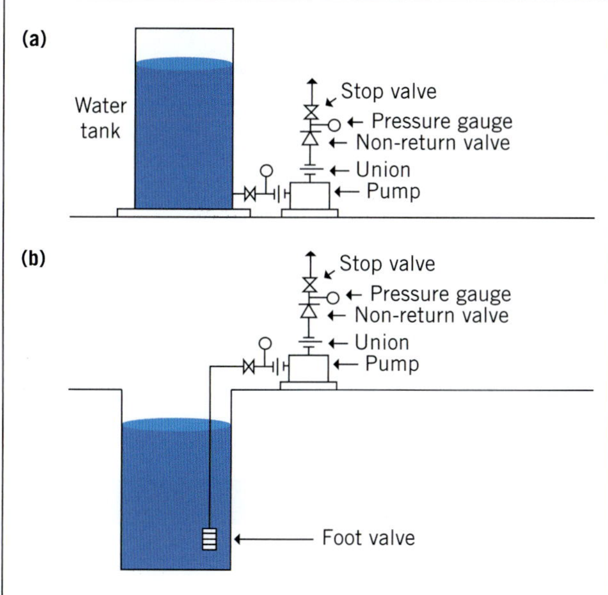

FIGURE 7.3 (a) Suction lift; (b) Suction head

Prepare for work

The plumber should be aware of what information is required by the pump manufacturer or supplier so that the right pump is chosen to suit the task. This may include the:

- source of water supply (such as a water main, storage tank or dam)
- temperature of the water being pumped
- purpose for which the pump assembly is required
- highest outlet above the pump
- pipework material and the type of valves being used
- material from which the pump is manufactured
- flow rate required at the point of use (number of outlets)
- pressure required at the point of use
- pressure losses to be overcome through the piping system
- type of power source required
- specific location of the pumps
- number of pumps required.

Typical examples of a manufacturer's pump selection sheets can be seen in Figures 7.4 and 7.5.

Quite often this information has already been catered for during the design process, particularly on larger installations, and is normally covered in the documentation. The plumber's role is to check the details on-site to ensure that all information is correct.

If there are any major variations it is always advisable to consult the client and obtain the necessary approvals before making any changes. Variations in the site conditions can also cause a variation to the installation of the pump assembly, so check all details relating to the pump before ordering and installing the unit/s.

Quality assurance

Following company policy for quality assurance can involve:

- carrying out WHS requirements such as JSA, SWMS and PPE
- keeping all tools and equipment in good working order
- carrying out the work in a professional manner
- communicating effectively with the client and other trades
- minimising material waste and time lost by writing accurate orders
- reusing and recycling material.

Tools and equipment

Take the time to prepare tools and equipment to ensure efficiency while the work is taking place. The typical tools and equipment used to assemble pipework, valves, unions and fittings to install and connect pumps are: Stillson's/footprints, spud wrench, shifting spanner, hacksaw, tube cutters, press-fit tool, electro-fusion welding machine and oxygen/acetylene.

Mechanical aids such as trolleys, winches or forklifts may be necessary to place heavy pumps into position.

Selecting the right pump

Once the pump supplier has been provided with all the necessary relevant information, they will select a pump using what is called a **pump curve**. A pump curve, also known as a **performance curve**, is a graph that pump manufacturers use to describe the performance of their products. This graph basically compares flow rate produced by the pump to the total head developed by the pump. From this graph, the supplier will provide options based on the best efficiency point (BEP) for specific pump types. For examples of pump curves.

Pump suppliers will readily provide pump curves to their clients to reinforce the decisions made when selecting the appropriate pump for the task at hand.

FROM EXPERIENCE

The experienced plumber will be familiar with pump curves and should be able to extract the relevant information from them to choose the best pump for the job.

Selecting materials for the pump components

When selecting a suitable material for a pump component, several factors are considered by the pump manufacturer. The choice for each component will

Source: © Grundfos Pumps Pty Ltd

What pump do I need?		1
Application		Household water supply
		Drainage
		Irrigation
		Water transfer
		Other
Water source		Above-ground tank
		Underground tank
		River
		Dam
		Other
Power supply		240 V single phase
		415 V three phase
Water requirement	Household	
		House only
		House and garden
		Showers (number)
		Sprinklers (number)
		Sprinklers (type)
		Evaporative airconditioner connected
	Irrigation	
		Sprinklers (number)
		Sprinklers (type)
		Automatic operation
		Manual operation
	Drainage and water transfer	
		Lift from pump (A) to point of discharge (B)
Details of existing pipeline		Size (mm)
		Type (polyethylene/PVC/copper/steel)
		Length (m)

How much flow (Q)?		2
Water pressure systems	✓	
Weekend cottage		10 to 20 L/min
Small home		20 to 30 L/min
Average home		30 to 50 L/min
Large home		50 to 90 L/min
Average water consumption	No.	
Standard shower head		15 L/min
Water saving shower head		6–7 L/min
Household standard tap		10–15 L/min
Tap with an aerator of flow restrictor		4–6 L/min
Lawn sprinkler		10–15 L/min

Calculate the flow rate	3
Q = ( ) L/min	

FIGURE 7.4 Selection sheet to help calculate flow rate

often depend on cost, availability and performance. Other factors affecting the materials selection include:

- the type of pump
- delivery (discharge) head in metres (how high the water needs to be pumped)
- flow rate in litres per minute (the amount of water required for all fixtures)
- liquid being pumped (consider issues such as temperature, corrosive qualities, abrasive qualities and the possibility of erosion)
- manufacturing method (casting or machining properties)
- initial cost
- economic life
- specific material properties such as strength, durability and flexibility.

General-purpose water pumps for low- and medium-pressure applications usually have a cast iron casing, high-tensile steel shafts, cast iron or bronze impellers and cast iron or bronze wearing parts. In some smaller pump units, certain plastic components may be used. For higher pressures, the casing is usually made from cast, forged or welded steel with the shaft, impellers and diffusers in stainless steel.

The plumber should be aware of the pump components and select pipe materials and ancillary equipment for connection to the pumpset that are compatible with the pump components and complement the specific application. When installing the pipework and ancillary equipment, care should be taken to ensure that there is no strain on the pumpset that could, in turn, cause the pump unit to malfunction.

Connection to the water utility's main system (towns and cities)

The designer and/or plumber must comply with the relevant water utility requirements when connecting a pump to the water supply system. This is important

How much pressure (P)? 4

P = pump head

Hd = height difference between the pump and the highest point of use

Hs = pressure already available at the pump level (tank example with positive suction head). If you pump water under the level of the pump (well, river, underground tank), contact your dealer in order to calculate the suction lift and to select the right pump.

Hf = friction loss or pipe resistance to water flow (see chart at right for poly pipe friction loss)

Pr = residual pressure, i.e. the required pressure at the tap, shower or sprinkler. As a guide, shower head, standard ½" tap or sprinkler requires approx. 150 kPa (15 m or 21 psi)

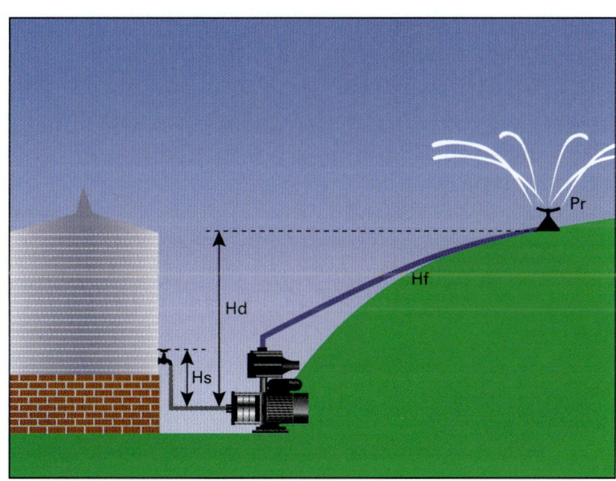

Flow Rate			Friction loss – PN12.5 high density polyethylene pipe (m/100 metres of pipe)					
L/min	m^3/hr	L/sec	20 mm	25 mm	32 mm	40 mm	50 mm	63 mm
12	0.7	0.2	10.9	3.7	1.1	0.4	0.1	–
24	1.4	0.4		13.4	3.9	1.3	0.4	0.1
36	2.2	0.6			8.3	2.8	0.9	0.3
48	2.9	0.8			14.2	4.8	1.6	0.5
60	3.6	1				7.2	2.4	0.8
72	4.3	1.2				10.1	3.3	1.1
84	5.0	1.4				13.5	4.4	1.5
96	5.8	1.6					5.7	1.9
108	6.5	1.8					7.1	2.3
120	7.2	2					8.6	2.8

Calculate the pressure 5

P = (   ) Hd – (   ) Hs + (   ) Hf + (   ) Pr

Example

Q (Flow rate) = **60 L/min** = **4** (sprinklers) × **15 L/min**

Hs = 2 m

Hd = 15 m

Hf = 3.6 m (50 m of 40 mm poly pipe — see friction loss chart above)

Pr = 15 m (150 kPa)

P = 15 – 2 + 3.6 + 15 = 31.6 m = 310 kPa = 44 psi

Useful conversion

From \ To	Flow conversion							
		litres per minute		litres per second		cubic metres per hour		gallons per min
1 L/min	=	1	=	0.017	=	0.06	=	0.22
1 m^3/hr	=	16.7	=	0.28	=	1	=	3.7
1 L/sec	=	60	=	1	=	3.6	=	13.2
1 gpm	=	4.5	=	0.076	=	0.27	=	1
Example								
60 L/min	=	60 L/min	=	1 L/sec	=	3.6 m^3/h	=	13.2 gpm
From \ To	**Pressure conversion**							
		Metre of head		Kilopascal		Pounds/inch^2		Bar
1 m	=	1	=	9.8	=	1.4	=	0.1
1 kPa	=	0.1	=	1	=	0.14	=	0.01
1 psi	=	0.7	=	6.9	=	1	=	0.07
1 bar	=	10.2	=	100	=	14.5	=	1
Example								
200 kpa	=	20 m	=	200 kPa	=	28 psi	=	2 bar

FIGURE 7.5 Selection sheet to help calculate pressure

because connecting a pump could have an adverse effect on the rest of the water supply system. If the water supply system cannot provide sufficient water for the pump to operate, the pump assembly will also be affected.

The minimum guaranteed water pressure in a water utility's main is 150 kPa (15 m head). It should be remembered that there will be a pressure loss brought about by friction losses through pipes, valves, fittings and other ancillary equipment – all of which have to be factored in when considering the pump installation.

The information required by the plumber is similar to that needed by the supplier when ordering the pump (see above), but would normally include extra information, such as:

- a written document from the water utility (water pressure enquiry) stating the available water pressure in the mains located near the building/property
- the site details, including nearest cross streets (for confirmation of the location).

Connection in other situations

Rural plumbers are often involved in installing water storage tanks and associated pump systems for drinking water and irrigation. With the move towards rainwater harvesting and recycled water in towns and cities for more efficient water usage, urban plumbers also require knowledge of such installations.

GREEN TIP

It is generally a standard requirement for new homes to store a minimum of 2000 L of rainwater for flushing toilets, irrigation and washing vehicles and equipment.

Whenever purchasing water tanks, a variety of suitable pumps are normally suggested to accompany them. The plumber should complete all necessary research before attempting to install the pump assembly. Consult *HB 230 – Rainwater Tank Design and Installation Handbook*.

An important issue to be aware of, particularly in urban situations, is that of cross connections. These may be direct (for example, a tank supply and a water main supply linked together by pipework so that the client can switch between the two supplies) or indirect (for example, where a client uses a hose to top up a rainwater tank from the main supply). In these situations, there is a potential for the tank water to enter the water main supply, creating a cross connection. Therefore, a suitable approved backflow prevention device may need to be installed to prevent cross connection.

Cross connections are classed into three hazard types:

1. *High hazard*: requires a testable device such as a registered break tank (RBT) or **registered air gap** (RAG) or a reduced pressure zone device (RPZD) is installed.
2. *Medium hazard*: requires a testable device such as a double check valve (DCV) or a pressure vacuum breaker (PVB) is installed.
3. *Low hazard*: requires a non-testable device such as a dual check valve (dual CV) or atmospheric vacuum breaker (AVB) or a hose connection vacuum breaker (HCVB) is installed.

Before installing or connecting in these situations, consult your local water utility for installation requirements. (Backflow prevention and tank water usage are covered in more detail in Chapters 8 and 9.) Also refer to the Plumbing Code of Australia (PCA) and AS/NZS 3500.1.

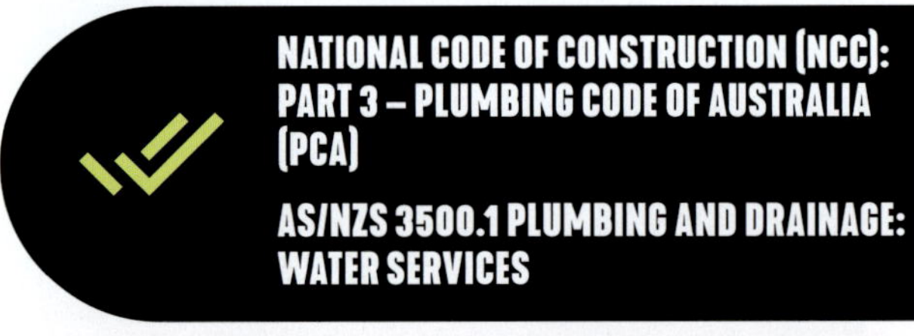

The plumber is directly responsible for identifying and preventing cross connection to avoid contamination of the community's drinking supply.

Types of pumps

Figure 7.6 shows examples of domestic pumps. The plumber's role is to install not only the pump but also the associated pipework, ancillary valves and equipment required for the pump assembly.

This chapter addresses the basic principles related to the more common types of water pumps and does not discuss the specifics about the many different pumps available for moving different fluids. If you wish to find out more about these pumps, the internet is a good place to start. Also see the References at the end of the chapter.

Rotodynamic pumps

A rotodynamic pump is a pump that uses the rotation of an impeller or propeller to create a pressure in the liquid it is moving. Pumps that use rotation to move a liquid are commonly referred to as centrifugal pumps.

Centrifugal pumps

Owing to its flexibility, the centrifugal pump is the most common and widely used pump in water supply systems, particularly where automatic

Self-priming
Built for long trouble-free life, the self-priming pumps with excellent suction capacity are suitable for a wide variety of water supply and transfer duties in home, garden and hobby applications as well as in agriculture and horticulture – wherever a reliable household water supply is needed.

Pressure systems
This range of Grundfos pumps are designed for any pumping applications involving clean and non-aggressive water in household applications and small-scale irrigation as well as in booster applications.

Submersible
Grundfos submersible pumps are easy to handle and suitable for operation for a variety of applications such as domestic water supply, irrigation and pressure boosting.

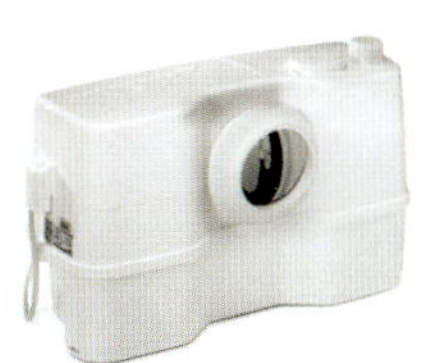

Wastewater and drainage
Grundfos wastewater and drainage pumps are designed to make pumping from household applications as simple and efficient as possible.

Hot water and heating
The Grundfos range of circulators for heating and hot water re-circulation are available in a variety of materials and finishes to cover even the most exceptional tasks.

Source: © Grundfos Pumps Pty Ltd

FIGURE 7.6 Examples of domestic pumps and their applications

control is necessary or where hot water is to be pumped (see **Figure 7.7**).

Larger pumps generally have a cast iron or cast metal casing. Domestic pumps are generally plastic. Centrifugal pumps have a volute, or spiral fin, that guides the water to the outlet of the casing. Inside the volute is an impeller with vanes or blades. The impeller, coupled with the pump driving mechanism, creates the energy to pump the water.

Vanes have a backward curve to the direction of rotation to minimise frictional resistance offered to the water as it passes from the centre (eye) of the impeller along the blade to its tip by centrifugal force.

How does a centrifugal pump operate?

This type of pump uses the principle of centrifugal force for its operation. The pump normally has a motor, attached to the pump, that drives the impeller to revolve at high speed. Water flows from the inlet pipe into the eye or centre of the impeller either by suction or by gravity. Water then passes through the eye of the impeller and is flung through to the tip of the vanes (blades), and into the volute casing by centrifugal force (this can be likened to a bucket of water being swung around 360°; no water is spilt because centrifugal force pushes the water to the bottom of the bucket). When it leaves the impeller blades, the water collects in the casing and is forced through the outlet, towards the point of discharge, at a constant pressure.

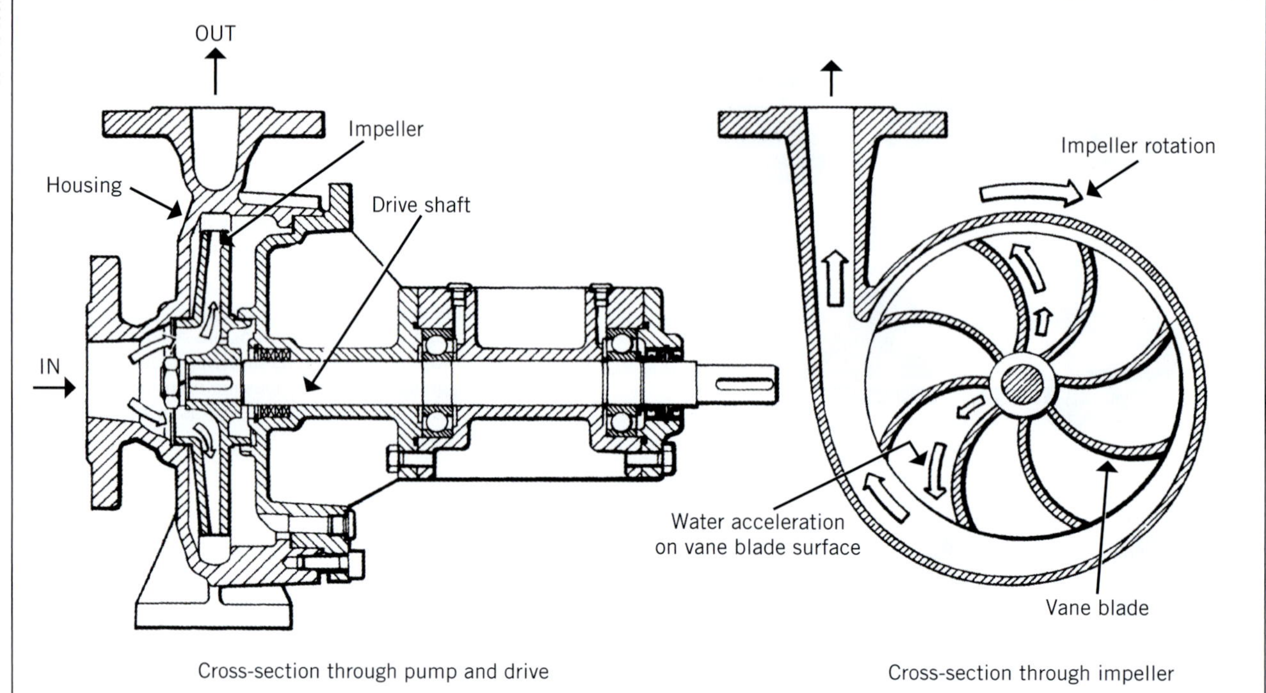

Source: *Basic Training Manual 11–3 Water supply 3*, Australian Government Publishing Service, from July 1970 to 1997.

FIGURE 7.7 Cross-section of a centrifugal pumpset

Pump connections

Centrifugal pumpsets are normally driven by a motor that is either close coupled or long coupled. As can be seen in Figures 7.8 and 7.9, a close coupled connection is where the pump is directly connected or bolted to the motor, and a long-coupled connection is where the pump and motor are a distance apart, each having its own shaft that is connected by a coupling. The advantage of a long-coupled motor is that it is easier to service and maintain, and therefore more common on larger installations, whereas the close-coupled motor is much more compact and is more commonly used in domestic applications. In some cases, centrifugal pumpsets also can be belt-driven.

Surasak_Photo/Shutterstock.com

FIGURE 7.8 Close coupled motor connection of a centrifugal pumpset

Pump efficiency

Because of its design and principle of operation, the centrifugal pump gives its best performance when used at the rate of discharge and head for which it is designed. The term used for this is 'best efficiency point' (BEP). The efficiency of the pump falls off when these conditions are varied.

For any given pump, the delivery, head and power input will increase as the speed increases. Thus, to make a pump deliver more water or pump against a greater head, the speed must be increased. However, it is undesirable to drive simple pumps of this type at very high speeds; therefore, a larger-capacity pump or several pumps should be installed.

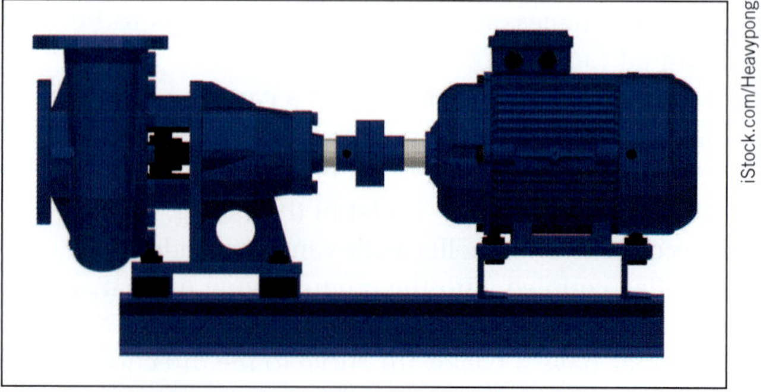

iStock.com/Heavypong

FIGURE 7.9 Long coupled motor connection of a centrifugal pumpset

Types of motors

Centrifugal and multi-stage centrifugal pumpsets may be driven by any form of motor or engine. However, the electric motor is the most common and generally the most convenient method of driving a pump.

In Australia, electricity is supplied in two voltages: 240 V and 415 V. Both have a frequency of 50 Hz. The electricity supply is often referred to as alternating current (AC) or direct current (DC). If the motor is wired up incorrectly, which sometimes happens, it can force the pump in the wrong direction. This can be dangerous and can damage the pump unit. If wiring has been done incorrectly, it must be corrected immediately to avoid serious damage.

It is important to have a licensed electrician connect any hardwire electric motor for a pump and be certain the pump is operating in the right direction.

Controlling the pump

The power source of the pump unit can involve simply plugging it into a 240 V power point (such as a pump unit connected to a domestic water storage tank) or hardwiring it through a control panel using either a 240 V or a 415 V power source (such as when pumping water from a low-level source to an elevated water storage tank).

The flow control of the water can also vary. The pump could simply operate each time a tap or valve is opened. This is controlled by an automatic pump control unit. But this can cause unnecessary wear and tear on the pump, as well as inconvenience to the user, because of noise transmission, hydraulic shock and water hammer. Another type of automatic control operation is the use of float switches or level controls in water storage tanks that are connected back to the pump control panel, or by flow switches or pressure switches located in a pipeline, which are again linked back to a pump control panel.

Float or level controls are generally used when filling storage tanks by means of a pump. The devices themselves can range from simple plastic floats, which incorporate a ball that acts as a switch, to more complex rods that incorporate sensors that control the flow of water.

To meet the varying demands that are required in certain situations, multi-stage pumps can be used. These pumps gradually increase in power and capacity as demand increases, thus building up in stages. This system helps to prevent undue stress on the pump system and ancillary pipework and equipment.

A typical example of pump information can be seen in Figure 7.10.

Positive displacement pumps

Positive displacement pumps have two main types: reciprocating and rotary. They have been used for a very long time and in fact were used extensively for water supply prior to centrifugal pumps becoming more popular. They were used to pressurise domestic services, domestic feeds to storage tanks, irrigation services and fire services. They were used to supply water from rivers and dams, or from underground to a storage tank in rural areas. The design of this type of pump allows it to obtain extremely high pressures and it should therefore be fitted with a safety valve on the outlet side to prevent over-pressurisation of the water service system.

Reciprocating pump

A reciprocating pump can be defined as a type of pump that operates with a constant motion, employing a straight-line path with a to-and-fro motion, moving backwards and forwards or up and down (as opposed to a circular motion). It may be of the piston, plunger or bucket type, unlike a centrifugal pump. Reciprocating pumps are generally made up of three moving elements for their operation and consist of:

- an inlet valve
- a piston or plunger
- an outlet valve.

One of the more common types of positive displacement pump can be seen on a windmill. The windmill makes effective use of the suction and lift pump, as can be seen in Figure 7.11. Multi-blade windmills traditionally pump water by directly operating a pump cylinder with a drive rod. The pump cylinder is submerged in the well attached to the end of the delivery pipe. It is a very simple pump, similar to a hand-operated bicycle pump. The drive rod is operated directly by the windmill rotor through the drive gearing, which translates the rotating motion to the up-and-down reciprocating motion. Two one-way valves in the pump direct water through the pump, as illustrated in Figure 7.12.

The bottom check valve is fixed in position and opens to allow water to be drawn into the bottom chamber during the upstroke of the plunger and valve, which is attached to the pump rod. As the pump rod moves down again, during the downstroke, the water in the lower chamber flows through the upper check valve into the upper chamber. This water is then lifted during the upstroke towards the point of discharge and so on (hence the term 'reciprocating', because it moves backwards and forwards).

The hydrostatic test bucket, which is used by plumbers when testing water services, operates on a similar principle to that mentioned above (see Figure 7.13).

Rotary pump

The rotary or geared pump relies on cogs (gears) turning together and displacing fluid. This type of pump is used to move highly viscous (thick and gooey) fluids such as oil, paint, ink and chocolate.

Source: © Grundfos Pumps Pty Ltd

Applications

The MQ booster pump is a self-priming compact unit designed for the distribution and boosting of clean domestic water. Its integral tank avoids the need for a cumbersome additional water tank and enables quick installation in confined spaces.

Features

- Quiet operation
- Protection against dry running
- Suction lift: up to 6.5 metres (see operating instructions for maximum suction pipe length)
- Built-in non-return valve
- Supplied with an electric cable (2 m) and plug
- Motor with thermal protection
- Discharge outlet adjustable by +/–5° to facilitate hose attachment

Construction

- Stainless steel pump housing
- Integrated 0.4 litres diaphragm tank
- IP54 motor
- Insulation Class B
- Voltage: 1 x 240 V – 50 Hz

Performance

Pumps	Discharge pressure (**kPa**/psi)					
	150	**200**	**250**	**300**	**350**	**400**
	22	29	36	44	51	58
	Output (L/min)					
MQ3-35	61	50	39	24	–	–
MQ3-45	66	58	49	39	27	13

As a guide 1 tap = 10 litres per minute

1 sprinkler = 15 litres per minute

Technical features

Pumps	P2 (W)	In (A)	Water temp	Connections Inlet	Connections Outlet	Weight (kg)
MQ3-35	550	4.0	0/35 °C	1″M	1″M	13
MQ3-45	670	4.5	0/35 °C	1″M	1″M	13

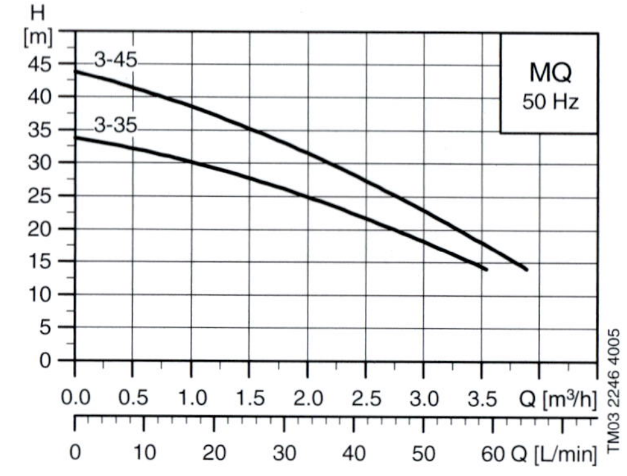

Dimensions / installation

240 G1 G1 320 192 190 4 x Ø10 570

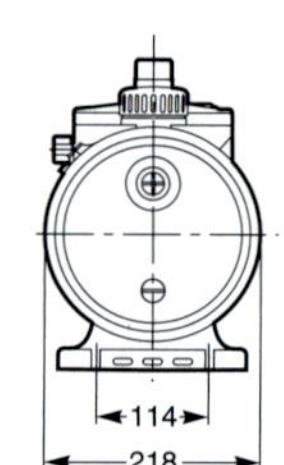

Dimensions in mm

The MQ booster pump's integrated control unit automatically protects against:

- dry running
- motor overheating
- motor overload.

In the event of a fault, the MQ stops automatically, then tries to restart after half an hour.

FIGURE 7.10 Typical example of pump information

As the gears turn, the cogs come apart on the inlet side, creating a vacuum that is then filled by the fluid and forced to the pump outlet by the rotating gears. At this point the fluid is displaced at a higher pressure (see Figure 7.14). The robust design with tight tolerances of the gears allows the ability to pump highly viscous fluids at very high pressures.

Material list

A typical material list for a pump installation could look something like this:

Material	Quantity
Pump (manufacturer recommended – required delivery head)	1
25 mm × M + F brass barrel union	2
25 mm × ball valve (water isolation)	2
25 mm line strainer	1
25 mm check valve (spring)	1
25 mm brass hex. nipples	6
25 mm M.I. connectors (no. 3s)	2

Source: *Basic Training Manual 11-3 Water Supply 3*, Australian Government Publishing Service, from July 1970 to 1997.

Gear box
Multi-blade rotor
Tail
Mill swivel
Pump rod swivel
Tower
Well seal (ground level)
Tower footing
Pump rod
Packer head
Stock tank
Water level
Drop pipe
Pump cylinder
Screen

FIGURE 7.11 Windmill driving a positive displacement pump

Source: *Basic Training Manual 11-3 Water Supply 3*, Australian Government Publishing Service, from July 1970 to 1997.

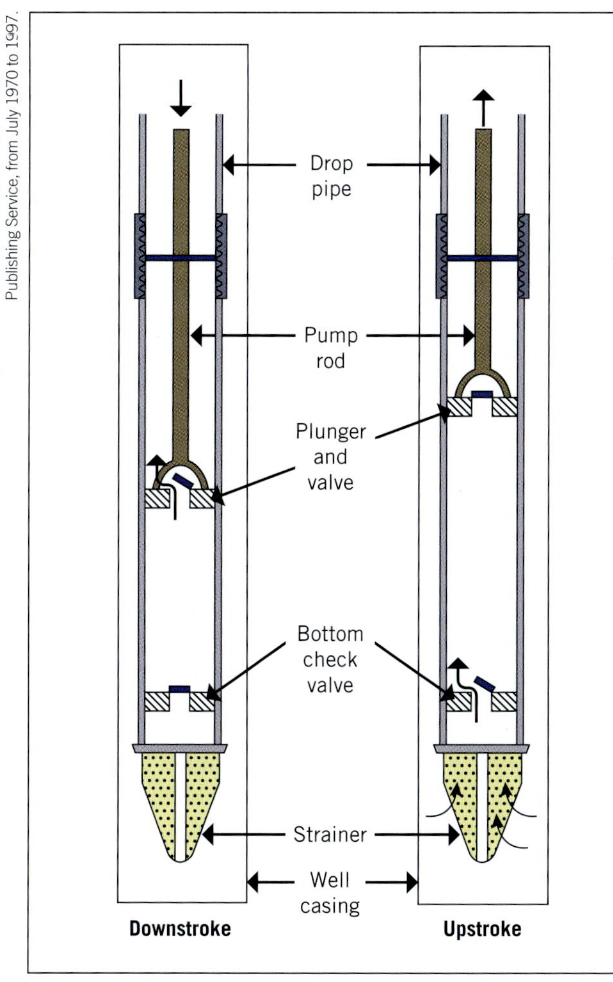

FIGURE 7.12 Operation of a typical windmill pump cylinder

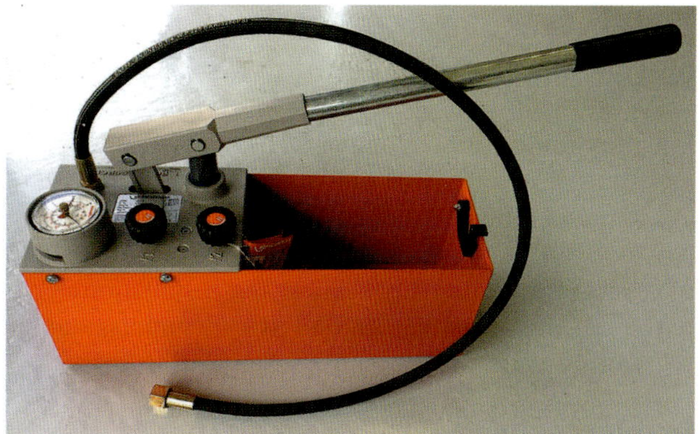

FIGURE 7.13 Test bucket – positive displacement (reciprocating) pump

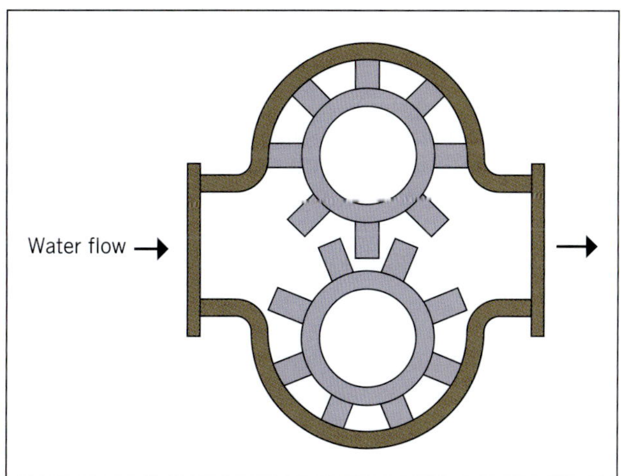

FIGURE 7.14 Rotary or geared pump

LEARNING TASK 7.3

1 Name the two main requirements necessary to choose the correct pump.
2 How may a potential cross connection occur with a pump installation?
3 Why is the centrifugal pump the most commonly used water pump?
4 What is the name of the casing in which the impellor rotates?
5 Name two devices that can control the pump operation.
6 Name a type of positive displacement pump.

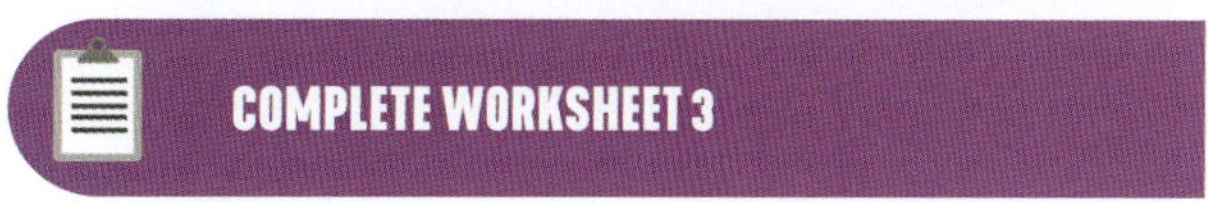

Install, test and commission water pumps

It is important for the plumber to have the knowledge and skills to determine the most suitable location for a pump and to install the correct valves, fittings and pipework required for efficient pump operation. This next section explains the location and installation requirements for pumps, as well as the commissioning and testing of pumps.

Location of pumps

Before installing a pump assembly, the plumber should check that the installation will meet the current standards and regulations. The pump assembly is normally located as close as possible to the source of water supply to minimise potential problems with the suction side of the pump.

In most situations, plumbers have the advantage of being able to locate a pump according to a plan. It is therefore vital that they can interpret the information provided on a plan with the provided specification as well as the manufacturer's specifications. For domestic installations, plumbers may have to use their own discretion to provide and install a pump (for example, for water supply from a rainwater tank), so they must have the skills and knowledge required to correctly locate and install the water pump.

FIGURE 7.15 Irrigation pump housing for weather protection

Installing the pump assembly

Pump assemblies, when located within a building, are usually located in a plant room or service duct. This room or duct must provide good accessibility and noise-proofing. Noise and vibration can be transferred from the pump assembly through the connecting pipework and/or the support structure into the rest of the building. This may not be too much of a problem in an industrial situation, but in residential or commercial buildings the noise can become quite disturbing. Continual vibration caused by the pump assembly not only may be a disturbance, but also may have a harmful effect on the connecting valves, pipework and surrounding equipment over a period of time. Pumps used for irrigation are usually located outside with some type of housing for weather protection and close to the water source, such as a storage tank, dam or bore (see **Figure 7.15**).

The pump installer should be aware of noise generation and take the necessary steps to prevent noise transfer and vibration. This can be achieved by installing anti-vibration flexible connections. Installed next to the pump unit on both the suction and the discharge sides of the pipework, flexible connections will prevent noise transmission and protect the pipework from being damaged when the pump motor starts and stops. If the pump is mounted on the floor area, it may also require rubber mountings or anti-vibration springs on the pump base.

On larger pumpsets (for example, long-coupled centrifugal pumps), the motor and pumpset are assembled in the factory and delivered to the site. The pump itself is connected to the motor shaft, sometimes with a combination of flexible couplings and spacer pieces. Because of its size, this pumpset must be mounted on either a frame or a specially designed pump base (sometimes called an inertia block). These pump bases must be carefully levelled, fixed and grouted into the building foundation if installed inside the building.

The foundation for the pumpset should preferably be a rectangular concrete block with a mass approximately equivalent to five times the weight of the complete pump unit. This foundation needs to support the pump and motor assembly and, if necessary, prevent vibration and noise transmission through to the building structure.

The final positioning of the complete pump unit is essential so that correct alignment of moving parts is maintained and the pump warranty is not voided. Therefore, the plumber should pay particular attention to following the accepted trade practice when grouting in and/or fixing the pump unit. This information can be obtained from the pump supplier or manufacturer if the plumber is unsure of what the acceptable method is.

FIGURE 7.16 Chilled water circulating pump. Note the protective guard over the drive shaft, the flanged flexible connections on the pipework and the pump mounted on springs (vibration pads)

When the pipework is installed for the pump assembly, consideration must be given to both the isolation and the disconnection of the pump so that it can easily be maintained or its parts replaced. This is usually achieved by installing an isolating valve on both the inlet and the outlet, and with flanges or unions for connection purposes. The type of isolation valve used depends on the type and size of pump assembly and is normally a full way type of valve (for example, a gate or ball valve). A non-return (check) valve must be installed on the outlet upstream of the isolation valve so that head pressure exerted on the pump outlet is avoided. The use of flanges or unions (for removing or maintaining parts of the pump assembly) depends on the size and type of pipe being used.

Assembly

A centrifugal pump is used in a reticulated (circulated) system, such as a water heating or water cooling system (see **Figure 7.16**). The assembly consists of:

- the suction side:
  - suction full way (gate or ball) valve
  - suction strainer
  - foot valve (check valve and strainer)
  - flexible coupling
  - suction gauge
  - unions/flanges
- the discharge side:
  - discharge gauge
  - flexible coupling
  - non-return (check) valve
  - full way (gate or ball) valve
  - unions/flanges.

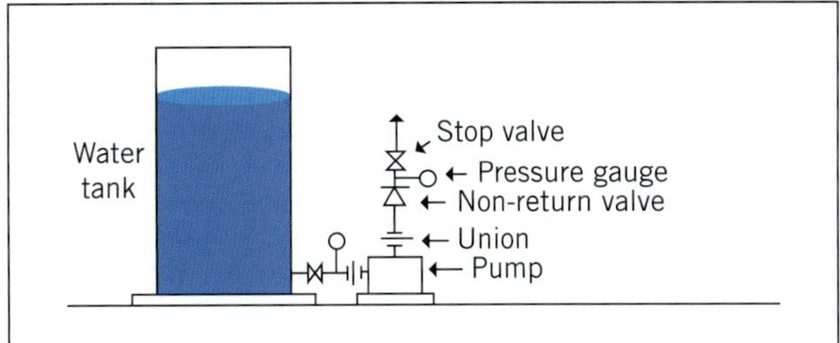

FIGURE 7.17 Positive suction head (water source above pump) installation and valves

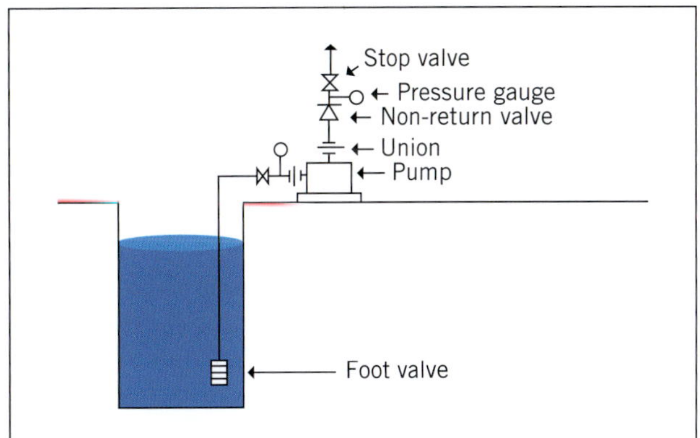

FIGURE 7.18 Negative suction head/lift (water source below pump) installation and valves

Pipe systems for pumps

Many of the problems experienced with pump installations are brought about by poor pipeline design or the improper application of the pump itself. If care is taken when designing the layout of the system and if good plumbing practice is used, there is normally a marked improvement in the pump's overall performance.

Piping systems for pumps, irrespective of the type of pump used, can be considered as two separate sections: the suction pipework and the discharge pipework.

Suction pipework

The suction side is the side between the source of the liquid to be pumped and the pump itself. As mentioned earlier, when the pump is below the liquid source it is said to have a positive suction head, which is preferred (see **Figure 7.17**). When it is above the liquid source it is said to have suction lift (negative suction head; see **Figure 7.18**).

Issues to consider with suction pipework include the following:

- Keep the suction pipe as short as possible.
- Where possible, try to provide positive suction head to the pump. (Install the pump below water source.)

- Ensure that air pockets are eliminated by allowing for a continual rise to the pump (if there is a potential for air pockets to form, consider redesigning the pipe layout or allow for an air release valve to be installed).
- When selecting the ancillary equipment, such as foot valves and strainers, select those that will provide minimal friction loss and resistance to flow.
- If a reducer needs to be installed in the pipework adjacent to the pump, use an eccentric (tapered) reducer rather than a concentric (bush) reducer to avoid the possibility of forming an air pocket and/or causing **cavitation (air bubbles)**.
- The suction pipe cannot exceed 6 m in length if the water source is below the pump.

Discharge pipework

Careful consideration should be given to the types of pipe, fittings and ancillary equipment used when installing the discharge pipework so that friction losses are kept to a minimum.

It is important to install a check valve (non-return valve) on the discharge side of the pump so that no undue backpressure is placed on the pump assembly, which could cause the pump to malfunction. It is good practice to install the check valve no closer than 15 times the pipe diameter to the pump outlet to prevent turbulence acting on the check valve.

When using reducers on the discharge pipework it is preferable to create a gradual taper using a tapered reducer piece, rather than using a reducer with a stepped shape, taking the larger diameter to a smaller diameter. This helps to prevent air being trapped and the possibility of the pump cavitating.

Ancillary equipment

Examples of ancillary equipment used with pumps include the following:

- **Check valve** – a non-return valve installed in the pipework. It is always installed on the discharge side and sometimes on the suction side of the pump. Avoid using a swing-type check valve as these have the potential to cause water hammer.
- **Foot valve** – a type of check valve with a built-in strainer. Used at the point of the liquid intake (suction side) to retain liquid in the system, it prevents the loss of prime when the liquid source is lower than the pump (generally when a suction lift is required).
- **Strainer** – a device installed in line on the suction side of the pump. The strainer is designed to prevent foreign matter from getting into the main body of the pump and possibly causing damage to the pump. This is incorporated into the foot valve.
- *Reducer* – pipe reducers are used for changing pipe sizes. It is important to use an eccentric reducer on the suction side of the pump so that air does not become trapped in the pipework and possibly contribute to cavitation.
- *Pipe supports* – these are a very important part of the pump assembly. Some pipe supports are designed only to take the load off the pipe assembly. Others are designed to support both the suction and the discharge line and may include noise suppressors as part of the support to help prevent pump noise being transmitted into the building.
- *Drains* – some larger pumps have seals that are designed to leak normally and may include a packing gland. For this type of pump, a drain must be installed to drain water away from the pump assembly to an approved point of discharge.
- *Pressure gauges* – these should be installed and can be used on both the suction and the discharge sides of the pump pipework assembly. They provide a ready check to see if the system is operating satisfactorily.
- *Air cells* – hydro-pneumatic pump assemblies include an air (pressure) cell (see **Figure 7.19**) on the discharge side. This cell has an internal bladder that is pressurised with air. When the pump cuts in, it initially fills the pressure cell with water, as well as pumping water through the system as it is required. The cell is designed to store water for intermittent (irregular) use so that the pump does not continually turn on and off (shunt) when there is a slight pressure drop. **Figure 7.20** shows a small hydro-pneumatic (air/pressure cell) installation.
- *Vibration eliminators* – installed in line on both the suction and the discharge sides of the pump and as close as possible to the pump itself. These are designed to prevent damage to both the pump and the pipework, particularly when the pump starts and stops.

LEARNING TASK 7.4

1. Why is the pump located as close as possible to the water source?
2. What valves and fittings are connected to the pump inlet (suction) side?
3. What valves and fittings are connected to the pump outlet (discharge) side?
4. Why is it important that no air enters the inlet (suction) pipe?

FIGURE 7.19 Pump with air pressure cell designed to boost water pressure

Testing the pump system

Once the water pump system has been installed, it is important to turn the system on and test for leaks. The manufacturer's instructions may also require that you test the flow rate and other factors to ensure the pump is working properly, and this information will be included as part of the commissioning process.

The plumber must check that the installation is free from leaks and defects, and that the required flow rate at all the outlets is correct. To check that the system is free from leaks, the plumber will need to hydrostatically pressure test the piping system in accordance with the job specifications or current relevant standards. Water services must be tested to the requirements of AS/NZS 3500.1.

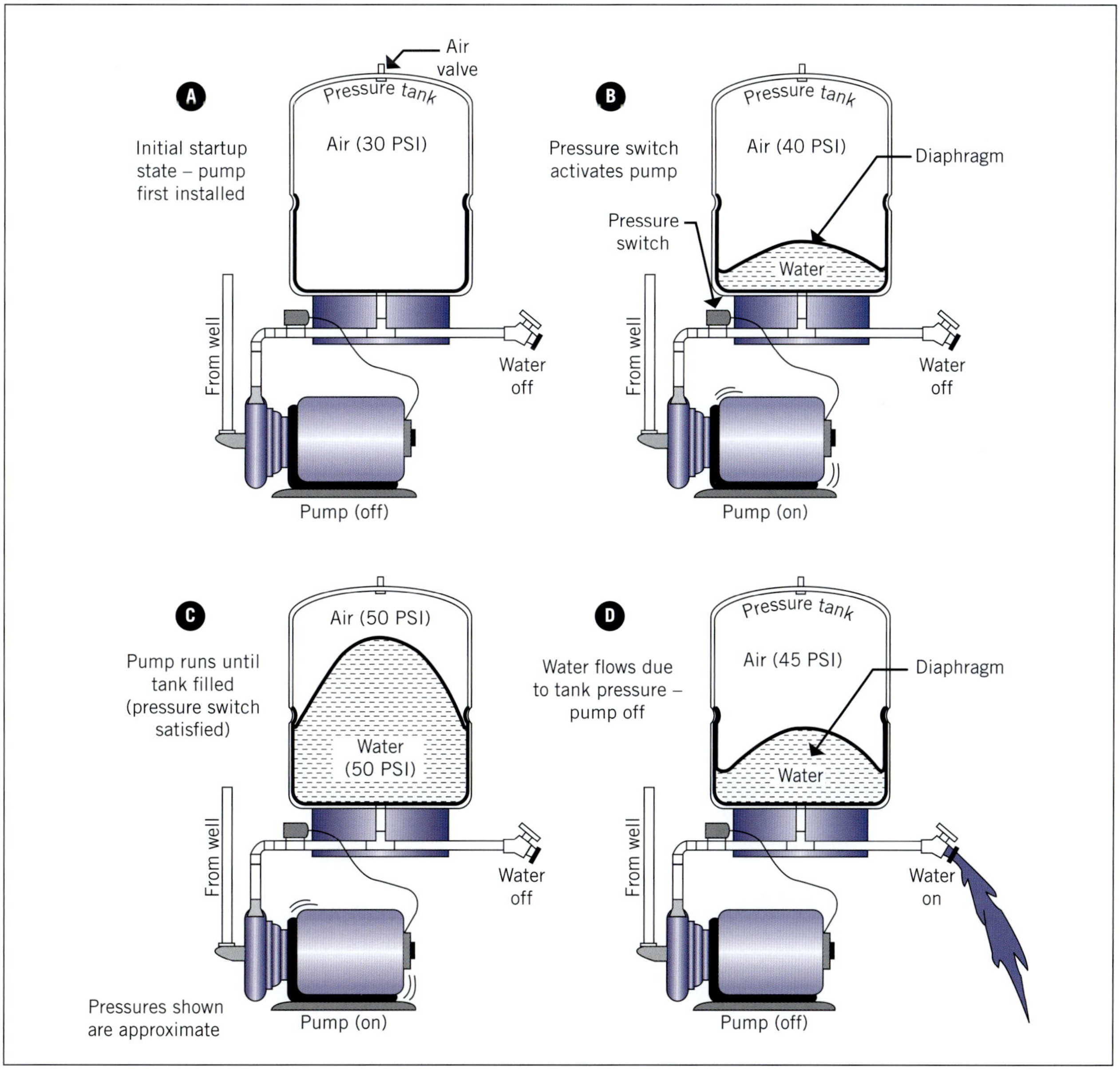

A
Initial startup state – pump first installed
Air valve
Pressure tank
Air (30 PSI)
From well
Water off
Pump (off)

B
Pressure switch activates pump
Pressure tank
Air (40 PSI)
Diaphragm
Pressure switch
Water
From well
Water off
Pump (on)

C
Pump runs until tank filled (pressure switch satisfied)
Air (50 PSI)
Water (50 PSI)
From well
Water off
Pump (on)
Pressures shown are approximate

D
Water flows due to tank pressure – pump off
Pressure tank
Air (45 PSI)
Diaphragm
Water
From well
Water on
Pump (off)

FIGURE 7.20 Diagram of a pump and bladder (air cell) assembly

AS/NZS 3500.1 calls for a test pressure of 1500 kPa to be maintained for a period of 30 minutes. Usually, a test bucket or an electric pump for large installations is used to hydrostatically test the piping.

AS/NZS 3500.1 PLUMBING AND DRAINAGE: WATER SERVICES

Testing procedure

When testing a piping system for leaks, plumbers need to take the following steps:

HOW TO

CHECK FOR LEAKS

1 Check the system components and find out the maximum rated pressure (plumbers may not exceed the lowest of these pressures unless the component is isolated from the pressure test).
2 Isolate the pump from the discharge end of pipework to be tested.
3 Connect the test bucket to the outlet point of the irrigation to be tested and fill the test bucket with water.
4 Plumbers need to remove any air in the system by leaving an open outlet to the system.
5 Close the drain valve and open the system valve on the test bucket.
6 Begin pumping, using the handle on the pump in an up and down motion, and carefully watch the system pressure gauge. Once all the air has been removed from the system, isolate the open end and continue pumping the water into the irrigation system.
7 Pump the system to the test pressure. This will be twice the working pressure of the system or 1500 kPa, as per AS/NZS 3500.1.
8 Check the pump and pipework for water leaks. If you find any, drain the system and repair them before continuing with the test.
9 Close the system isolating valve on the test bucket and record the pressure reading on the pressure gauge.
10 Observe this pressure for 30 minutes and record any fall in pressure. If the pressure falls, it means a leak is detected. A visual inspection is the most accurate observation.
11 Flush out the system if the pressure test is completed correctly.
12 Once the piping system is complete, open all valves that were closed before the pressure test.

To test the performance of the pump for correct flow and operation, plumbers will need to record the:

- operating pressures at both the suction and discharge sides of the pump
- flow rate required (i.e. l/sec or l/min)
- pump data from the compliance plate located on the pump unit.

A flow cup or a digital flow meter is an efficient way to check the flow rate.

Commissioning the pump system

Commissioning occurs when responsibility for an irrigation system or any other water supply system is handed over from the designer and installer to the owner/operator, and may have implications for insurances, maintenance programs and compliance. Commissioning is the final phase in the installation process and is undertaken by the installer to ensure the pump system and controls are operating correctly. Where a designer has been involved, they are often involved in commissioning, either at the system testing phase or providing input on how to correct performance issues.

The pump system should now be ready to start so that the plumber can check whether it is operating correctly. Before starting the pump, however, there are several points that must be checked (failure to do this could result in damage to the pump):

1 *Electric wiring* – this must be checked by a qualified electrician and must comply with all current and relevant codes.
2 *Direction of rotation* – depending on the type of pump this can be checked by removing the coupling between the motor and the pump and briefly starting the motor. Confirm that the direction of rotation is the same as that indicated on the pump casing. If it is incorrect, the wiring needs to be checked by a qualified electrician.
3 *Pump gland* – there are two types of shaft seals used on pumps: mechanical seals and packed glands. Mechanical seals do not drip, whereas packed glands (stuffing boxes) require a drip rate of approximately 20 drips per minute. If it does not drip, the gland instantly overheats and destroys itself and the pump shaft.
4 *Priming the pump* – do not operate a pump dry. Before operating the pump, the system must be flushed to remove any debris and then it must be **primed** (filled) with clean water. The pump and the suction line must be full with water before the pump is switched on (see following instructions).
5 *Other considerations* – carefully read the manufacturer's installation and starting instructions before switching on the pump.

Priming a water pump

Remove the prime airlock plugs as shown in Figure 7.21. Fill a bucket with water and pour water into the pump prime airlock plugs casing until the suction line and pump casing are full. Reinstall the priming plug loosely to allow air to bleed from the pump. Turn on the pump and monitor the sound of air bleeding from the priming plug. When air stops bleeding, turn off the pump. Then tighten the plugs and connections. The pump is now primed and ready to use with all the air removed.

All the relevant information should then be recorded by the plumber for their records. Depending on the nature of the installation, this same information is usually given to the owner or the facility manager for maintenance purposes. It is also usual for this information to be permanently affixed next to the pump assembly or the pump control panel. Record the result to a test datasheet.

Commissioning and testing documentation will consist of the following:

- a commission report
- a pump selection datasheet
- as-built plans
- operation and maintenance manual
- inspection and testing of backflow prevention device – submitted to water supply regulator
- handover of warranty and operation instructions to the client.

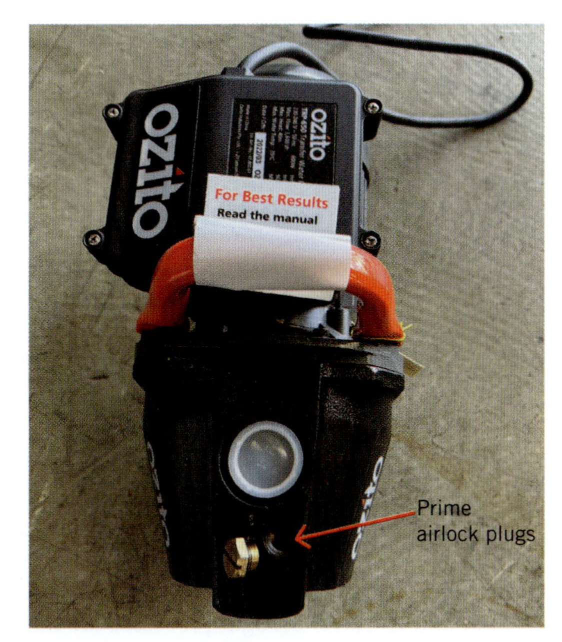

FIGURE 7.21 Priming a water pump

FROM EXPERIENCE

It is important to have a thorough knowledge of pressures and flow rates to ensure the correct pumps are selected for a particular job.

LEARNING TASK 7.5

1. A packing gland on a shaft seal is meant to leak. What is the recommended drip rate?
2. Explain why pumps need to be primed.
3. Why is it important to check the rotation direction?
4. Name two hazards that could be present when installing pumps.

Troubleshooting

In most cases, troubleshooting should be left to those who are well experienced in pump installations. Different pump installations may have different problems. Before making any alterations to the pumps or the pump installation, it is advisable to consult with the pump manufacturer so that any warranty is not voided.

There are, however, some simple checks that can take place when faced with a problem onsite, as shown in Table 7.1.

Troubleshooting is, of course, not limited to the items listed in Table 7.1. Where there is a problem, it is advisable to contact the manufacturer of the pump. They, in turn, may:

- refer you to the literature provided with the pump
- arrange for their representative to meet you onsite
- refer you to their website to access the necessary information to assist you.

Good planning and design of a water pump installation minimises the long-term costs by:

- ensuring that the pump operates efficiently
- ensuring that the right type of pump is installed to achieve the required flow rates and pressures
- ensuring that the life of the pump is maximised
- minimising hydraulic (friction) losses
- helping minimise energy costs
- ensuring easy pump operation
- ensuring that foundation and structural requirements are met.

Figure 7.22 shows a typical pump assembly with dual pumps, a pressure cell and pump control panel.

TABLE 7.1 Troubleshooting for pumps

Problem	Possible causes and/or remedies
No water is being pumped or the pump is not delivering enough water.	The suction line and pump casing are not properly primed (air in the system).
	Ensure that the system is completely flushed prior to commissioning the installation.
	The delivery head is too high (check the friction head).
	The mechanical seal may be worn and leaking air or water. Check the weep hole.
	Suction lift is too great. If pumping from a well or pit the maximum practical distance between the water level and the pump inlet is about 6 m (see the discussion on suction lift earlier in the chapter).
	The impeller, suction pipe or foot valve is clogged (also check the strainer in the suction line, if fitted).
	There is an air leak in the suction line.
	The suction and/or discharge piping is too small in diameter.
The pump operates for a while then stops pumping.	The pump is not properly primed.
	There is an air pocket in the suction line.
	There is an air leak in the suction line.
	The suction strainer is blocked.
The pump is noisy when operating.	The pump and motor are out of alignment. Check the pump alignment.
	Foreign matter is rotating with the impeller. Visually inspect the pump suction inlet port.
	The pump is cavitating which can be caused if the pump suction lift is too great or the pipework is incorrectly installed. Cavitation occurs when the liquid in a pump turns to a vapour at low pressure. It occurs because there is not enough pressure at the suction end of the pump. When cavitation takes place, air bubbles are created at low pressure, preventing the liquid from being pumped.
The pump vibrates during operation.	The pump and motor are out of alignment.
	The foundation is insufficiently rigid.
	The suction or discharge piping may be unsupported.
	The stuffing box becomes hot. The packing is too tight or not dripping enough water (approximately 20 drips per minute are required on packed gland seals).
The pump does not prime properly.	Make sure the pump casing is filled with water.
	Look in the suction line or fittings. Check to see that all fittings are tight in the suction line and make sure there is no leak in the pipe itself.
	The mechanical seal may be worn and leaking air.
	The inlet valve rubber may be frozen to the seat.
	The pump may be running too slowly.
	Suction lift may be too high. Keep the pump as close to the water source as is safely possible.

FIGURE 7.22 Completed pump assembly with control panel

Clean up

Having a general tidy-up in the work area at the end of the day is a good practice. The work area is more productive when clean and tidy. It is easier and safer to move about and workers' thought processes are more direct without the distraction of clutter.

Clients often associate cleanliness with quality. If they see a job site that is a mess, they might think less of the quality of work being carried out.

A clean workspace will result in happy customers, who are likely to call next time they need work done or will be happy to refer others. Part of the job is not just the hands-on labour, but providing a great customer experience. Taking a few minutes to clean up could result in more leads from referrals.

Any leftover material, pipe and fittings should be stored and reused on another job. All recyclable rubbish should be disposed of in the proper bins.

All tools and equipment must be cleaned and stored so they are ready for use next time. Check all tools and equipment for any damage and report any faults to the supervisor. Remember to tag any faulty tools and equipment so they are not used until repaired.

All company quality assurance procedures must be followed. These include the clean-up process, the pack-up and storage process, and issuing any required final documentation (as mentioned in 'Commissioning').

LEARNING TASK 7.6

1 What could cause the pump to not deliver water or to stop pumping?
2 What could prevent the pump from priming properly?
3 What should happen to leftover material?
4 You have been asked to install a water pump to connect a rainwater tank to a tap so the customer can water the garden. The tap location is approximately 4 metres away from the pump. The pump will sit next to the already installed water tank. Develop a materials, tools and PPE list for the task.

COMPLETE WORKSHEET 4

SUMMARY

- Water pumps have three purposes: to increase water pressure, to move water from a lower place to a higher place and to circulate water in a continuous loop.
- The two major groups of pumps are rotodynamic and positive displacement.
- Centrifugal pumps are the most commonly used and come under the rotodynamic group.
- There are several different mechanisms that control the pump operation, including float or level controls and pressure controllers.
- It is preferable that the water source is above the pump (positive suction head) because the pump will self-prime and it doesn't have to draw (suck).
- A pump cannot draw water that is more than approximately 6 m below it, regardless of its size.
- It is preferable to install the pump as close to the water source as possible for pump efficiency.
- Always install a full-way isolation valve, line strainer and union on the pump inlet.
- A foot valve must be installed on the suction pipe when the water source is below the pump (negative suction).
- Always install a union, non-return (check) valve and full-way isolation valve on the pump outlet.
- Installing an air cell gives the pump a longer life because it reduces the pump's operating time and saves energy.

REFERENCES

Davey: **http://www.davey.com.au**

Grundfos: **http://www.grundfos.com**

NSW Environment Protection Authority: **http://www.epa.nsw.gov.au**

Standards Australia: **http://www.standards.org.au**

GET IT RIGHT

1 Discuss the reasons a pump may not operate after installation.

WORKSHEET 1

To be completed by teachers	
Student competent	☐
Student not yet competent	☐

Student name: ______________________

Enrolment year: ______________________

Class code: ______________________

Unit competency code/title: CPCPWT3025 Install water pumpsets and CPCPIG3022 Install and commission domestic irrigation pumps

Task: Review 'Identify installation requirements' and answer the following questions.

1 Name the three main functions of a water pump.

____

____

____

2 Describe at least four situations where pumps may be used within a water supply system.

____

____

____

____

3 What is the definition of a water pump?

____

____

4 State the two basic categories that pumps fall into.

____

____

5 List two examples of each of the pump types from the two categories above (see **Figures 7.1** and **7.2**).

Type 1

____

____

Type 2

____

____

6 Which type of water pump is the most commonly used?

____

7 Name the main component in the rotodynamic pump that accelerates the fluid.

____

8 Name three sources where water for irrigation can be pumped from.

____

____

____

WORKSHEET 2

To be completed by teachers	
Student competent	☐
Student not yet competent	☐

Student name: ______________________________

Enrolment year: ______________________________

Class code: ______________________________

Unit competency code/title: CPCPWT3025 Install water pumpsets and CPCPIG3022 Install and commission domestic irrigation pumps

Task: Review 'Identify installation requirements' and answer the following questions.

1 When selecting and installing pumpsets for particular applications, the plumber has a number of responsibilities. List four of these responsibilities.

____

____

____

____

2 On large pump installations it is normal for more than one pump to be installed. Give two reasons why this is done.

____

____

3 Explain the term 'delivery head'.

____

____

4 Explain the term 'flow rate'.

____

____

5 Explain the term 'total dynamic head'.

____

____

6 Explain the formula $P = H \times 9.81$

where P = ______________

H = ______________

9.81 = ______________

7 Using the above formula, determine the pressure supplied to outlets on the 8th, 18th and 26th floors of a high-rise building that has a water tank located on the roof of the 31st floor that is 2 m in height. (Note that each floor is 3 m high.) What delivery pressure would be required of a pump at ground level to fill the water tank?

8th floor:

18th floor:

26th floor:

Pump delivery:

8 a In relation to the term 'suction head', what is the theoretical maximum height that a pump can be above the water source?

b Due to friction loss, what is the maximum height a pump should be positioned above the water source?

c Give a reason why the maximum theoretical height cannot be achieved.

9 Define the terms 'positive suction head' and 'negative suction head'.

10 Why is it preferable to have positive suction head?

11 Explain the term 'discharge head'.

12 Explain the term 'friction head' and state examples of situations that cause friction head.

WORKSHEET 3

To be completed by teachers	
Student competent	☐
Student not yet competent	☐

Student name: ______________________

Enrolment year: ______________________

Class code: ______________________

Unit competency code/title: CPCPWT3025 Install water pumpsets and CPCPIG3022 Install and commission domestic irrigation pumps

Task: Review 'Prepare for work' and answer the following questions.

1 Before contacting the pump manufacturer or supplier, the plumber must obtain certain facts about the pump installation so that the right pump is chosen. Name five pieces of information that may be required in selecting the right pump.

____

____

____

____

____

2 What is the function of the pump (performance) curve?

____

____

____

3 Name the three hazard types for cross connection.

____

____

____

4 On what basic principle does a centrifugal pump operate?

____

____

____

5 Name two major parts of a centrifugal pump.

____

____

6 Name two fuel types available to run pumps.

7 What is the basic principle of operation for a multi-stage pump?

8 When testing water services for soundness, plumbers use a test bucket. Name the type of pump that is used on the test bucket.

9 Explain the differences between a close-coupled pump and a long-coupled pump.

10 What type of fluids are rotary pumps designed to move?

WORKSHEET 4

To be completed by teachers	
Student competent	☐
Student not yet competent	☐

Student name: ______________________________

Enrolment year: ______________________________

Class code: ______________________________

Unit competency code/title: CPCPWT3025 Install water pumpsets and CPCPIG3022 Install and commission domestic irrigation pumps

Task: Review 'Install, test and commission water pumps' and 'Clean up' and answer the following questions.

1 Why is it necessary to use flexible connections and vibration eliminators when installing pumpsets?

____

____

____

2 Why should a non-return valve be installed on the discharge side of a pump?

____

____

____

3 Explain how the hydro-pneumatic pumping system works.

____

____

____

____

4 Why should a pump never be run dry?

____

____

5 What does the term 'cavitation' mean?

____

____

6 What does 'priming' a pump mean?

____

____

7 Name the two functions of a foot valve.

____

____

8 Create a material list for the valves and fittings required to connect the pump from the diagram below. Assume the pump has 25 mm F.I. connections.

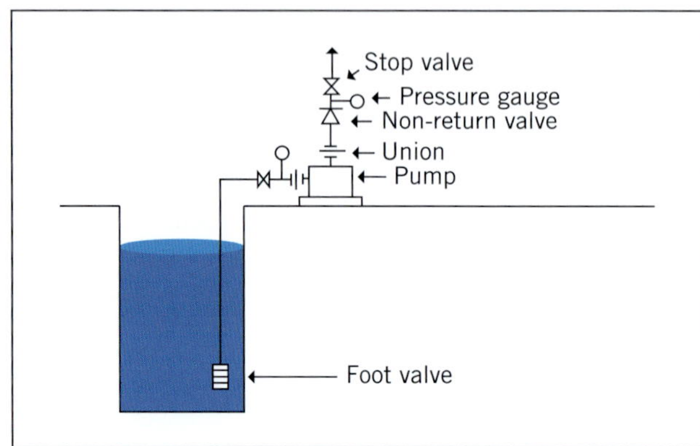

QUANTITY	MATERIALS

8

CONNECT AND INSTALL STORAGE TANKS TO A DOMESTIC WATER SUPPLY

Chapter overview

This chapter looks at the connection and installation of storage tanks to a reticulated water supply pipe system on new or existing sites.

The content of this chapter relates to the basic principles of the connection and installation of above-ground water storage tanks that are 'open' to the atmosphere. The water supply is connected to the tank via a physical air gap and is therefore subjected to atmospheric pressure, so it is known as a gravity feed supply system,

All water connections made to fixtures, tanks or service pipes must have the relevant authority's permission. Strict regulations apply, whether the fixtures are connected directly to the drinking water supply or indirectly through storage tanks or flushing devices.

Water from any fixture, storage tank or flushing device cannot flow back into the supply pipes. Cross connection must always be prevented and is to be achieved by providing a minimum air gap of at least 20 mm between the lowest level of the water inlet and the highest part of the overflow pipe or rim of the fixture or by fitting a backflow prevention device.

Learning objectives

Areas addressed in this chapter include:

- identify installation requirements
- prepare for work
- install and test storage tank
- clean up.

Identify installation requirements

Before commencing any work in relation to connecting or installing storage tanks, it is important to comply with the Plumbing Code of Australia (PCA), Australian Standards, plans, specifications, the relevant authority's requirements and the manufacturer's installation instructions.

Access codes and standards, plans and specifications

So that proper planning for the task can take place, the following steps are important:

- Refer to the National Construction Code (NCC), specifically Volume 3 – Plumbing Code of Australia (PCA) for compliance. The PCA can be freely accessed online.
- Refer to the Australian Standards – AS/NZS 4020 Testing of Products for Use in Contact with Drinking Water; AS/NZS 3500.0 Plumbing and Drainage – Glossary of Terms; AS/NZS 3500.1 Plumbing and Drainage, Part 1: Water Services for 'Deemed to Satisfy' compliance as stated in the PCA. The Australian Standards can be purchases online.
- Plans and specifications relating to the task must be obtained and carefully reviewed to ensure that the system complies with current rules and regulations. These are accessed from the client/builder.
- All tasks must be planned and sequenced in conjunction with others involved in or affected by the work.
- Sustainability principles and concepts must be applied to work preparation and application.
- Safety and workplace environmental requirements associated with connecting static storage tanks to a water supply system must be followed.

The information in this chapter will show the vast range of storage tank installations. It will explain some of the complexities related to the installation of storage tanks so there is a better understanding of what is seen on various work sites. It is also recommended to visit various websites relating to tanks and tank manufacturing to supplement the contents of this chapter.

Work health and safety (WHS) and environmental requirements

Legislation requires that work health and safety (WHS) requirements be observed and adhered to. At the minimum, a job safety analysis (JSA) must be completed on all jobs – identifying the hazards, assessing the risks and applying control measures to minimise the risks.

Working at heights, working in confined spaces and hot works are considered high-risk activities involved with the installation of storage tanks.

Manual lifting and handling techniques must always be followed according to WHS requirements. A correct size-up of the load to be lifted is important to determine whether a mechanical lifting device, a crane or a two-person lift is required. Try to avoid excessive manual handling and double handling of materials wherever possible. Do not place yourself or others in a dangerous position and ensure that all the necessary safety precautions have been taken.

Safe Work Australia requires a safe work method statement (SWMS) to be completed for any high-risk activity, as well as a JSA, before starting these activities.

Environmental requirements involve taking the appropriate measures to reduce excessive noise and dust when drilling, cutting and sawing different materials.

Workers must be appropriately trained and accredited to work in confined spaces. Working in a tank is a confined space, so the correct safety procedures are mandatory. Workers have died working inside tanks due to contaminated air and poor ventilation. Watch the video: 'Worker killed inside Braintree water tank'. WCVB Channel 5 Boston (1.27mins): https://www.youtube.com/watch?v=N3-2wiVu7Dw

Storage tanks

A storage tank is defined as a container or vessel open to the atmosphere at the top, connected to a drinking (potable) water supply and filled with water to be used for various purposes. It is hard to find a clear definition of what constitutes small- and large-capacity storage tanks, but in this chapter small-capacity storage tanks are deemed to be those up to 500 L and large-capacity tanks are those more than 500 L.

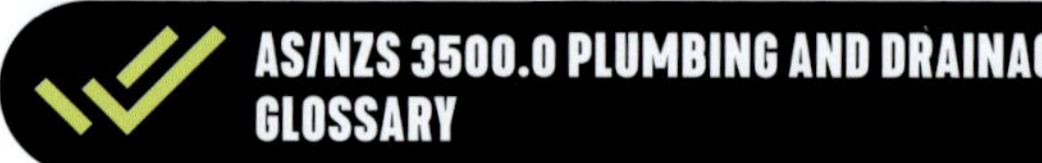

The following illustrate the range of storage tanks available:

- For a simple cistern for a water closet (see Figure 8.1), the plumber ensures that the cistern is installed properly, connects it to the water supply and checks that it operates correctly. With these installations the cistern is manufactured with an approved air gap and the water level is controlled by a float valve and an overflow, which is integral to the unit.

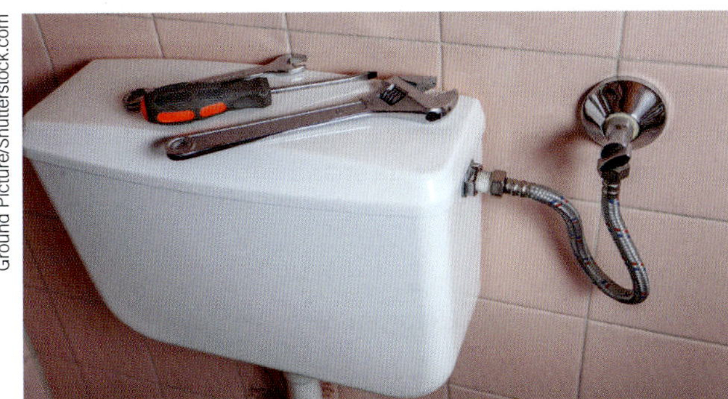
Ground Picture/Shutterstock.com

FIGURE 8.1 Water supply to a cistern

- A header tank, or make-up tank (see Figure 8.2) or similar, involves a considerable amount of planning to manufacture the tank, locate it onsite and carry out the installation. The tank is sized according to the specific site requirements. It is manufactured with the necessary connection points and placed in its nominated position, and then all connections are made to and from the tank.

FIGURE 8.2 Header tank or make-up tank, showing the outlet connection from the tank, the sludge valve, safe tray and safe waste, and tank support (note the bracket holding the tank and support material)

- A large-capacity water storage system is an extremely large tank or several tanks that may contain several thousand litres of water, required for a variety of reasons, on a building site (see Figure 8.3). The tank is sized according to the specific site requirements. Obviously, more detailed planning and approval must be carried out before installation can take place, as it requires input from other trades. This type of system may use tanks that are either prefabricated or built in situ. Some installations require a dedicated supply of water at all times, such as water storage for firefighting. In this situation, it is advisable to have more than one tank available for supply (see Figures 8.27 and 8.28 later in this chapter). This helps avoid major disruptions to supply or possibly unsafe situations due to breakdown or maintenance of the system.

Source: Shaun Kristen.

FIGURE 8.3 Large-capacity storage tank dedicated for a fire service

Storage tanks may be used to serve the following purposes:

- *To provide a supply of water that is physically disconnected from the mains supply* (e.g. see Figures 8.1, 8.2 and 8.3). This includes all items that have the potential to cause a cross connection and where a registered air gap (RAG) or registered break tank is required for containment, zone or individual backflow protection purposes. Other examples of services that may be coupled with these include, but are not limited to, sanitary flushing (flusherette supply), air-conditioning services, 'make-up' water, contingency reserve, ablutions and combined systems (e.g. rainwater harvesting, where rainwater storage is backed up by a mains pressure supply).
- *To store and distribute water in buildings at a level higher than that which the supply authority can provide through its water mains.* This is generally the case for multistorey buildings where pump sets would be required to supply water to a storage tank located at or near the top of the building.
- *To provide a large reserve of water (such as for fire services) where the potential demand exceeds the available mains supply.* Storage tanks are also required where required flow rates are insufficient or there is a lack of suitable water pressure available. Pumps are used in conjunction with these tanks.

GREEN TIP

It is common practice and compulsory for new houses (depending on local council regulations) to install rainwater tanks with mains water back-up to supply toilets, washing machines and garden taps to save drinking water.

Prepare for work

Licensed plumbers must consider many factors when selecting and installing a storage tank for specific purposes. The two main considerations are:

1. Basic design requirements for locating a storage tank, including:
   - tank materials
   - the shape, construction method and location of the tank.
2. Basic tank installation requirements, including:
   - water supply piping both to and from the tank
   - parts of the storage tank
   - support for the storage tank.

Other factors may also have to be considered, but these are not discussed here as they relate more to design than installation. An example would be whether wind and earthquake design considerations are required to enable water tanks to survive seismic and high-wind activity.

An important consideration is that the work area should be prepared to allow for an accessible connection and installation. Again, this is extremely important, whether the site is a domestic residence or a larger building site, as storage tanks are very bulky, which can cause difficult installation access.

Create a material list and collect materials

Quantities of required materials should be calculated from plans and specifications. Materials and equipment should be ordered and collected according to workplace procedures. The size of the task will determine when the items are ordered and collected, and whether there is a suitable secure location to store them.

Tank manufacturers can provide tanks in all shapes and sizes and generally have a range of tanks from which to select to suit a specific installation. In most cases, tanks can be ordered directly from the manufacturer, but in some cases they need to be ordered through the local plumbing supply company. Tanks can usually be delivered directly to the site as and when required.

When placing a special order for a tank (that is, one designed to suit a particular installation), it is essential to provide specific details relating to the tank's connection points. Most times, a sketch indicating any specific requirements will be helpful for the supplier.

Materials and equipment should be checked for compliance with relevant Australian Standards, docket and order forms, and for acceptable condition. If any items do not comply with the above, they should be noted and returned to the supplier for replacement.

Tank materials

The tank material chosen should comply with plans, specifications and current Australian Standards, such as AS/NZS 4020 Testing of Products for Use in Contact with Drinking Water. Materials may include, but are not limited to:

- cast iron (sectional)
- concrete (cast in situ or precast)
- copper (both in sheet form and corrugated)
- fibreglass
- plastic
- stainless steel
- steel (galvanised, Colorbond® or sectional).

Shape and construction method

Storage tanks come in a variety of shapes and sizes, and can be custom-made to suit a particular size and shape. The main shapes used are either circular or rectangular in shape. The shape chosen generally complements the tank's location with the available space during construction and after completion, as space is a valuable resource on any premises.

FROM EXPERIENCE

Tanks can be custom-made to suit specific spaces. It is important for the plumber to measure the dimensions accurately and allow enough room for access to maintain the tank and connect the pipework.

Tanks must be fitted with a close-fitting cover to prevent the entry of dust, birds and animals. Figure 8.4 indicates the various parts of a storage tank installation, including the cover.

Circular tanks

Circular tanks are available in a range of approved materials. The size of the tank may influence the material chosen and the choice of construction method. The advantage of round tanks is that there are fewer corners and crevices for mould and mildew to form in. Also, they are easier to transport. The main disadvantage is that round tanks use more space. See Figure 8.5, which indicates a tank with a close-fitting (removable) cover. Other points to note in this photograph are:

- the water supply pipe to the tank
- the supply pipe from the tank
- that each pipe is connected with a demountable joint (threaded joint and union) to enable easy removal if required
- the timber support for the tank
- the safe tray.

Smaller tanks are generally made of lighter materials; for example, thin sheet metals (such as copper and Colorbond®), fibreglass and UV-resistant plastics.

In larger applications, there is greater demand for site space, and along with this increase in size come heavier construction materials. Larger tanks quite often incorporate a different range of materials from those used for smaller tanks. They can be manufactured

Demountable joint (union coupling)
Close-fitting cover
Float
Air gap
Inlet orifice
Water inlet valve with spindle in vertical position
Overflow pipe (min 40 mm)
Overflow pipe 40 mm (alternative connection)
Signage indicating contents of tank
Contents
Water outlet isolating valve (compulsory if tank exceeds 50 L capacity)
75 mm min
12 mm min gap
50 mm min
Gate valve
Minimum gap 25 mm for air circulation
Safe tray
Additional isolating valve (when tank is in roof space)
Safe waste (min 50 mm)
Water inlet

Source: *Basic Training Manual 11-1 Water Supply 1*, Australian Government Publishing Service from July 1970 to 1997.

FIGURE 8.4 Parts of a storage tank installation

FIGURE 8.5 A corrugated copper tank with a close-fitting cover

off-site using materials mentioned in the tank materials list, and then delivered to the site. They can also be assembled onsite in sectional panels or cast in situ using reinforced concrete for the floor and walls and sheet metal for the roof.

Rectangular tanks

Rectangular tanks are available in similar materials to those mentioned earlier, such as concrete, sectional steel or cast iron, and have similar influences on material type.

Figure 8.6 shows a bolted together, sectional tank construction. Such a design allows for variations in shape and size if site conditions alter, although standard four-sided shapes are the preferred economical choice. Sectional tanks are obviously made up in sections – the more common sizes generally being 600 mm × 600 mm square and 1000 mm × 1000 mm square – which are bolted together onsite. One of the main advantages of sectional tanks is that they can be put together within a confined area; for example, a tank built in situ in a plant room or car park area. They can be of the internally or externally bolted type of construction, depending on specific needs.

Source: Shaun Kristen.

FIGURE 8.6 Rectangular (sectional construction) storage tank; note the bollards located near the corners to help provide protection

Select and check serviceability of tools and equipment

The tools required for the installation of static water tanks will depend on the type of tank being installed and the circumstances relating to the installation. Typical tools would include levelling equipment when setting up the base of the tank platform and shifting spanners or Stillson's to make pipe connections to the tank.

Tools and equipment used to install static water tanks should be maintained in accordance with the manufacturer's instructions.

There are specialised tools and equipment necessary for different jointing systems when installing storage tanks with the associated pipework, valves and fittings.

Crimp and press-fit tools used on copper and plastic pipe and fittings need calibrating regularly to ensure the tool is pressing or crimping to the correct tolerances, so the joint is sound. Depending on the tool, recalibration should take place after 50 000 cycles. Some tools have a warning light that alerts when recalibration is due.

Electro-fusion welding machines used for jointing polyethylene pipe also need to be regularly calibrated to ensure they are heating to the required tolerances.

All electrical tools and leads must be tagged for safety regularly to the local authorities' rules. The general requirement for building and construction is to inspect and tag electrical equipment every three months because of the harsh environment the electrical equipment is exposed to.

As with all plumbing hand and power tools, personal protective equipment (PPE) and other associated equipment, it is essential that the correct tool is selected for the task and checked to ensure it is working in accordance with the manufacturer's instructions and used correctly.

Any damage that is found during inspection should be reported to the supervisor immediately and the tool/equipment needs to be tagged out and removed from circulation until the issue has been resolved.

Personal protective equipment

The risk of injury can be reduced by the appropriate use of PPE. The types of PPE include:

- *Safety boots* have a steelcapped toe to protect your feet from falling objects, and a hard-wearing sole that sharp objects cannot pierce.
- *Safety glasses or goggles* will protect your eyes from injury when drilling, cutting or sawing with hand or power tools.
- *Ear plugs or earmuffs* will protect your hearing when using power tools and working on noisy work sites.
- *Dust masks or respirators* will protect your respiratory system from breathing in dust and fine particles when drilling, cutting or sawing and working in a dusty environment.
- *Gloves* will protect your hands from cuts, abrasions and burns when working with different materials, tools and equipment.
- *Overalls* will cover your skin and protect you from exposure and different conditions.
- *High-visibility clothing* allows workers onsite to easily identify where others are and helps to avoid accidents.

Tank location

The location of a storage tank influences the choice of material that is most appropriate for the design and construction of the tank. Factors affecting this may include atmospheric conditions, surrounding equipment or storage, and pedestrian or vehicular traffic. The tank shape is an important consideration, as is the access available to get the tank into position.

Current regulations state that storage tanks must be constructed and installed entirely above ground level, *unless* permitted otherwise by the relevant authority.

The location of any tank will be dictated by the site conditions and the required usage. As stated above, the manufacture and location of the tank depend upon the tank's environment – that is, indoors, outdoors, above ground or below ground (considering the temperature of the area where water will be stored with regard to freezing). Suitable access to the tank and its controls must also be considered (see Figure 8.7). If a tank is located internally, sufficient headroom above the tank and space around the tank should be provided to allow access for maintenance. Another point to consider is whether a pressurised (pumped) supply is required to deliver water to the tank and therefore what type of controls are necessary for the installation.

FIGURE 8.7 A concrete storage tank located in a plant room at a hospital; note where the water supply enters (top left-hand side), the water level indicator (clear plastic tubing on left), the panel to gain access to the inside of the tank (with a danger sign indicating a confined space) and the warning sign indicating the contents of the tank and that no chemicals are to be added

Tanks located externally may have some restrictions placed upon them. Some of these restrictions include, but are not limited to:

- a minimum distance between the tank and surrounding buildings or structures; for example, fire brigade requirements
- minimum clearances from heavy vehicular traffic to avoid possible damage to the tank
- tanks that store drinking water cannot be located directly beneath any sanitary plumbing or pipes carrying non-drinking water.

Note: These restrictions are normally mentioned in the specifications for the project.

Storage tanks can be placed in a variety of places around a building or property. In a domestic residence, they may be placed in a ceiling space, on a roof, on the outside ground or concrete, or on the floor within a building.

On larger building sites, tanks may be placed in plant rooms, on rooftops, in car park areas or in other approved locations.

Remember to carefully consider any manual handling issues. Excessive lifting can cause permanent spinal damage.

Plant rooms

The location of a plant room will govern the source of supply and connection to the tank/s. The plant room may be located at a level where the tank/s can be directly fed from the authority's water main and reticulated supply and so may not require pumps for the inlet supply, or it may be located at a high level with a pumped water supply as part of the reticulation system and therefore use large-capacity storage tank/s to provide water for the rest of the building.

A plant room may contain small-capacity and/or large-capacity storage tanks.

- Small-capacity storage tanks (see **Figures 8.5**, **8.8** and **8.9**) may serve a variety of purposes, including heated water, sanitary flushing, air-conditioning services, ablutions and make-up water.
  - If the water supply to the plant room is fed directly from a water main and reticulated supply, then each individual tank may be connected directly to that system in accordance with the relevant authority's requirements.
  - If the water supply is from a larger storage tank, then a separate pump system, located within the plant room, is normally installed to provide water to these points. The pump system used for this purpose is normally a hydro-pneumatic system (see **Figure 8.9**).
- Large-capacity storage tanks located in a plant room within a high-rise building often will have a greater need for a dual water supply. These tanks generally consist of separate units, but may be a single tank partitioned to contain both the domestic and fire supply (hydrants and hose reels) in the one tank. *Note:* Tanks supplying fire sprinkler systems generally require a separate tank.
  - Tanks located in these areas are generally square or rectangular in shape due to the limited space available. Materials used may vary (such as concrete construction; see **Figure 8.7**), but they are normally made from either cast iron or steel and are generally of a sectional construction.
  - These larger tanks are normally built on engineered elevated platforms, helping to provide adequate clearance and strength in supporting their weight (see the section 'Support for storage tanks' later in the chapter).

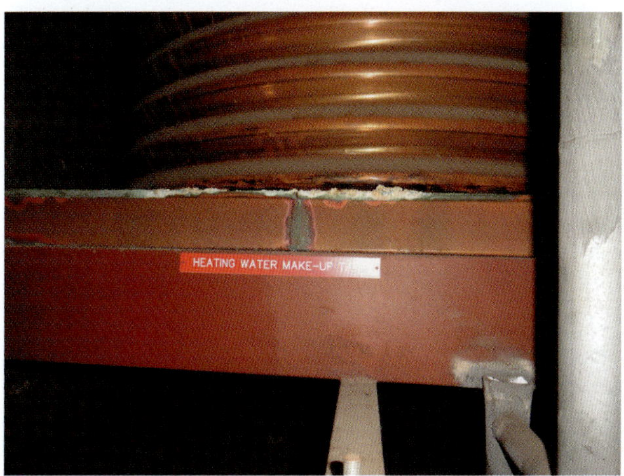

FIGURE 8.8 A small-capacity tank with signage indicating that the tank is used for heating water make-up supply

FIGURE 8.9 Storage tanks and a hydro-pneumatic system located in a plant room

Rooftops and car park areas

Tanks located on rooftops or in car park areas are designed similarly to those located in plant rooms and are manufactured from the wide range of approved materials. Both small- and large-capacity storage tanks can be placed in these locations. Tanks placed within car park areas may require additional protection, such as bollards placed at the corners to prevent vehicular damage (see Figure 8.6); consideration should also be given to eliminating vandalism and mechanical damage (see Figure 8.10).

FIGURE 8.10 This tank installed within a car park is fenced within a recessed area, providing protection from vehicular and pedestrian traffic and vandalism, and displaying efficient use of a limited space

Suitable access must be provided to any tank in any location, and some may require a ladder, trafficable walkways or provisions to eliminate fall hazards (see Figures 8.27, 8.28 and 8.30 later in the chapter).

LEARNING TASK 8.1

1 What is a bollard?
2 Explain the purpose of a sectional tank.
3 What is the purpose of a tight-fitting cover on a water tank?
4 Name four materials that tanks are made from.
5 List two hazards involved in working inside a tank.
6 At what height in a tank is the water capacity of a tank calculated?

 **COMPLETE WORKSHEET 1**

Install and test storage tank

It is important that plumbers have the skills and knowledge required to install, connect piping and test a storage tank. Selecting the correct approved valves and fittings ensures a professional job. Having a water tank installation fail inside a building could cause serious flooding and costly damage.

Install storage tank components and associated pipework

Connections to the inlet and outlet of a storage tank are much the same regardless of the tank's size. However, the type and size of piping materials, tank construction materials and installation procedures for the tank may vary considerably depending on the requirements of the client, various authorities, manufacturers and suppliers.

The materials used for water supply systems to static storage tanks throughout Australia will vary depending on the area. The plans and specifications generally identify the materials chosen to supply water to the storage tank and state any specific requirements for the installation. Although the materials are selected from those listed in the Standards, they must also comply with the local authority's codes and regulations. These details would be dealt with at the development application stage. It is essential that an approved air gap is provided on the inlet supply. This air gap may have to be calculated, recorded and checked on a regular basis once the tank has been installed (for example, RAG or registered break tank). AS/NZS 3500.1 explains how the air gap is calculated.

The connections to and from the tank will use either union couplings or flanges adjacent to the tank itself to allow for easy disconnection.

All piping materials must be adequately supported according to current regulations.

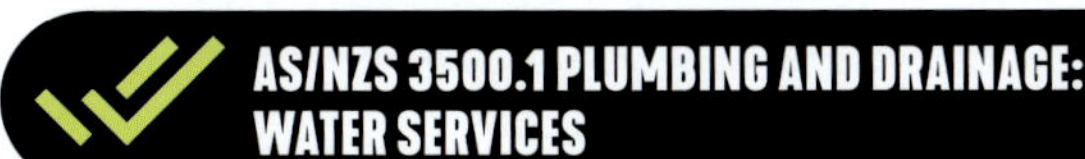

Water supply piping materials

The most common piping materials used for water supply to and from storage tanks are plastics and copper. Other materials – such as ductile iron, cast iron–cement lined or galvanised mild steel – may be used depending on the purpose of the tank and its specific water supply source.

Plastics must be installed and joined according to the manufacturer's recommendations, and may include:

- acrylonitrile-butadiene-styrene (ABS)
- cross-linked polyethylene (PE-X)
- polybutylene (PB)
- polyethylene (PE)
- polypropylene (PP-R)
- unplasticised polyvinyl chloride (PVC-U).

Copper must also be installed in approved locations and according to the manufacturer's recommendations. Various methods may be used for jointing copper pipe, including:

- silver solder
- push-fit
- press fit
- flanged
- roll-grooved.

FROM EXPERIENCE

It is important for the plumber to have the knowledge and skills to choose the correct material and jointing method for a specific application.

Valves and ancillary items

There is a variety of valves and ancillary items that may be required for tank installation, depending on the size, location and purpose of the tank itself. Different valves may be used for the inlet supply, outlet supply and other parts of the tank (for example, sludge valves).

Inlet supply to a water storage tank

The more common valves for the inlet supply are approved stop taps/valves; for example, loose jumper valves (see Figure 8.11), gate valves (see Figures 8.12 and 8.13), float valves (see Figure 8.14) and ball valves (see Figure 8.15). *Note:* It is recommended that the control valves be installed with the valve stem in a vertical position.

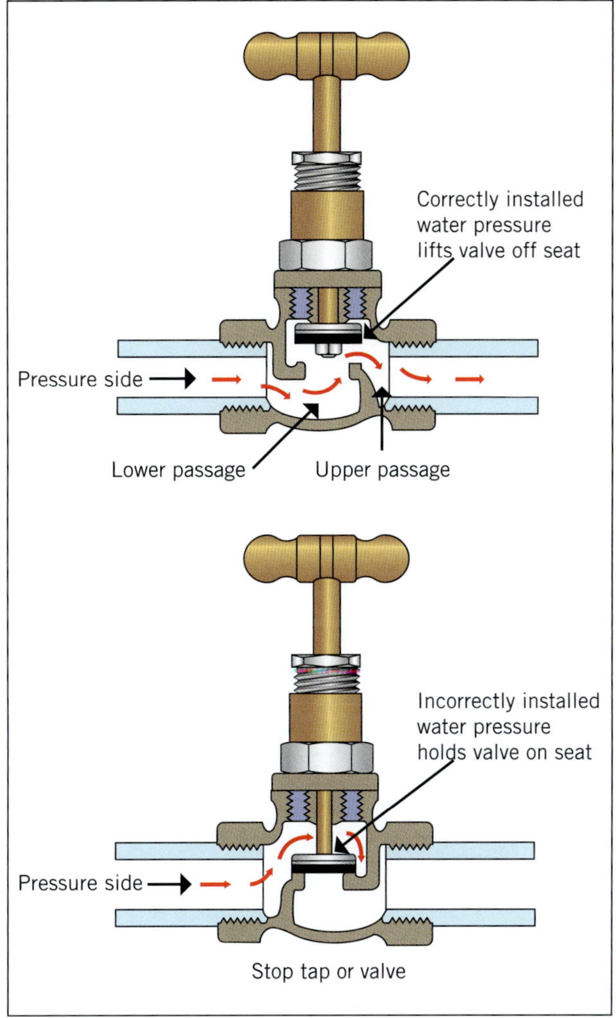

Source: *Basic Training Manual 11-2 Water supply 1*, Australian Government Publishing Service from July 1970 to 1997.

FIGURE 8.11 Stop tap with a loose valve (jumper valve)

- *Loose valves (jumper valves), control valves.* These are designed with a loose valve (jumper valve) inside and fitted loosely to the spindle, thus making them a one-way flow valve. (Check the direction of flow carefully when installing this kind of valve, as no flow will occur if these are installed back to front.)
- *Gate valves.* Designed as a full flow valve, they open and close by having a sliding gate that operates laterally inside the valve.
- *Float valves.* The level of water within the tank is normally controlled by an approved float valve or other flow control device. The float valve dictates the operational level of the water tank. To check the operational level is correct and aligns with the job plans and specifications, the float valve may need to be adjusted to either reduce or increase the operational water level of the tank. The size of the inlet float valve affects the size of the overflow outlet (refer to AS/NZS 3500.1). For some installations (such as a pumped supply) the flow

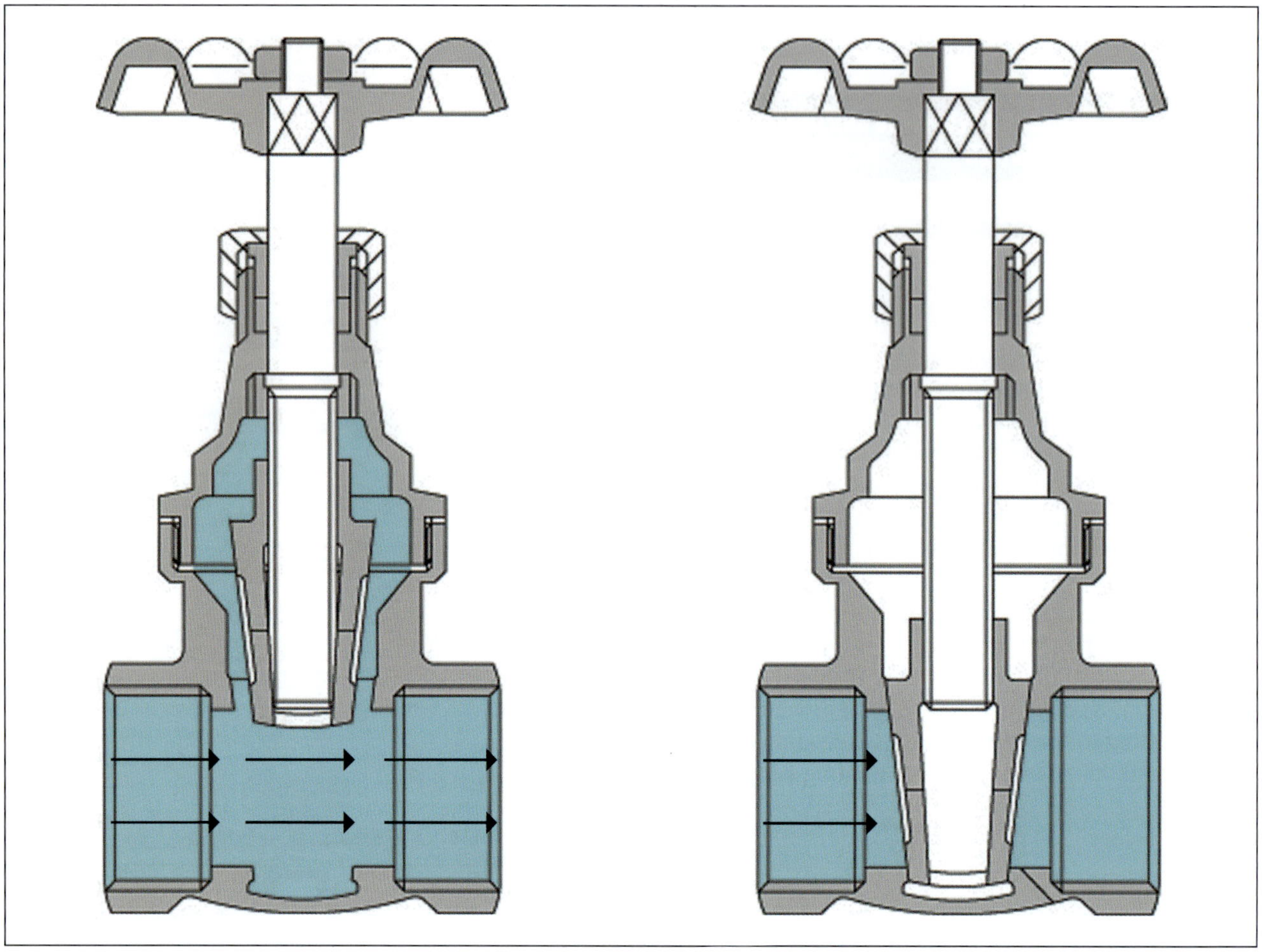

Source: Zetco Valves Pty Ltd.

FIGURE 8.12 A sectional view showing a gate valve in both the open and closed position (refer to Figures 8.13, 8.16 and 8.18, which show a gate valve installed)

control may be regulated by a float control device, or devices, coupled with a control panel and pumps.

- *Ball valves* are the most used and part of the rotary movement family of valves, which are often called quarter-turn valves and have a full flow function.

AS/NZS 3500.1 PLUMBING AND DRAINAGE: WATER SERVICES

Outlet supply from a water storage tank

According to current regulations, any storage tank with a capacity of more than 50 L requires an isolating valve on the outlet, and the outlet connection is to be situated at least one pipe diameter (that is, of the outlet pipe itself) from the bottom of the tank.

The actual connections to the tank outlet are normally full-way flow valves such as gate valves or ball valves (see Figures 8.12 and 8.15). These are used so as not to restrict the flow of water from the tank. These valves may be used as sludge valves where the tank capacity is more than 500 L, and in these cases it is advisable to plug or cap the open end of the valve to help prevent any wastage of water.

Source: Shaun Kristen.

FIGURE 8.13 A gate valve used on a fire service (notice the '003' padlock for authorised access only)

Source: Terry Bradshaw.

FIGURE 8.14 A float valve inside a tank

Source: Zetco Valves Pty Ltd.

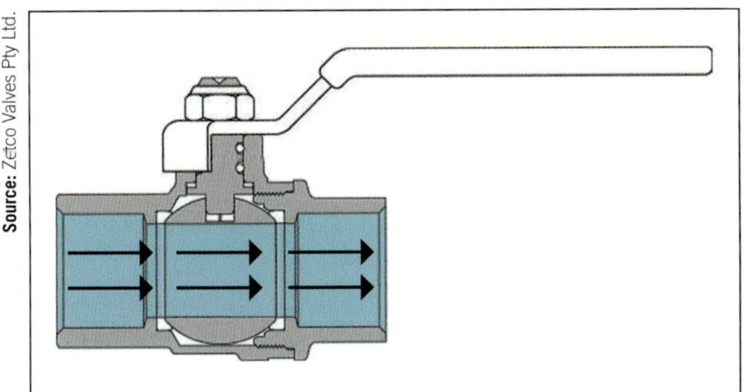

FIGURE 8.15 A sectional view showing a typical ball valve in the open position

In cases such as fire service supply, the valve may have to be locked in the open position to ensure that supply is always maintained. Figure 8.16 shows a fire service with an outlet using a 150 mm gate valve strapped in the open position, with an anti-tamper device fitted to the isolating valve.

The tank shown in Figure 8.17 appears to have a capacity greater than 50 L. The photo shows that it does not comply with current regulations because it does not have an isolating valve on the outlet. It would have to be completely drained to allow any work to take place on the outlet side, causing a large amount of water being wasted.

Source: Shaun Kristen.

FIGURE 8.16 Outlet from a fire service tank

FIGURE 8.17 The outlet pipe from a storage tank without an outlet valve

Ancillary items

Several ancillary items are used on the actual connection of the water service to the tank. They may include, but are not limited to, check valves (non-return valves), union couplings and flange connections.

In some situations, an additional valve such as a check valve (non-return valve) may be required downstream of the outlet valve (see Figures 8.18 and 8.19). These are designed to prevent the reversal of flow from the downstream section of pipe. Depending on the design of the non-return valve, the position of its installation may affect its performance. Always check the manufacturer's recommendations for installation requirements.

Regulations state that the inlet supply and outlet connection to a water storage tank must have an allowance for disconnection. This can be in the form of a union coupling or a flange, and is dependent on the size of the service connecting to the tank.

Source: Terry Bradshaw.

FIGURE 8.18 Threaded union coupling connection, gate valve and check valve on the outlet of a tank

Source: Zetco Valves Pty Ltd.

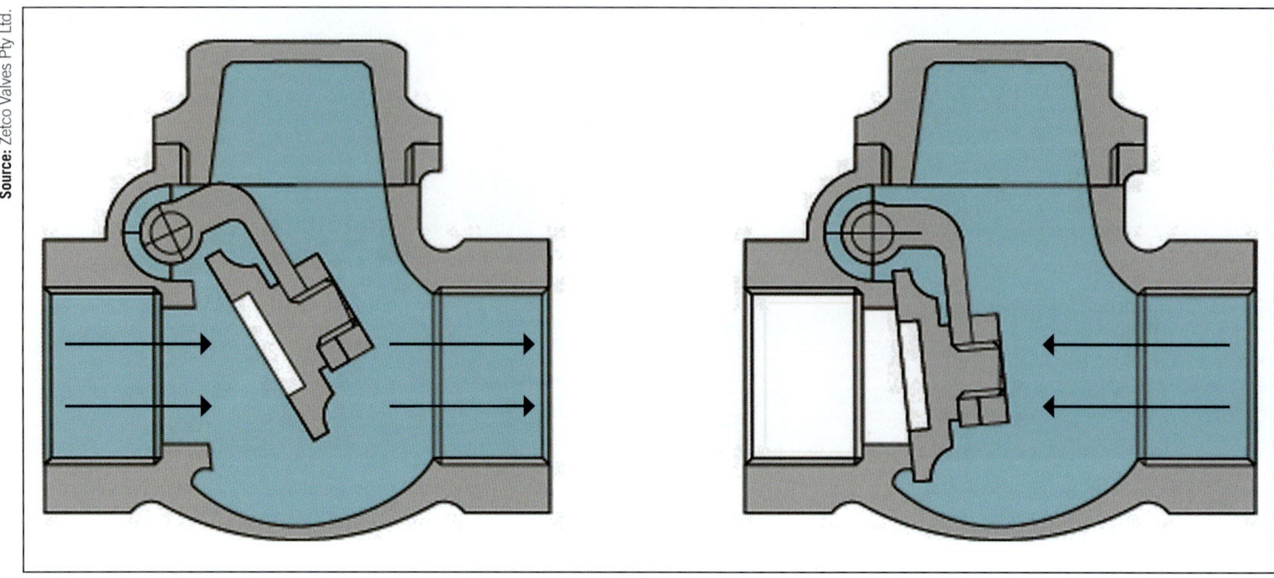

FIGURE 8.19 A sectional view showing a swing check valve in both the open and closed position – these must be installed in a horizontal and upright position (refer to Figure 8.18)

- *Union couplings* are defined as a demountable connection between two pipes, or fittings, that can be separated without damaging the pipework or components for reassembly. Union couplings are normally used on pipe sizes up to and including 50 mm. Unions also have the advantage of allowing some minor deflection between the hard piping and the tank connection, which can assist during installation.
- *Flanges* are also a demountable joint but have much less flexibility than a union coupling. Flanges are normally used on pipe sizes greater than 50 mm. The type of flange will vary but must always match the valve/tank connection to which it will be joined. Appropriate gaskets are also used to ensure that the joint is watertight. It is important when installing flange connections to ensure that the alignment of the faces of the flanges and their bolt holes are accurate, as there is no margin for error if flanges are fixed.

Other examples of ancillary items that may be used (but are not discussed here) include:

- alarm valves
- butterfly valves
- flow switches
- pressure-limiting valves
- pressure switches
- strainers.

Parts of the storage tank

Plumbers need to understand several key parts of the storage tank.

The overflow pipe

The overflow pipe should be designed according to current regulations such as AS/NZS 3500.1 and the specification (if provided). The minimum size for any storage tank overflow is DN 40. Sizing the overflow outlet is dependent on the outflow rate (determined by the inflow rate in l/s), spill level (mm) and overflow type (horizontal, vertical or weir) based on AS/NZS 3500.1.

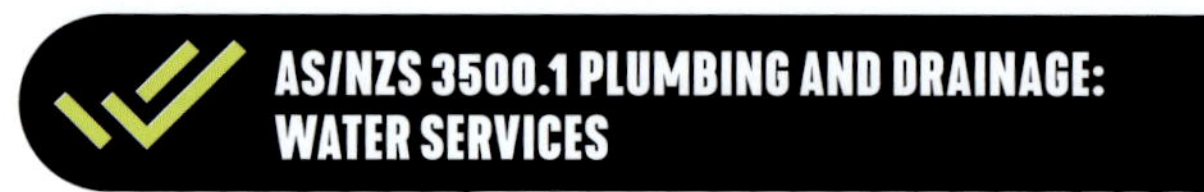

Materials used for the overflow must match the specification requirements and be of a type that conforms to current regulations. The overflow pipe will normally discharge into the safe tray, but on occasions may discharge onto the floor or other approved area.

Figures 8.20 and 8.21 show an overflow pipe on a storage tank located externally. This can drain to the stormwater drainage system (if approved).

Safe tray and safe waste

The safe tray and safe waste must be designed according to current regulations such as AS/NZS 3500.1 and the specification (if provided).

The safe tray is a watertight tray that is fitted under a storage tank, where required; it drains through an outlet to a safe waste (see Figures 8.2 and 8.4). The safe tray should be constructed from approved materials, generally sheet metal, with specific measurements that must comply with current regulations. The minimum height of the sides of the tray is 50 mm, and the minimum clearance between the side of the storage tank and the safe tray is 75 mm. Timber supports should be placed both beneath and inside the safe tray; the storage tank is placed on these supports. The main purpose of these supports is to help to evenly distribute

FIGURE 8.20 Typical overflow connection to a storage tank

FIGURE 8.21 Dual storage tank overflow connection

the weight of the tank and its contents and to allow air movement beneath the base of the tank.

The safe waste must be of an internal diameter greater than that of the tank overflow but not less than a diameter of 50 mm. It must be constructed from approved materials in accordance with current regulations. The safe waste must discharge to a point where it is noticeable but does not cause any inconvenience.

Sludge valve

The sludge valve is primarily designed to allow the tank to be cleared of silt/sediment build-up that has settled out of the water and deposited on the bottom of the tank. Sludge valves are located at the bottom of the tank with a minimum size of half the outlet diameter or DN 40 minimum. They must be fitted to any tank that exceeds 500 L in capacity. As stated above, it is advisable to plug or cap off the open end of the valve to help prevent any wastage of water. An example of a sludge valve is shown in Figure 8.22.

Approved cover

The cover should enable easy access to the various controls for the tank (float valves, level controls, etc.) and be either removable (see Figure 9.5) or of an approved design and fixed in position (such as approved roofing material or a specially constructed cover that is compatible with the tank). In the case of larger-capacity tanks, approved access must be allowed both onto and into the tank itself for maintenance purposes. The removable access cover must be a minimum of 0.5 m^2 in size as per AS/NZS 3500.1. Refer to Figure 8.23 and the section 'Access to all parts of the tank' later in the chapter.

Support for storage tanks

The structural supports for water storage tanks must be designed according to current regulations such as AS/NZS 3500.1 and the specification (if provided).

When support for a tank is required, the material used is generally determined by the location, size and volume of the tank. As you can imagine, a smaller tank for residential and domestic supply located externally to the building may require only a concrete pad to support the base of the tank. Larger tanks may require additional support, such as the installation of a concrete or timber plinth to support the tank (see Figures 8.24 and 8.25). These plinths may have a dual application as they also allow ventilation under the tank, thus reducing the likelihood of corrosion and thereby increasing the tank's longevity.

FIGURE 8.22 Sludge valve and drain

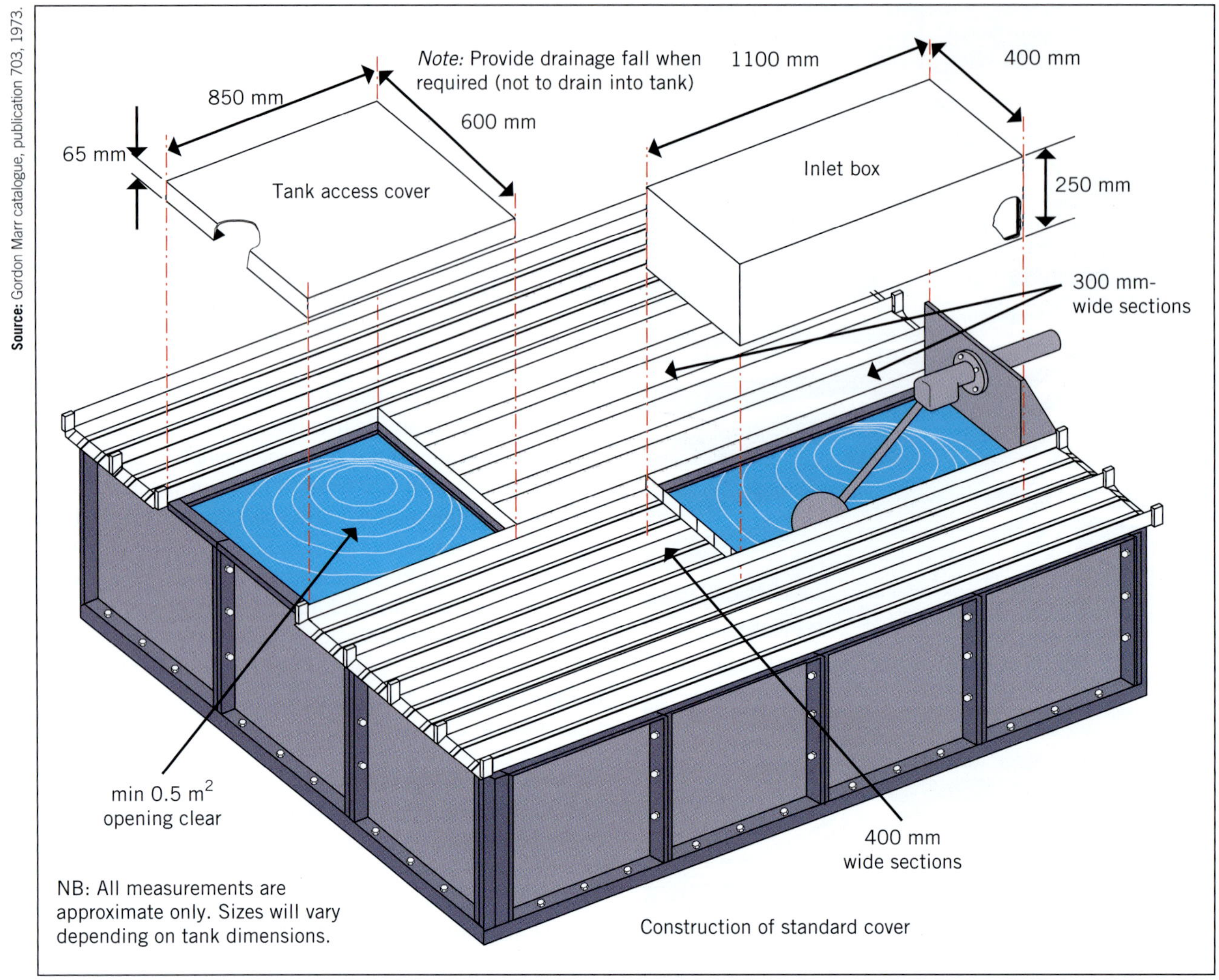
Source: Gordon Marr catalogue, publication 703, 1973.
Note: Provide drainage fall when required (not to drain into tank)
850 mm
600 mm
65 mm
Tank access cover
1100 mm
400 mm
Inlet box
250 mm
300 mm-wide sections
min 0.5 m²
opening clear
400 mm
wide sections
NB: All measurements are approximate only. Sizes will vary depending on tank dimensions.
Construction of standard cover

FIGURE 8.23 Allowance for access to a large-capacity storage tank

Source: Gordon Marr catalogue, publication 703, 1973.
RSJs on 610 mm centres
(internal dimensions of plates)
Outside bolted floor plates
Tar/cork strips
500 mm
min
300 mm
min
Mild steel flats
Intermediate wall(s)
(if required)

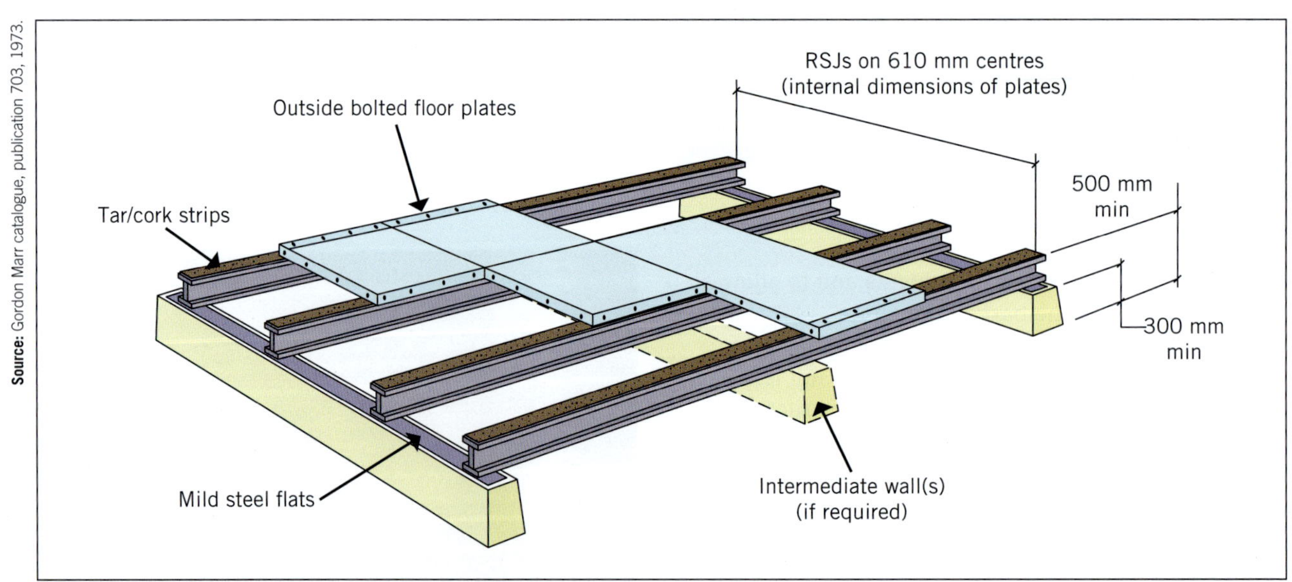

FIGURE 8.24 The type of support that may be required for a large-capacity storage tank

Tanks placed in roof spaces or in elevated locations may require the design input of an engineer to enable additional support structures to be put in place. The main thing to consider, regardless of where the tank is located (internally, externally, on the ground or elevated), is the mass of the tank, its contents and any material used to support the tank.

A structural engineer can calculate and make provision for the structural requirements for large-capacity tanks. The engineer must take into consideration the mass of the tank material itself plus the support material – for example, a cast iron tank with reinforced concrete supports – and then design the structure to support the entire installation.

EXAMPLE 8.1

HOW TO CALCULATE TANK CAPACITY AND MASS

A water storage tank is 10 m long, 6 m wide and 4 m high. Determine the volume of the tank in cubic metres (m^3) and litres (L), and the mass in tonnes (T).

Note: 1 L of water has a mass of 1 kg and 1 m^3 contains 1000 L, so 1000 L has a mass of 1000 kg or 1 T.

Volume in cubic metres

We can calculate the volume in cubic metres by using the formula:

$$V = L \times W \times H$$

where:

V = volume in cubic metres

L = length in metres (10)

W = width in metres (6)

H = height in metres (4)

$$V = 10 \times 6 \times 4$$
$$= 240\ m^3$$

Volume in litres

We know that there are 1000 L in a cubic metre and that this tank contains 240 m^3 of water.

Thus, multiplying 240 (m^3) by 1000 will give the capacity in litres (and also in kilograms).

$$\text{Volume in litres} = 240\ (m^3) \times 1000$$
$$= 240\ 000\ L$$

Mass in tonnes

Mass is 240 000 kg; that is, 240 T.

Understanding the above method to determine a tank's water capacity and total mass is important to plumbers in everyday work activities related to the installation of tanks. For example, if a tank is to be situated in a ceiling space in a domestic residence, careful consideration should be given to the location of that tank. It is essential to locate the tank over corners of the wall below, and preferably spanning two or three walls, such as near a bathroom or hallway area.

Source: Tank Industries, A Division of Hunt Engineering & Staff Pty Ltd.

FIGURE 8.25 This sectional tank has been placed on a roof with additional lateral support being provided to the timber slats, giving the system the ventilation and corrosion control it needs

Personal injury and structural building damage could occur if a tank was installed in the ceiling space in the middle of a room without adequate support.

Other points to be considered regarding tank installation

There are a few other points to be considered regarding tank installation, including:

- access to all parts of the tanks
- internal corrosion protection
- use of an approved membrane
- sustainable identification markings.

Access to all parts of the tank

As mentioned earlier, it is important to have suitable, approved access to the tank for maintenance purposes. An access ladder is required if the internal tank depth is 2 m or more. Any access ladder should be designed to specifications (as per AS 1657), incorporating safe work practices at heights and allowing for access to both the tank cover and access opening for periodical maintenance and cleaning purposes. All access openings must have signage adjacent to the opening stating 'Warning – confined space'.

Figures 8.26 and 8.27 illustrate various forms of access.

Additional 'working at heights' protection may be required to minimise hazards related to working at heights.

Internal corrosion protection

Some storage tanks are required to have internal corrosion protection. One of the reasons for accessing some larger tanks is to check and possibly resurface

Source: Shaun Kristen.

FIGURE 8.26 Trafficable access provided to a tank's maintenance cover

FIGURE 8.27 Storage tanks with approved ladder access

the inside of the tank to help prevent corrosion. It is therefore important to have suitable access to the tank and for the personnel working on this process to have current confined-space qualifications.

Use of an approved membrane

In some situations, regulations and/or specifications may state that a membrane of non-corrosive insulating material may be required for all metallic tanks or other tanks as directed. It is advisable to check all documentation to determine what is required.

Suitable identification marking

In some situations, regulations and/or specifications may state that piping materials, valves, specific tanks or parts of the tank installation may require identification. It is advisable to check all documentation to determine what is required.

An example of a water storage tank installation

Currently, it is very common for the installation of a storage tank system for rainwater harvesting where the stored roof water is used for activities such as:

- flushing toilets
- topping up a swimming pool
- washing clothes
- watering the garden
- washing cars.

When this type of system is used, the water storage can be supplemented by topping up the tank water from the reticulated system, and therefore an approved backflow prevention device must be installed on that inlet supply to prevent cross connection. The water from the tank system is distributed by means of a pump unit to the various points of usage. Refer to Figures 8.28 and 8.29.

COMPLETE WORKSHEET 2

LEARNING TASK 8.2

1. What is the minimum size of a removable access cover?
2. Name three uses of storage tank water.
3. What does 1 L of water weigh?
4. When must a sludge valve be fitted?
5. What type of isolation valve must be installed on the tank outlet?

Testing and commissioning the storage tank

The process for testing a storage tank as per AS/NZS 3500.1 is the tank must be filled with water until it overflows for a period of at least one minute. This is to ensure the overflow is operating correctly and the air gap is maintained as calculated.

FIGURE 8.28 Multiple rainwater storage tanks, a service from the drinking (potable) supply – complete with a backflow prevention device and its drain – the pumping unit and the supply from the tanks to the points of usage

FIGURE 8.29 The roof water drainage pipe to the tanks, the drinking (potable) water supply to the tank system and the outlet supply to the points of usage; note the approved identification stickers indicating flow

The installed system should be pressure tested and commissioned in accordance with the relevant Australian Standards and job specifications. All tanks should be tested for soundness prior to, and after, installation so as not to cause any inconvenience to the user once installed.

Test data should be recorded in the format required by the job specifications and quality assurance procedures.

Cleaning and maintaining the storage tank

Storage tanks must be cleaned and disinfected before use.

HOW TO

CLEAN AND MAINTAIN THE STORAGE TANK

Storage water tanks must be cleaned and disinfected before initial use and when the tank is being repaired or serviced.

This involves draining the tank so all the sludge can be removed and all the internal surfaces can be cleaned with a high-pressure water jet, scrubbed and swept.

After the cleaning is complete, the tank is disinfected with a chlorine solution. There are two methods that can be used, and they have retention times that must be adhered to. Refer to AS/NZS 3500.1 for this information.

AS/NZS 3500.1 PLUMBING AND DRAINAGE: WATER SERVICES

Clean up

Upon completion, the work area must be left clean and tidy. Any material left over should be stored and reused on another job or recycled, which will reduce waste.

Tools and equipment need to be cleaned before storing, and kept in good working order. Any tools or equipment on hire should be returned in good condition.

All relevant documentation must be submitted to the regulatory authorities, principal contractor (builder) and client (owner).

Figure 8.30 shows a tank installation site with the clean-up completed.

Source: Picture courtesy of Altanks Pty Ltd.

FIGURE 8.30 A tank installation site with the clean-up completed

LEARNING TASK 8.3

1. What needs checking when the tank is under test?
2. What chemical is used to disinfect a water storage tank?
3. What are some main considerations when installing a water storage tank?

COMPLETE WORKSHEET 3

SUMMARY

- Water storage tanks have three main purposes:
  1. to provide a water supply that is physically disconnected from the mains supply to prevent cross connection
  2. to store and distribute water in buildings at a level higher than that which the supply authority can provide through its water mains (high-rise buildings)
  3. to provide a large reserve of water (such as for fire services) where the potential demand exceeds the available mains supply.
- The advantage of round tanks is that there are fewer corners and crevices for mould and mildew to form in, but they use more space.
- The advantage of rectangular tanks is that they can save space and can be built in sections, but they have more corners and crevices for mould and mildew to form in.
- It is important to have isolation valves on the inlet and outlet connections, with unions or flanges for disconnection.
- Adequate tank support is critical to support large tanks.
- Sufficient access must be allowed to maintain tanks.

REFERENCES

AS 2845.2 Water Supply – Backflow Preventions Devices, Part 2: Registered Air Gaps and Registered Break Tanks

AS/NZS 3500.0 Plumbing and Drainage, Part 0: Glossary of Terms

AS/NZS 3500.1 Plumbing and Drainage, Part 1: Water Services

National Construction Code (NCC)

AS/NZS 4020 Testing of Products for Use in Contact with Drinking Water

Plumbing Code of Australia (PCA)

GET IT RIGHT

1 Which photo has a correct connection?

____

____

____

2 Why is it correct?

____

____

____

WORKSHEET 1

To be completed by teachers	
Student competent	☐
Student not yet competent	☐

Student name: ______________________________

Enrolment year: ______________________________

Class code: ______________________________

Unit competency code/title: CPCPWT3020 Connect and install storage tanks to a domestic water supply

Task: Review 'Identify installation requirements' and 'Prepare for work', then answer the following questions.

1 Name the codes and standards that apply to water storage tanks.

____

____

____

2 Name two high-risk activities associated with installing water tanks.

____

____

3 What is meant by the term 'open type' in relation to water storage tanks?

____

____

____

4 State three main purposes of water storage tanks.

____

____

____

5 Give three examples of what a storage tank may be used for.

____

____

____

6 Give a definition of a water storage tank.

____

____

____

7 List two basic design requirements for locating a storage tank.

____

____

8 List six different materials that may be used to construct a storage tank.

____

____

____

____

____

____

9 List the two main shapes of storage tanks in use today.

____

____

10 Describe the main advantages of a circular tank.

____

____

11 Describe an advantage of a rectangular storage tank.

____

____

12 Describe the difference between a 'small capacity' storage tank and a 'large capacity' storage tank.

____

____

13 Why is it important to have more than one large-capacity storage tank available for a water supply in some instances?

____

____

____

____

14 Identify the various items shown in the diagram of a storage tank below.

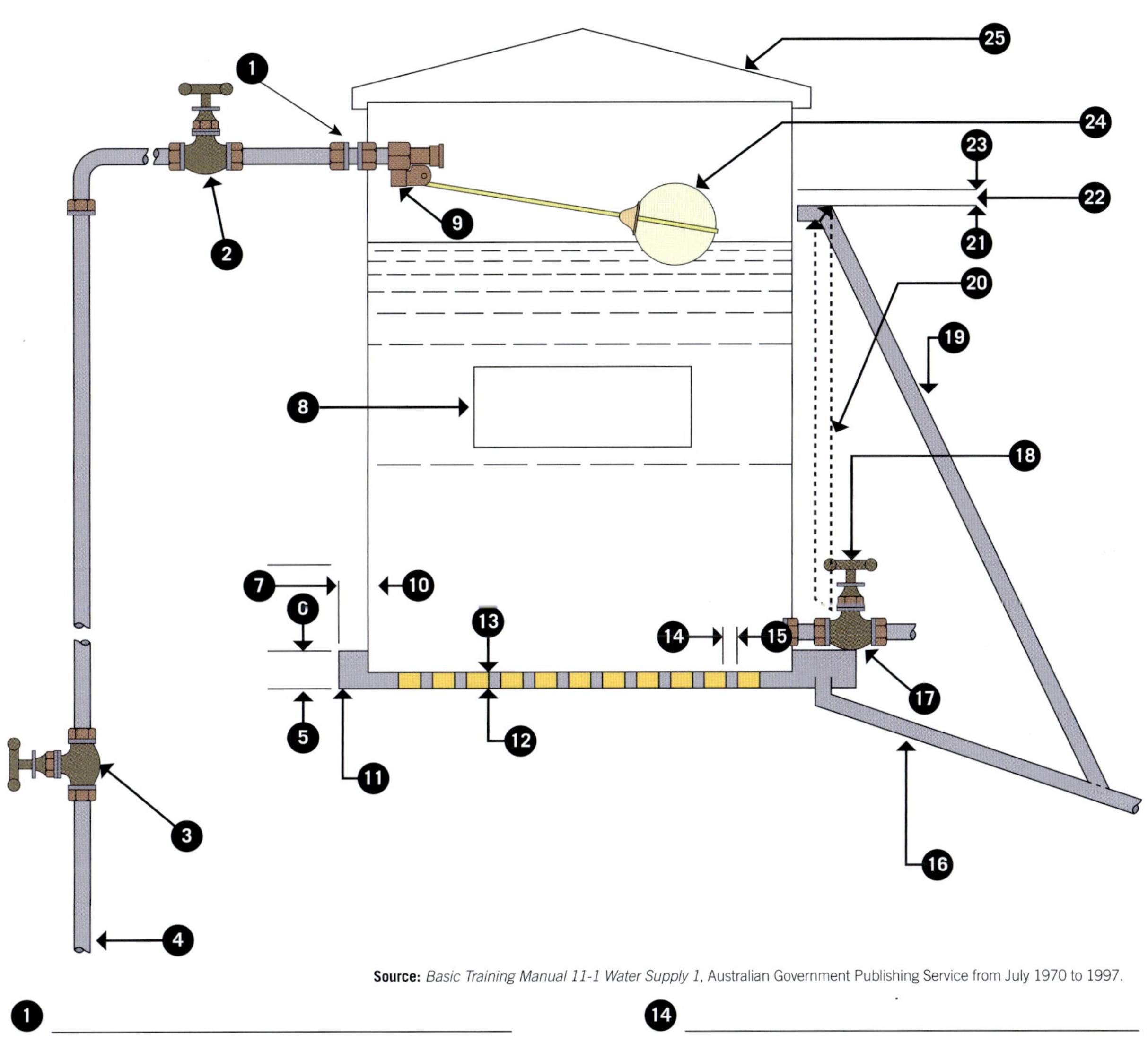

Source: *Basic Training Manual 11-1 Water Supply 1*, Australian Government Publishing Service from July 1970 to 1997.

1 ______________________

2 ______________________

3 ______________________

4 ______________________

5 ______________________

6 ______________________

7 ______________________

8 ______________________

9 ______________________

10 ______________________

11 ______________________

12 ______________________

13 ______________________

14 ______________________

15 ______________________

16 ______________________

17 ______________________

18 ______________________

19 ______________________

20 ______________________

21 ______________________

22 ______________________

23 ______________________

24 ______________________

25 ______________________

15 List three possible locations in which storage tanks can be positioned on large building sites.

16 Describe how a storage tank located in a public car park area could be protected from vandalism and mechanical damage.

17 What is recommended for you to do when placing an order for a tank with a supplier?

18 State the main WHS points to be considered when carrying out the work related to installing and testing a storage tank for a work activity.

WORKSHEET 2

To be completed by teachers	
Student competent	☐
Student not yet competent	☐

Student name: ______________________________

Enrolment year: ______________________________

Class code: ______________________________

Unit competency code/title: CPCPWT3020 Connect and install storage tanks to a domestic water supply

Task: Review 'Install and test storage tanks' and answer the following questions.

1 Why is an air gap important when connecting a potable water supply to a storage tank?

____

____

2 Name the two most common piping materials used for water supply to and from storage tanks.

____

____

3 State two methods that may be used for jointing copper pipe.

____

____

4 Where would a loose jumper isolation valve be installed on a water storage tank?

____

5 Describe, in your own words, why the connections to the tank outlet are normally gate valves or ball valves.

____

____

6 Describe how a float valve works.

____

____

____

____

7 Complete the following statement: 'According to current regulations, any storage tank with a capacity in excess of _____ litres requires an isolating valve on the outlet.'

____

8 Describe the method used to allow for disconnection of a water service from a storage tank.

9 When would flanges be used instead of unions?

10 State the minimum size for any storage tank overflow.

11 As per AS/NZS 3500.1, what three considerations determine the overflow pipe size?

12 State the minimum size of a safe waste pipe from a safe tray.

13 Describe the purpose of a sludge valve on a storage tank.

14 What is the required distance that an outlet is located from the bottom of the tank?

15 What is the required distance that an outlet is located from the top of the tank?

16 What is the minimum size of a removable access cover?

17 What are three considerations to determine adequate support of a water storage tank?

18 State four uses for water contained in a tank used for rainwater harvesting.

____

____

____

____

19 Complete the following calculation and fill in the missing parts of the formula and other statements with the correct response.

A water storage tank is 6 m long, 3 m wide and 2 m high. Determine the volume of the tank in cubic metres (m^3) and litres (L), and the mass in tonnes (T).

Volume in cubic metres

We can calculate the volume in cubic metres by using the formula:

$V = L \times W \times H$

where:

V = ______________

L = ______________ (___)

W = ______________ (___)

H = ______________ (___)

V = ___ × ___ × ___

= ___ m^3

Volume in litres

We know that there are 1000 L in a cubic metre and that this tank contains ______________ m^3 of water.

Thus, multiplying ______________ (m^3) by 1000 will give the capacity in litres (and also kilograms).

Volume in litres = ______________ (m^3) × 1000

= ______________ L

Mass in tonnes

Mass is ______________ kg; that is, ______________ tonnes.

WORKSHEET 3

To be completed by teachers	
Student competent	☐
Student not yet competent	☐

Student name: ______________________

Enrolment year: ______________________

Class code: ______________________

Unit competency code/title: CPCPWT3020 Connect and install storage tanks to a domestic water supply

Task: Review 'Install and test storage tank' and 'Clean up' and answer the following questions.

1 How would a tank installation be tested?

____

____

____

2 Why is this necessary?

____

____

____

3 How is a tank cleaned?

____

____

4 How is a tank disinfected?

____

____

5 Why is it important to reuse and recycle left over material?

____

____

6 Have a class discussion about each of the installations shown in the two photographs below and note any items that do not comply with AS/NZ 3500.1 Plumbing and drainage, Part 1: Water services. (For example, look closely at the jointing method used on the cold water supply inlet to the tank in photograph (a).) Note the specific clause number that relates to the non-compliant item and write it down in the space below.

(a)

(b)

INSTALL BACKFLOW PREVENTION DEVICES

9

Chapter overview

This chapter provides the knowledge and skills that a plumber needs to protect the potable water supply by installing backflow prevention devices.

The drinking (potable) water supply for a community should never be taken for granted and should always be protected from contamination. If the water supply becomes contaminated in any way at all, the health and hygiene of the community are potentially at risk.

Procedures involved in determining the size of the connection to the water service – and, more specifically, what precautions must be taken to protect the drinking water supply from contamination that could occur due to a cross connection between the drinking water supply and any alternative water supply, service, system or fixture – must be adhered to as per the Plumbing Code of Australia (PCA) and AS/NZS 3500.

The avoidance of cross connection is called backflow prevention. This is an area of utmost importance to the water utilities, as they make every effort to provide good-quality drinking water to the community. It is for this reason that the water utilities insist that once the water leaves their supply main, through a control valve and then onto a property, it must not re-enter that water main.

Learning objectives

Areas addressed in this chapter include:

- identify installation requirements
- prepare for work
- connect and test system
- clean up.

Identify installation requirements

Backflow is managed by the installation of a backflow prevention device. This can be a mechanical device, a registered air gap or a break tank, which is designed to automatically ensure the disconnection or separation of the drinkable (potable) water supply from a potential source of contamination or backflow.

A cross connection is any connection, physical or otherwise, between any drinkable (potable) water supply system either directly or indirectly connected to a water main and any fixture, storage tank, receptacle, equipment or device through which it may be possible for any contaminated water or substance to enter a water system. Backflow can be caused by either *backpressure* or *back siphonage*. Cross connection may take many forms, including:

- hoses left in pools
- hoses with chemical injectors attached
- outlets lower than the flood level of the fixture (submerged inlet)
- irrigation systems
- cooling towers
- bidets
- dental cuspidors
- photographic laboratories and other industrial and commercial processes.

Backpressure is the difference between the pressure of the water supply and a higher pressure within any vessel or pipework to which it is connected. If the water outlet on a property is much higher than a low-pressure water main, it is possible for backpressure to occur.

Back siphonage occurs when the water supply pressure falls below atmospheric pressure – in other words, a vacuum or partial vacuum is created. Reductions in drinkable (potable) water supply pressure occur whenever the amount of water being used exceeds the amount of water being supplied – such as during water mains flushing, firefighting or breaks in water mains – which creates back siphonage. In a water-utility potable water system, the effect of back siphonage is similar to drinking water through a straw. Watch the video called *Introduction to Backflow Prevention* by the Sunshine Coast Council on YouTube (4:46 min).

Access, read and determine backflow prevention device installation requirements

Access to the regulatory requirements for the installation of backflow prevention devices are as follows:

- *ABCB Cross Connection Handbook* (abcb.gov.au)
- National Construction Code (NCC)
- Plumbing Code of Australia (PCA)
- AS/NZS 3500.0 Plumbing and Drainage, Part 0: Glossary of Terms
- AS/NZS 3500.1 Plumbing and Drainage, Part 1: Water Services
- AS/NZS 2845.1 Water Supply – Backflow Prevention Devices, Part 1: Materials, Design and Performance Requirements
- AS 2845.2 Water Supply – Backflow Prevention Devices, Part 2: Registered Air Gaps and Registered Break Tanks
- AS/NZS 2845.3 Water Supply – Backflow Prevention Devices, Part 3: Field Testing and Maintenance of Testable Devices.

It is important to read the manufacturer's specifications when installing backflow prevention valves. Some backflow prevention valves have a considerable pressure loss between the inlet and outlet, and this must be considered to supply the required water pressure.

Protection required

The type of backflow protection required within properties is defined into three different areas, as per the Plumbing Code of Australia (PCA) and AS/NZS 3500.1 (see Figure 9.1).

Containment protection

Containment protection is installed immediately downstream of the meter assembly for a property that may have an area of risk. Generally, this means all commercial and industrial properties. It should be noted that the highest identified hazard onsite will also determine the hazard rating applied to the containment device.

Zone protection

Zone protection is when hazards with the same rating are identified and a device is installed on the supply line to those areas, such as a particular building in a complex, so as to protect the rest of the property from backflow or contamination of the drinking supply.

Individual protection

Individual protection is when a single hazard is identified, and a backflow prevention device is installed for protection of individual fixtures, appliances or apparatus.

Selecting backflow prevention devices

The type of backflow prevention device required for the installation is determined by the hazard rating (see Table 9.1) as per AS/NZS 3500.1. The information required to select the appropriate backflow prevention device for the hazard rating is determined by:

- referring to the Plumbing Code of Australia (PCA)
- identifying the hazard rating from the determination method in the PCA
- selecting the valve most appropriate for the installation in the determined hazard rating from AS/NZS 3500.1.

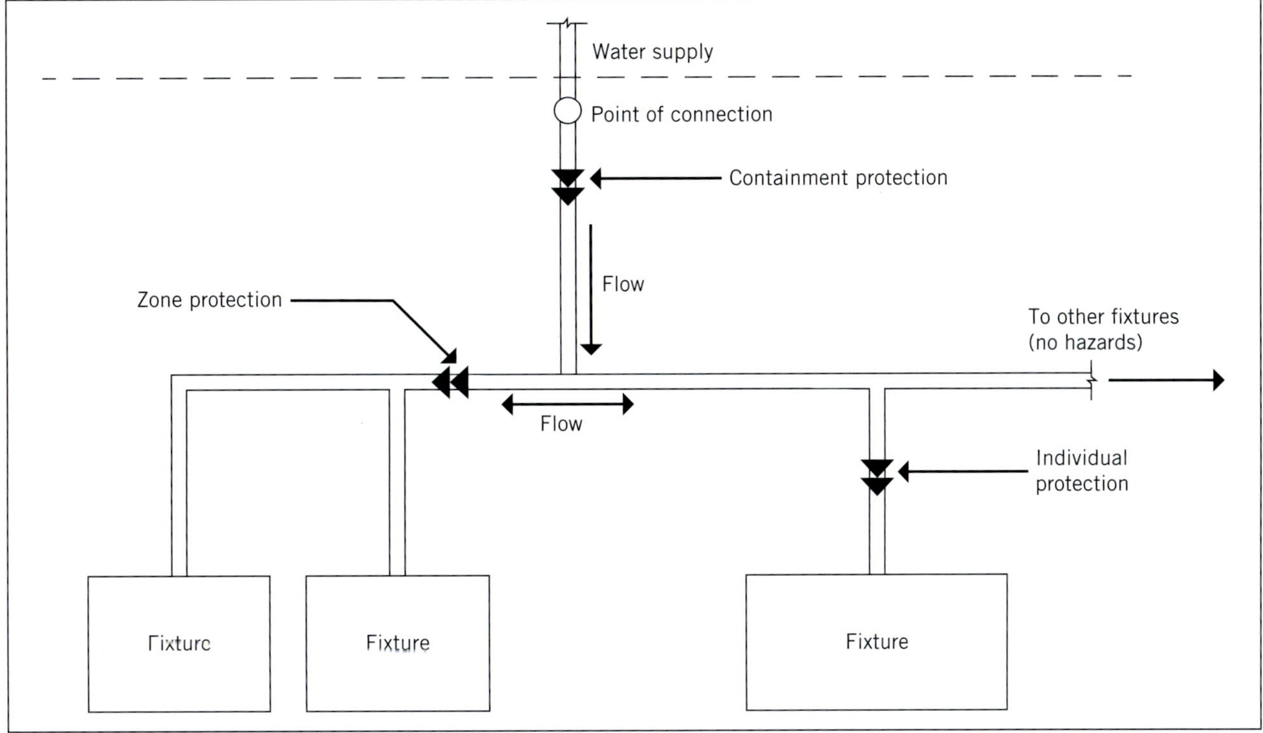

FIGURE 9.1 Individual, zone and containment protection

TABLE 9.1 Hazard ratings for water supply systems

Hazard rating	Definition
High hazard	Any condition, device or practice within a water supply system and its operation that could potentially cause death.
Medium hazard	Any condition, device or practice within a water supply system and its operation that could potentially endanger health.
Low hazard	Any condition, device or practice within a water supply system and its operation that could potentially constitute a nuisance but not endanger health.

NCC: PART 3 – PLUMBING CODE OF AUSTRALIA (PCA)

Backflow prevention devices

Backflow prevention devices must be manufactured in accordance with AS/NZS 2845.1 Water supply – Backflow prevention devices, Part 1: Materials, design and performance requirements. Air gaps and break tanks used as backflow prevention devices must be installed in accordance with AS 2845.2 Water supply – Backflow prevention devices, Part 2: Registered air gaps and break tanks and AS/NZS 3500.1 Plumbing and drainage, Part 1: Water services.

When the plumber tests any backflow prevention device, whether it is after the initial installation or as part of general ongoing maintenance, the testing must be done by an accredited backflow licensed plumber in accordance with AS/NZS 2845.3 Water supply – Backflow prevention devices, Part 3: Field testing and maintenance.

FROM EXPERIENCE

Only licensed plumbers with backflow accreditation may test and commission backflow prevention devices.

The role of the plumber in backflow prevention is important, and that is why it is necessary to undergo extensive training. It is the plumber's responsibility to help protect the community's water supply.

Apply workplace, work health and safety (WHS) and environmental requirements

Legislation requires that work health and safety (WHS) requirements be observed and adhered to. At the minimum, a job safety analysis (JSA) must be completed on all jobs – identifying the hazards, assessing the risks and applying control measures to minimise the risks.

Safe Work Australia requires a safe work method statement (SWMS) to be completed for any high-risk activity, as well as a JSA, before starting these activities. Working at heights, working in confined spaces, excavation and hot works are considered high-risk activities involved with the installation of backflow prevention devices.

If a backflow prevention device is installed in a confined space, use a gas detection device to ensure the area holds no poisonous or explosive gases. A harness with a tripod must be set up and a rescue plan in place for if a dangerous situation arises.

If a backflow prevention device is installed in an elevated position, the use of scaffolding reduces risk. If the backflow prevention device is fitted to a fire service, a drop in water pressure might set off fire alarms. Be sure to notify the fire brigade and area wardens before work begins on fire services.

Large diameter backflow prevention valves are heavy, and care must be taken when lifting them. Manual lifting and handling techniques must always be followed according to WHS requirements. A correct size-up of the load to be lifted is important to determine whether a mechanical lifting device, a crane or a two-person lift is required. Try to avoid excessive manual handling and double handling of materials wherever possible. Do not place yourself or others in a dangerous position and ensure that all the necessary safety precautions have been taken.

Environmental requirements involve taking the appropriate measures to reduce excessive noise and dust when drilling, cutting and sawing different materials.

Electrical safety

Electrical conductivity and stray electrical currents are also potentially fatal when dealing with metallic water services and associated valves. Before removing or isolating a backflow prevention device, or cutting into a metallic water service, check for stray current with a neon tester.

Ensure no stray current is present before fitting bonding straps. Check the bonding straps have been applied correctly with proper contact to ensure any stray current that occurs while working on the water service is bridged. Also have rubber matting or timber platforms to kneel on.

Don't let water discharge from a test point onto any nearby electrical equipment.

FROM EXPERIENCE

Plumbers should always be aware of all current rules and regulations, and safe work practices, and be aware of any new technology related to their work.

Prepare for work

It is the responsibility of the licensed plumber to prevent any cross connection with the drinking water supply, and install the appropriate backflow prevention device. Low hazard backflow prevention devices installed on water services are not testable devices. Medium- and high-hazard backflow prevention devices are testable devices and are registered with the local water authority to ensure annual testing is carried out.

Create a material list and collect materials

It is important to write out a detailed list of the materials required to carry out the installation of a backflow prevention device for the job to go smoothly. Always check the order upon delivery to ensure that the correct materials have been supplied, as well as the correct quantities.

A typical list to install a reduced pressure zone device (RPZD) on a 20 mm water service for containment protection may look like this:

- 1 m of 20 mm copper piping
- 1 × 20 mm RPZD
- 2 × 20 mm resilient seated ball valves
- 1 × 20 mm line strainer
- 2 × 20 mm brass M + F unions
- 4 × 20 mm brass hex nipples.

Select and check serviceability of tools and equipment

Some of the basic tools and equipment required for installing backflow prevention devices are:

- shovels and spades for excavation
- mattock, pick and bar for breaking up ground
- tape measure
- hacksaw and tube cutters for cutting pipe
- press-fit tool and oxygen/acetylene kit for pipe and fitting jointing
- pipe-grips (Stillson's, footprints and multigrips), shifter-spanner and spud wrench for screwed joints.

All electrical tools and leads must be tagged for safety regularly to the local authorities' rules. The general requirement for building and construction is to inspect and tag electrical equipment every three months because of the harsh environment the electrical equipment is exposed to.

As with all plumbing hand and power tools, personal protective equipment (PPE) and other associated equipment, it is essential that the correct tool is selected for the task and checked to ensure it is working in accordance with the manufacturer instructions and used correctly.

Any damage found in the tools or equipment must be reported to the supervisor immediately and the item should be taken out of circulation and tagged as damaged to prevent anyone using the item.

Personal protective equipment

The risk of injury can be reduced by the appropriate use of PPE. The types of PPE include the following:

- *Safety boots* have a steelcapped toe to protect your feet from falling objects, and a hard-wearing sole that sharp objects cannot pierce.
- *Safety glasses or goggles* will protect your eyes from injury when drilling, cutting or sawing with hand or power tools.
- *Ear plugs or earmuffs* will protect your hearing when using power tools and working in noisy work sites.

- *Dust masks or respirators* will protect your respiratory system from breathing in dust and fine particles when drilling, cutting or sawing, and working in a dusty environment.
- *Gloves* will protect your hands from cuts, abrasions and burns when working with different materials, tools and equipment.
- *Overalls* will cover your skin and protect you from exposure and different conditions.
- *High-visibility clothing* allows workers onsite to easily identify where others are and helps to avoid accidents.

Types of backflow prevention devices

There are many types of backflow prevention devices designed for different hazard ratings. The low-hazard devices are not testable, but the medium- and high-hazard devices must be tested annually by an accredited licensed plumber.

Atmospheric vacuum breakers

An anti-siphon or **atmospheric vacuum breaker (AVB)** is designed to protect against back siphonage of contaminated water into the drinkable supply by allowing air into the pipe upstream of the device, so preventing a vacuum from forming. AVBs should be installed vertically and at least 150 mm higher than the highest downstream outlet. No isolation valve is fitted downstream of the AVB, to avoid a continuous pressure situation. Atmospheric vacuum breakers are for low-hazard situations such as Type B irrigation installations, which are not subject to continuous pressure. (see Figure 9.2).

FIGURE 9.2 Atmospheric breaker valve (ABV) – typical installation

Hose connection vacuum breakers

Hose connection vacuum breakers are used on hose taps to prevent back siphonage when hoses are connected for low hazards. The breaker has a single check valve, which is spring-loaded to hold it in a closed position with an atmospheric vacuum breaker vent (see Figure 9.3). These devices are generally attached to a hose tap and can also be used in other applications, including laboratory sinks, service sinks, wall hydrants, and commercial dishwashers and soap dispensers. They must be installed on all drinking water hose taps in dual water areas.

FIGURE 9.3 Hose connection vacuum breaker valve

Dual check valve

A **dual check valve** is a low-hazard device and is installed on the outlet side of the water meter where dual water services are installed in a domestic residence to prevent the backflow of polluted water into the water utility's service. The dual check valve assembly consists of two in-line spring-loaded poppet check cartridges that require a 7 kPa pressure difference to open. If a backflow situation occurs, the two valves will close, preventing reverse flow, which protects against backpressure and back siphoning (see Figure 9.4).

Dual check valve with atmospheric port

A dual check valve with atmospheric port is a low-hazard device made for smaller supply lines. It is ideally suited for laboratory equipment, processing tanks, sterilisers, dairy equipment, dental consoles and similar applications with a low hazard rating where the device is under continuous pressure. It stops back siphonage or backpressure occurring and allows air into the pipe via the atmospheric port, preventing a vacuum from forming, and discharges any reverse water flow to the atmosphere (see Figure 9.5).

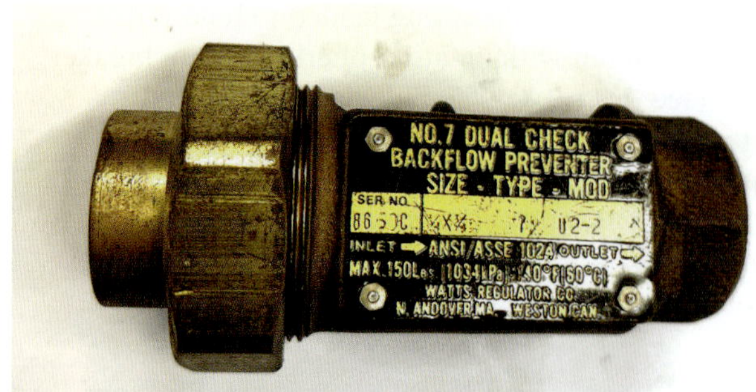

FIGURE 9.4 Dual check valve

FIGURE 9.5 Dual check valve with atmospheric port

Double check valve

The **double check valve** provides protection to the drinkable (potable) water supply from contamination in medium-hazard applications, such as Type C irrigation systems, where the device is under continuous pressure and backpressure. It is a testable device that must be tested annually. Double check valves consist of two spring-loaded check valves that require a 7 kPa pressure difference to open. In a non-flow condition the check valves hold 7 kPa differential minimum in the direction of flow. In a flow condition the check valves are open, proportional to the flow demand. In a backflow condition both checks will close, preventing reverse flow until the resumption of normal flow, when they will reopen. They are designed to protect against backpressure and back siphoning (see **Figure 9.6**).

FIGURE 9.6 Double check valve

Pressure vacuum breaker

Pressure vacuum breakers (PVBs) are used on medium-hazard installations, such as Type C irrigation systems, where the device is under continuous pressure. PVBs stop back siphonage occurring and allow air into the pipe via the atmospheric port. This prevents a vacuum from forming and discharging any water backflow to the atmosphere. To allow them to operate and break siphon conditions, they must be installed at least 300 mm above the highest hazard they protect and must be tested annually (see **Figure 9.7**).

Reduced pressure zone device

Reduced pressure zone (RPZ) devices RPZDs are designed to protect against backpressure, back siphoning or a combination of both. They consist of two independently operating, spring-loaded, Y-pattern check valves and one hydraulically dependent

FIGURE 9.7 Pressure vacuum breaker

differential relief valve. These devices provide protection to the drinkable (potable) water supply from contamination in high-hazard applications and are testable devices. Under flow conditions the check valves are open with the pressure between the checks, called the zone, being maintained at least 35 kPa lower than the inlet pressure and the relief valve being maintained closed at 14 kPa. Should abnormal conditions arise (no flow or reversal of flow), the relief valve will open and discharge its contents through the relief port to maintain the zone at least 15 kPa lower than the supply (see **Figure 9.8**). When normal flow resumes, the zone's differential pressure will resume, and the relief valve will close. The minimum height of the relief port on a reduce pressure zone (RPZ) valve is 300 mm from ground level.

A significant pressure loss of up to 100 kPa can occur through an RPZD, so this needs to be factored into the installation to achieve the required pressure at the most disadvantaged outlet. An RPZD is required to be tested annually by a registered backflow accredited plumber.

FIGURE 9.8 Reduced pressure zone device

Detector assemblies

On jobs where large diameter backflow protection devices are installed with large diameter water meters in the containment zone, a detector assembly is installed to measure small quantities of water being used. As a large water meter will not detect low flow, a smaller water meter connected on a bypass will detect and measure the low flow (see **Figure 9.9**). Usually, a 20 mm water meter with a 20 mm backflow prevention device equivalent to the main backflow prevention device is fitted for the detector assembly. The same commissioning and testing procedures apply to the detector assembly as required for the main backflow prevention device by a registered backflow accredited plumber.

FIGURE 9.9 Double check valve detector assembly

Registered break tanks

A **registered break tank (RBT)** is a storage tank that incorporates an air gap that is specifically calculated from the orifice size of the inlet and the size of the overflow. This information is set out in AS/NZS 3500.1. It is a high-hazard testable device that is registered by, or on behalf of, a regulatory authority for the purposes of inspection and maintenance to ensure its functional requirements are maintained (see **Figure 9.10**).

Registered air gap

A registered air gap (RAG) is an air gap in a storage tank or the air gap over a fixture, vat tanker filling point or drum of non-drinkable (non-potable) liquid

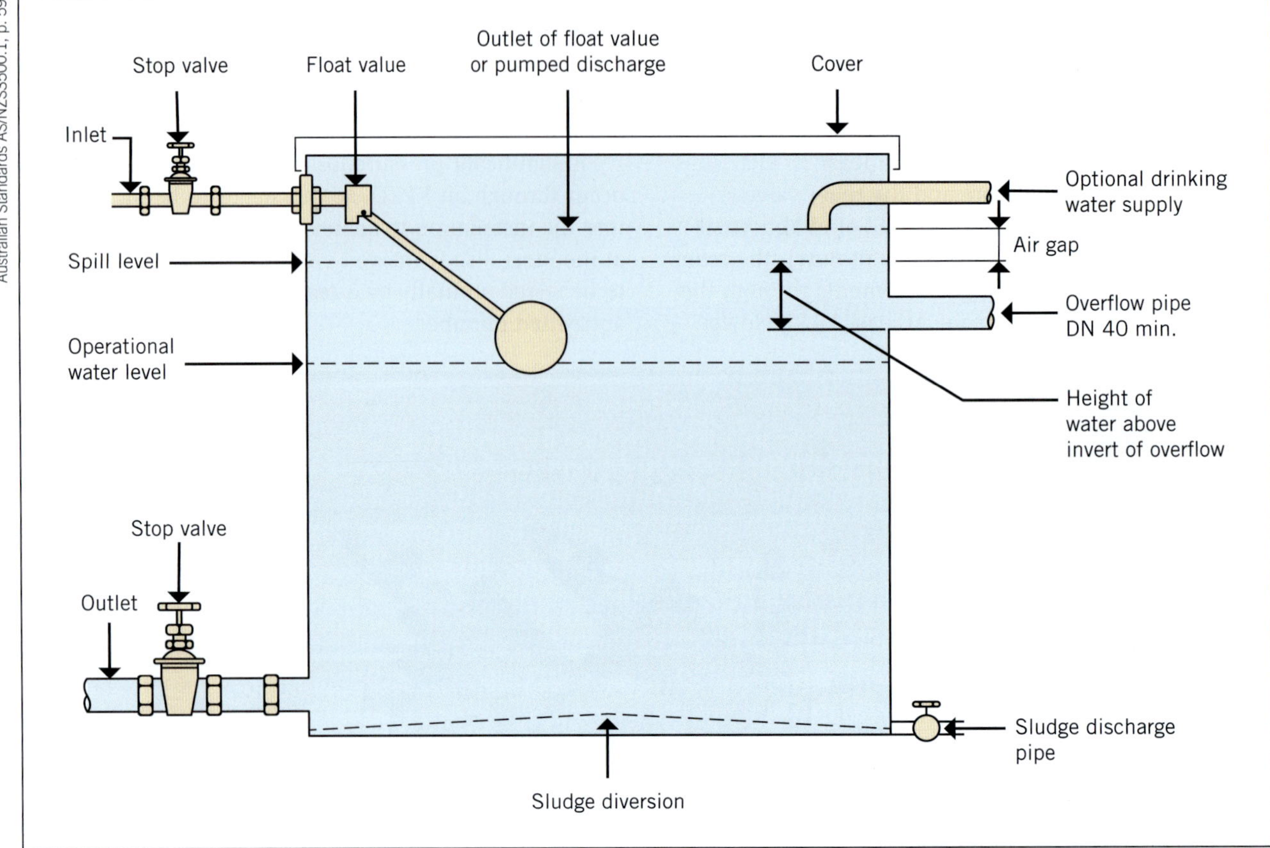

Source: Adapted from Figure 8.4.1 Air gap on a registered break tank. Australian Standards AS/NZS3500.1, p. 59.

FIGURE 9.10 Air gap on a registered break tank

(see **Figure 9.11**). It is a high-hazard testable device that is registered by, or on behalf of, a regulatory authority for the purposes of inspection and maintenance to ensure its functional requirements are maintained. The minimum air gap is 20 mm as per AS/NZS 3500.1.

Registered air gap must be 2 × pipe inside diameter (ID)

Source: Backflow Awareness Course © http://www.tich2o.com, 1 January 2004. All backflow materials are used by permission from CMB Industries, Inc.

FIGURE 9.11 Registered air gap

Ancillaries

Most valves have limitations regarding their temperature and pressure capabilities. These factors must be considered when selecting backflow prevention devices. The appropriate valve for the application should be ordered.

When installing backflow prevention devices, it should be remembered that no heat should be applied to them. Depending on their type, and whether they are testable or not, they should be provided with a line strainer, except when used on a fire service. Testable devices must have a resilient-seated isolating valve installed immediately upstream of the line strainer or immediately upstream of the device in cases where no integral line strainer is fitted (see **Figure 9.12**). A resilient seated valve must also be installed on the outlet of the device. It is essential to flush the pipework before connecting the device to clear any foreign matter, which will prevent any debris from getting into the valves and damaging areas such as the valve seats.

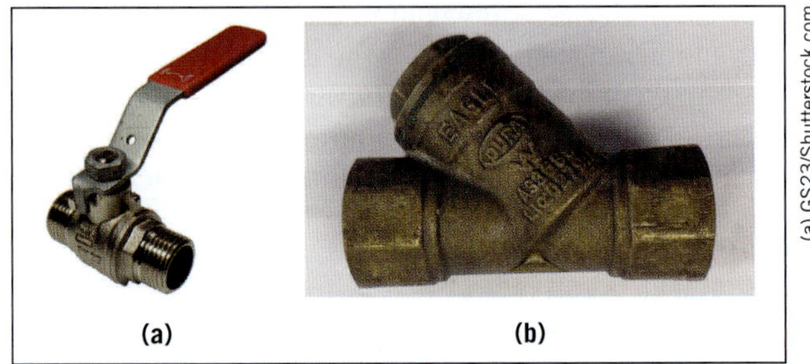

(a) GS23/Shutterstock.com

FIGURE 9.12 Isolating valves and line strainer installed on testable backflow prevention devices

LEARNING TASK 9.1

1 What is the difference between backpressure and back siphonage?
2 Where would a pressure vacuum breaker (PVB) valve be installed?
3 Why is it important to flush the pipework before connecting a backflow valve and line strainer?
4 Where is the backflow valve installed when containment protection is required?
5 Why are isolating valves required to be installed immediately upstream and downstream of a testable backflow prevention device?
6 What information is required to calculate the air gap for a registered break tank?

COMPLETE WORKSHEET 1

Connect and test system

When installing backflow prevention devices, there are several issues that specifically need to be considered. As a guide, consider the following (although the research required is not limited to these points alone):

- the hazard rating – low, medium or high (refer to the PCA)
- the backflow prevention device to suit the hazard rating (refer to AS/NZS 3500.1)
- access to install the backflow prevention device
- the pressure loss across the backflow prevention device
- water pressure and volume requirements for the water supply to the water supply system (flow rate)
- use of an alternative water supply; will there be an alternative water source for an irrigation system/s and, if so, what effect will this have on the overall system, including the drinking water (refer to Chapter 11 'Set out, install and commission irrigation systems')?

Some of the information required for the above points can be gathered from the plans and specifications, but it is best to visit the site prior to commencing work to understand the requirements.

GREEN TIP

Using an alternative water supply, such as rainwater or bore water, for irrigation saves drinking water and cost.

Set out and install backflow prevention devices according to relevant specifications

AS/NZS 3500.1 makes specific reference to the installation of backflow prevention devices. The rules are based on current 'best practice' principles and must be observed when installing the devices as well as following manufacturer's installation instructions.

AS/NZS 3500 stipulates regulations related to the installation of the backflow prevention device, as do the manufacturer's instructions. It is important to be familiar with both sources of instruction before attempting to install the device and any associated valves and pipework.

There are several different types of backflow prevention devices that can be installed in certain locations. As mentioned, the plumber should be aware of the current rules and regulations relating to the location and installation of the backflow prevention device. The plumber should also be aware of the current categories and terminology related to each of the categories and devices. Some of the terminology used in relation to the devices can be found in AS/NZS 3500.0 Plumbing and Drainage, Part 0: Glossary of Terms and in the Glossary of this book.

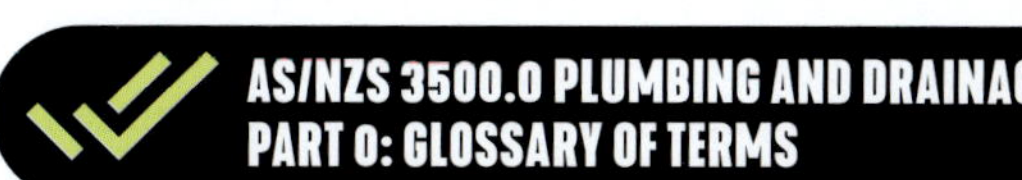

HOW TO

INSTALL A BACKFLOW PREVENTION DEVICE

1 Do not apply heat to any backflow prevention device during installation. This is to prevent distortion or weakening of the valve body and internal workings through the application of heat; for example, from an oxygen/acetylene flame.
2 Before installing the device, fit it with an approved line strainer. This is to help prevent any debris from fouling the internal workings of the actual backflow prevention device.
3 Place an approved isolating valve upstream of the strainer and the device as well as downstream of the device. These isolation valves are required so that the device can be isolated for maintenance and repair, and to prevent draining the piping system of water during any maintenance and repair.
4 Flush all piping before connecting the device (and strainer). This is to help prevent any debris from getting into the valves and damaging areas such as the valve seats.
5 Do not install unprotected bypasses around the device. This is to protect the supply system from cross connection. If a bypass is installed, then the same type of backflow prevention device must be

>>

>>

installed on the bypass to prevent contaminated water from flowing through the bypass back into the drinking water supply.

6 Install the device according to the current codes and standards. Follow the manufacturer's written instructions to ensure safe/best practice. If unsure about the correct way to install the device, or the connecting pipework, it is advisable to contact the manufacturer so that any warranty is not voided. The local water authority should also be consulted to determine whether there are any rules or regulations specific to the local area.

7 Protect the device from damage such as freezing. Regulations state the actual installation requirements for the device, but if it is installed in a location where it can be damaged, or tampered with, necessary steps need to be taken to protect the valves. This should be considered when planning for the project and priced into the tender submission price.

8 If continuous water supply is essential, install another valve set-up 'in parallel' to the first, so that the latter can be shut down. This would be dependent on the purpose of the system, and if continuous supply is essential, then it must be catered for. This allows a valve or fitting on the line to be serviced or replaced while supply of water is maintained for the system. What it means is that if a particular backflow prevention valve is sized for a project, then another complete valve set-up of the same size should be installed next to – that is, 'in parallel' with – the other valve set-up. The valve configuration should be exactly the same. This can be achieved by rising up out of the ground with a supply pipe, fitting a 'tee' branch, completing the duplicate valve arrangements – complete with isolating valves, strainers, unions or flanges and the backflow prevention device – then connecting to another 'tee' branch and onto the pipework system (see Figure 9.13).

9 Ensure that in-line devices can be removed or replaced. This can be interpreted to mean that unions or flanges are required on certain backflow prevention device installations. This is to allow for the removal or replacement of a device or other valve or fitting without damaging the rest of the installation.

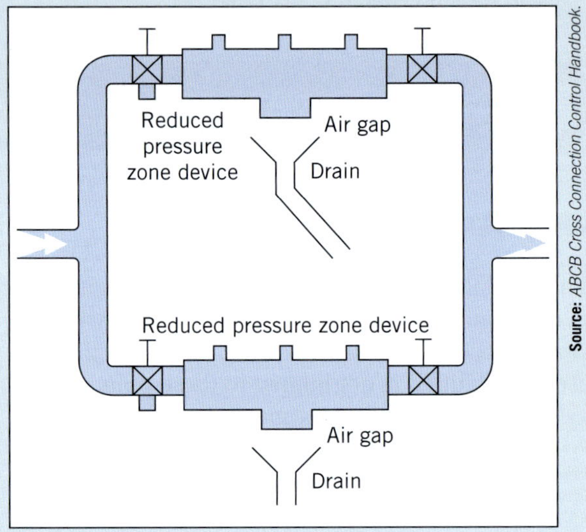

Source: ABCB Cross Connection Control Handbook.

FIGURE 9.13 Valve set-up 'in parallel'

Once installed, all testable backflow devices must be commissioned and registered by a licensed plumber with an accredited backflow commissioning and testing qualification. All testable backflow devices are registered with the local water authority so the annual testing of the devices can be monitored and regulated. The devices are tested to ensure the check valves are tight to prevent any flow reversal and the relief valve (if fitted) is operating correctly.

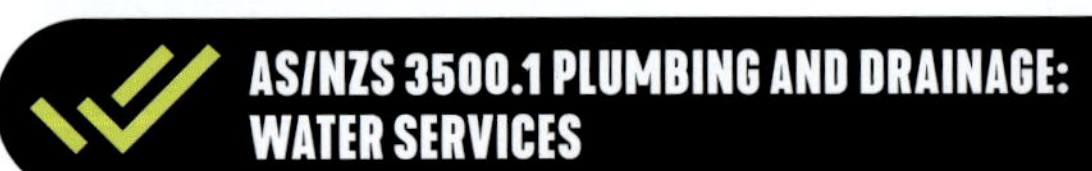

AS/NZS 3500.1 PLUMBING AND DRAINAGE: WATER SERVICES

Remember when cutting into a metallic water service to check for stray current and that bonding straps should always be connected.

Water pressure and volume requirements

As discussed earlier, the determination of water pressure and volume requirements for the water supply system is important and must be checked onsite before work begins. Provision must be made for the correct flow of water through the pipes. The installation may need to be ring-fed to ensure adequate water supply.

Should pressure, volume and flow be a problem for the water supply system, a pumped system, a combination of a pump and tank supply system or an alternative means of water supply may be required.

FROM EXPERIENCE

Proper planning leads to a safe and efficient installation, ensuring customer confidence and satisfaction.

Flow of water through pipes

While static pressure is a guide to the volume of water that might be expected to flow through a pipe system, other factors govern the volume that will affect the flow. The most important factors are as follows:

- *The diameter of the pipe* – the larger the bore of the pipe, the more water it can deliver.
- *The length of the pipe* – the longer the pipe, the greater the friction between the water and the inside surface of the pipe, causing pressure loss.

- *Changes in direction* – sharp bends cause eddies that, in turn, cause friction and restriction of the water flow. Frictional loss varies with the angle of the bend and the radius of curvature. The sharper the angle and the shorter the radius, the greater the frictional losses will be. A radius of five times the pipe bore on the centre line gives no more friction than the same straight length of pipe.
- *The quality of jointing* – burred pipe ends, excess sealing compound and gaps between pipe ends restrict the flow of water.
- *Passage through valves* – each time water passes through a valve, it is restricted and usually forced to change direction once or twice, depending on the design of the valve. Allowance for pressure loss through backflow prevention valves must be considered. For isolation, full-way valves (such as ball valves) are preferred.
- *Enlargement or contraction in size* – to gain maximum flow, enlargements and contractions of pipe diameter should be made as gradually as possible.

De-burring the pipes

When the pipes have been cut, the burrs must be removed from both edges of the cut. A burr left on the outside of any pipe may foul the mating socket of the next piece of pipe or fitting. A burr left on the inside causes friction when water passes through the pipe. Outside burrs should be removed with a file. Inside burrs should be removed with a reamer, a rat-tail file or a rotary burr remover.

The water pressure and flow rate

The two categories are static pressure and dynamic pressure.

- *Static pressure* is the pressure available when there is no water flowing through the pipework system. This pressure alone does not indicate whether there will be enough water flow for the water supply system.
- *Dynamic pressure (working pressure)* is the pressure available when the water is flowing, which helps determine the sufficient flow rate required.

Determining the static pressure can be done by connecting a pressure gauge to a hose tap and taking an approximate reading while there is no flow. Although static pressure does not indicate whether there will be enough water flow, it does help to determine if the adequate pressure is available. When pressure and volume of water are essential, greater care and accuracy must be used to determine flow rates. Determining the dynamic pressure will indicate whether there is enough pressure and water flow combined to supply the water system requirements.

If these pressure readings are low, it may be necessary to take some other readings at various times throughout the day.

Flow rate and dynamic (working) pressure depend on the pressure available in the water main, the size of the water meter assembly and the size and length of the drinking (potable) water service. Elevation (height) above the water supply point also needs to be considered, as this will affect the dynamic pressure.

When pumps and systems that use set pressures to operate are required, a dynamic pressure/flow measuring device will have to be connected (see, for example, **Figure 9.14**). This will measure the dynamic pressure.

There are different methods used for calculating flow rates and pressure readings depending on the nature of the plumbing work involved. A sample of the readings taken could be recorded, as shown in **Table 9.2**.

FIGURE 9.14 Pressure reading device

HOW TO

ESTABLISH THE FLOW RATE

1. Connect the device to a hose tap with the hose tap adapter.
2. Ensure that the ball valve is turned off.
3. Turn the water on (slowly) at the hose tap and take the static pressure reading at the pressure gauge; take note of this pressure.
4. Carefully open the ball valve and take note of the pressure reading on the pressure gauge as the water is flowing through the device. This will give the dynamic (working) pressure.
5. While the water is flowing through the device, direct the water into a bucket and, using a stopwatch, time how long it takes to fill the bucket. This will help to determine the flow rate (generally in litres per minute or litres per second). The size of the bucket should be a minimum of 10 L capacity. For a more accurate reading of the flow rate, a larger bucket should be used, or a purpose-made flow cup could be used (see **Figure 9.15**).

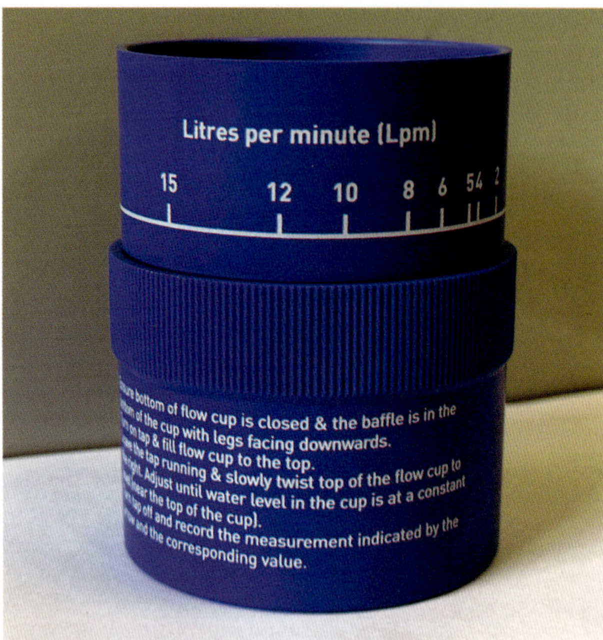

FIGURE 9.15 Flow cup

TABLE 9.2 A worked example of pressure and flow information for a water supply system

Client details	Name: Apprentice plumber Address: Supply Road, Waterview
Time of test	10.00 am
Location of hose tap	Outside laundry
Static pressure reading	400 kPa
Dynamic pressure reading	200 kPa
Capacity of bucket	10 L
Time taken to fill bucket	20 seconds
Flow rate (capacity of bucket ÷ time taken to fill it)	0.5 L/second
Flow rate in litres per minute (litres per second × 60)	30 L/minute

Remember that there will also be a pressure and flow loss through the backflow prevention device (up to 100 kPa, depending on the device) and any other valves and fittings that are installed as part of the connection, and this must be factored in when considering any installation.

Once these factors have been considered, it can be decided whether the available water supply is sufficient, or whether a pump *or* a storage tank and pump are required to provide both volume and pressure for the system. If a pump and storage tank are required, then if set up correctly, the storage tank would fulfil any backflow prevention requirements by using a registered break tank or registered air gap.

FROM EXPERIENCE

It is important for the plumber to be aware of the pressure loss across the backflow prevention valve, especially in low-pressure areas.

The effect on the consumer of closing down the water service for the connection

During the planning phase, the plumber should be aware of, and plan for, any effect that the closing down of the water service will have on the client and the client's property. This will depend on:

- the type of establishment (residential, commercial or industrial)
- the timing of the shutdown (e.g. whether it can be done in normal working hours)
- how long the system will be closed down
- what needs to be done to the rest of the water service after the connection is satisfactorily completed and the water is turned back on (e.g. the effect on old joints, the hot water service and air in the line).

FROM EXPERIENCE

Effective communication with the client is important to avoid conflict when disrupting the water supply.

LEARNING TASK 9.2

1. What procedure is carried out to a testable backflow prevention device after it is installed?
2. What is the difference between water pressure and water flow?
3. How is flow rate measured?
4. What is the difference between static pressure and dynamic pressure?
5. What can cause pressure loss in a water supply system?

Using an alternative water supply

When using an alternative water source, possibly for an irrigation system, consideration must be given to the quality of the water and the possibility of any cross connection with the drinking water supply.

Consideration must also be given to the size of the main water supply service and the pressure and volume of water available in that main. For this reason, on larger projects a pumping system is required. This may be combined with a storage tank that feeds water to the pumps. To meet BASIX requirements, it is compulsory in most areas for new homes to harvest rainwater in storage tanks to supply water for toilet flushing, irrigation, car washing and, in some areas, clothes washing (see Figure 9.16). These systems usually have a mains water back-up in case the rainwater supply becomes low. A backflow prevention device is installed to prevent the rainwater cross connecting with the mains water supply.

In some cases, there may be an alternative water supply for the system, such as a dam, a creek, a river, an artesian bore or some other source. This alternative supply may either supplement the supply from the water main or be supplemented by the water main. In these cases, it is very important to protect the main water supply from contamination because there is a high danger for a cross connection to occur from the non-drinking water supply to the drinking water supply (see Figures 9.17 and 9.18).

Testing installed pipework according to relevant job specifications

Once the backflow prevention device has been installed, the water supply is turned back on and a visual check for leaks is carried out on all the connections, valves and fittings.

As mentioned earlier in this chapter, if the backflow prevention device is testable, then it must be commissioned. This is the initial service of the backflow prevention device, which is done by an accredited backflow licensed plumber with a calibrated test kit (see Figure 9.19). This is when the backflow prevention device is registered with the local water authority.

FIGURE 9.16 Storage tanks being used to supply water to a domestic irrigation system

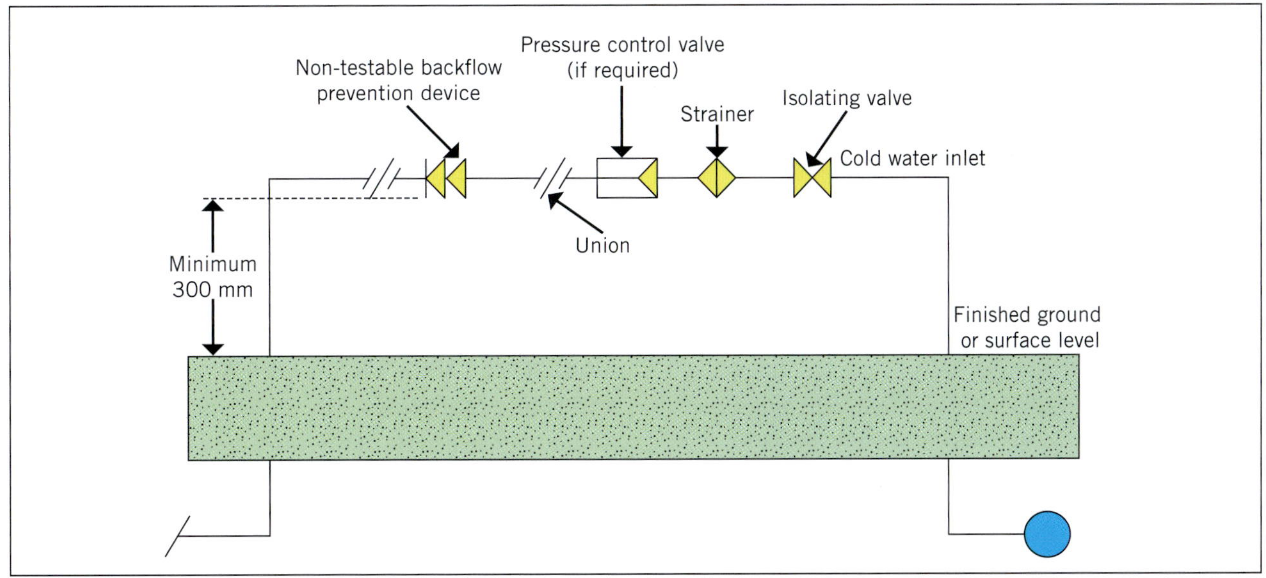

FIGURE 9.17 Standard irrigation system connected to a drinking or non-drinking water service with non-testable backflow prevention device

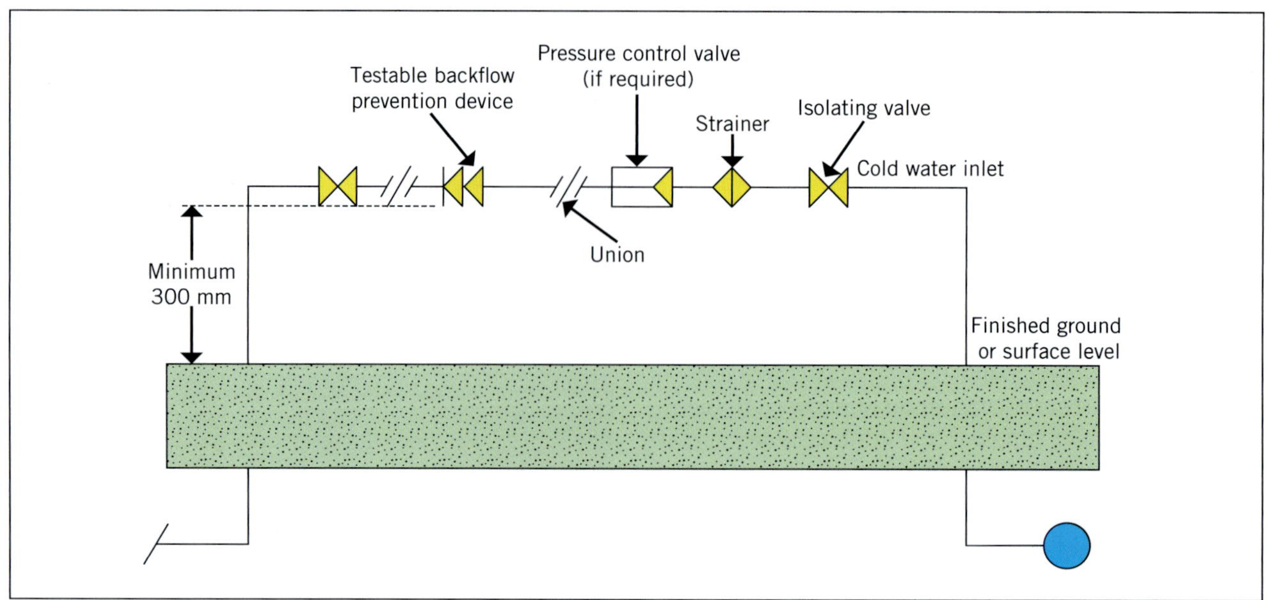

FIGURE 9.18 Standard irrigation system connected to a drinking or non-drinking water service with testable backflow prevention device

FIGURE 9.19 Calibrated backflow valve test kit

Testable backflow prevention devices are those that include features to enable the device to be tested in-situ; that is, without the need to remove the device. According to AS/NZS 3500.1, the following types of backflow prevention devices are testable devices:

- reduced pressure zone device
- reduced pressure detector assembly
- double check valve assembly
- double check detector assembly
- spill-resistant pressure-type vacuum breaker
- pressure-type vacuum breaker
- single check valve (testable)
- single check detector assembly (testable).

Commissioning and testing a backflow prevention device in-situ requires the water supply to the device to be shut off. Therefore, prior to conducting the test, it is important to alert the owner and/or occupants of the property that their water supply will need to be shut off for a period of time, until the test is complete. In situations where it is not viable to shut off a water supply, such as hospitals or industrial premises that operate 24 hours a day, the backflow prevention device can be duplicated by installing a second equivalent device in a bypass, configured in a parallel arrangement with the primary backflow prevention device. This is called a protected bypass. Figure 9.13 shows how to configure a protected bypass.

Clean up

A clean workplace reduces the chance of accidents occurring, such as trip hazards and falling objects. Another reason is that the build-up of waste and debris on a job site creates a problem for storage and access. A clean and tidy work area helps the job run more smoothly, which helps to increase profit.

Clear work area, remove waste and reuse material

A thorough sweeping of the work area helps remove dirt, dust and rubbish. At the end of the job, there is usually a large amount of cardboard and paper packaging. This should be placed in the recycling waste so it can be reused and reduce landfill.

Offcuts and leftover material should be appropriately stored and used on another job. This not only prevents waste but also enables savings in cost and increases in profit.

Clean and store tools and equipment

Tools and equipment should be packed away after they have been cleaned and maintained in proper working order. Any tools and equipment on hire must be returned as soon as possible to avoid extra cost. Any faulty tools must be tagged out of use and reported to the supervisor. Blades and cutting tools must be kept sharp or replaced if necessary to be effective in use.

All final documentation, such as the certificate of compliance, must be submitted to the regulatory authorities and client. Delivery dockets, invoices and order sheets should be reconciled as per company policy, and time sheets accurately completed.

LEARNING TASK 9.3

1 When would a backflow prevention device be required on a rainwater tank?
2 Who commissions a testable backflow prevention device?
3 Where would a protected bypass be installed?

COMPLETE WORKSHEET 2

SUMMARY

- Plumbers have a major responsibility to protect the community's drinking water.
- Any potential cross connection must be identified and prevented.
- Hazard ratings must be determined as per the Plumbing Code of Australia (PCA) so the appropriate backflow prevention device can be identified.
- Backflow prevention devices must be located and installed as per AS/NZS 3500.1.
- Testable backflow prevention devices must be commissioned and tested annually by an accredited backflow licensed plumber.
- Available water pressure and flow rate is a major consideration for water supply systems.
- Materials and equipment must comply with the relevant Australian Standards.
- The pressure loss across a backflow prevention device must be considered.
- Alternative water supplies must be installed so there is no possibility of cross connection with the drinking water supply.
- The interruption of the water supply to a premises must be considered.

REFERENCES

ABCB Cross Connection Handbook: **NCC-2022-Cross connection-control-Handbook-FA.pdf (abcb.gov.au)**

AS 2845.2 Water supply – Backflow Prevention Devices, Part 2: Registered Air Gaps and Registered Break Tanks

AS/NZS 2845.1 Water Supply – Backflow Prevention Devices, Part 1: Materials, Design and Performance Requirements

AS/NZS 2845.3 Water Supply – Backflow Prevention Devices, Part 3: Field Testing and Maintenance of Testable Devices

AS/NZS 3500.0 Plumbing and Drainage, Part 0: Glossary of Terms

AS/NZS 3500.1 Plumbing and Drainage, Part 1: Water Services

NCC: Part 3–Plumbing Code of Australia (PCA): **https://ncc.abcb.gov.au/**

GET IT RIGHT

1 What are the issues with this RPZD set-up?

____

____

____

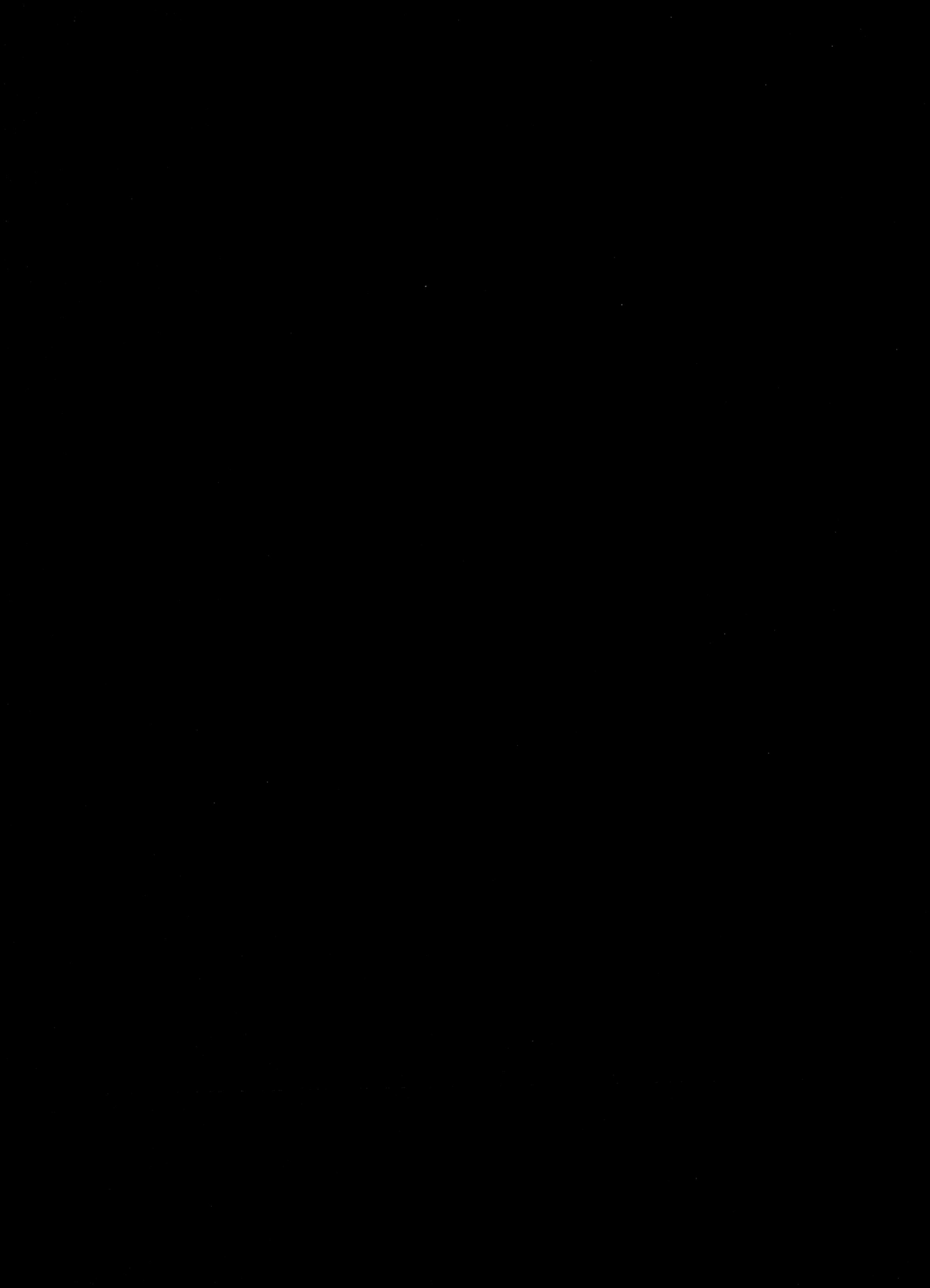

WORKSHEET 1

To be completed by teachers	
Student competent	☐
Student not yet competent	☐

Student name: ______________________

Enrolment year: ______________________

Class code: ______________________

Unit competency code/title: CPCPWT3027 Install backflow prevention devices

Task: Review 'Identify installation requirements' and 'Prepare for work' and answer the following questions.

1 List the two causes of backflow.

____

____

2 Provide two examples of a cross connection.

____

____

3 Describe the term 'backpressure'.

____

____

____

4 Describe the term 'back siphonage'.

____

____

____

5 Name the three different areas where backflow protection may be required on a property.

____

____

____

6 What code defines the protection required?

____

7 Name the three hazard ratings for backflow prevention devices and explain what they mean.

8 Who is permitted to commission and test backflow prevention devices?

9 What is the definition of a high hazard?

10 What does the abbreviation 'RBT' mean?

11 Break tanks used as backflow protection must comply with which Australian Standard?

12 Provide an example of a submerged inlet.

13 Describe the term 'zone protection'.

14 What is the minimum air gap size?

15 What does the abbreviation 'RAG' mean?

16 Where should isolating valves be installed on a testable backflow prevention valve?

17 What backflow device is used to prevent backpressure and back siphonage on a medium-rated hazard?

18 When and where is a line strainer required to be installed on a backflow prevention device?

___

___

___

19 What is the minimum height of the relief port on an RPZ valve from ground level?

___

20 Name two situations where it is recommended that a dual check valve with atmospheric port be installed for protection.

___

___

21 What pressure difference is required across the spring-loaded valves of a dual check valve to operate correctly?

___

22 Explain why backflow protection is important to water utilities.

___

___

23 What is the difference between a dual check valve and a double check valve?

___

___

24 Registered backflow devices should be tested at intervals not longer than ________ months.

WORKSHEET 2

To be completed by teachers	
Student competent	☐
Student not yet competent	☐

Student name: ______________________________

Enrolment year: ______________________________

Class code: ______________________________

Unit competency code/title: CPCPWT3027 Install backflow prevention devices

Task: Review 'Connect and test system' and 'Clean up' and answer the following questions.

1 What is recommended when installing backflow devices if continuous water supply is required?

____

____

2 Why are testable backflow devices registered with the local water authority?

____

____

3 Why is it important that no heat is applied to any backflow prevention device during installation?

____

____

4 Name three ways to reduce friction loss in a piping system.

____

____

____

5 Explain the difference between static pressure and dynamic pressure.

____

____

____

6 Why is it important to establish both the static and dynamic water pressures for a water supply system?

____

____

____

7 What type of isolation valves are used to minimise friction loss?

8 Why is determining the pressure loss across an RPZD important? How would you determine the pressure and flow loss across a device?

9 What method is used to establish the flow rate of a water supply system?

10 What unit of measurement is used to express the flow rate?

11 Why is it important to notify the client if the water supply is going to be disrupted?

12 How can the water pressure be increased if the mains supply is too low?

13 What backflow prevention devices can be used in a high hazard situation?

14 Does a Type A irrigation system require backflow prevention?

WORKSHEETS 9

15 List three alternative water supplies.

16 Which backflow prevention device is recommended for a Type D irrigation system?

17 When should the backflow prevention device be commissioned?

18 What is the purpose of a detector assembly?

19 What is the purpose of a protected bypass?

20 Who registers the testable backflow prevention device with the water authority?

INSTALL HOME FIRE SPRINKLER SYSTEMS 10

Chapter overview

The first automatic fire sprinkler system was developed by John Carey in England in 1806. He patented an apparatus for the 'extinguishment of fires in gentlemen's apartments and warehouses'. Combustible cords were attached to weighted valves so that when the cords burnt, the valves opened, spraying tank-supplied water onto the fire.

In modern times, a fire sprinkler service combined with smoke alarms will detect and help to control fires in a home, aiming to prevent loss of life and reduce property damage. A correctly installed sprinkler system can delay and possibly prevent flashover (that is, a fully established fire) in a room, improving the chances of safe evacuation from the building. The system components are generally located in unoccupied spaces, such as ceilings.

A typical home sprinkler system comprises piping from a permanent water supply connected to automatic sprinklers located in the cooking, living and sleeping areas. The liquid (alcohol) inside the sprinkler bulb expands when exposed to extreme heat, causing the bulb to break and permitting water to flow freely from the sprinkler to the fire-affected area.

This chapter addresses the skills and knowledge required to install a home fire sprinkler system inside a single-storey Class 1a building to the relevant standards and using approved materials and techniques. It does not address bushfire protection.

Learning objectives

Areas addressed in this chapter include:

- identify installation requirements
- prepare for work
- install, test and commission system
- clean up.

Identify installation requirements

The general requirements for installing home fire sprinkler systems for a Class 1a building can be obtained from job specifications, relevant Australian Standards, codes, manufacturers' instructions and jurisdictional requirements. Chapter 4 Carry out interactive workplace communication' in *Basic Plumbing Skills* contains information on how you can access and review these standards as required.

Access, read and determine Australian codes and standards

The requirements for the installation and testing of fire sprinkler protection systems for homes must be in compliance with AS 2118.5 Home Fire Sprinkler Systems, as well as the National Construction Code (NCC).

Australian Standards

The associated Australian Standards for automatic fire sprinkler systems are:

- AS 2118 Automatic fire sprinkler systems
- AS 2118.1 Part 1: General requirements
- AS 2118.2 Part 2: Wall wetting sprinklers (drenchers)
- AS 2118.3 Part 3: Deluge
- AS 2118.4 Part 4: Residential
- AS 2118.5 Part 5: Home fire sprinkler systems
- AS 2118.6 Part 6: Combined sprinkler and hydrant
- AS 4118 Fire sprinkler systems
- AS 4118.1.1 Part 1.1: Components – Sprinklers and valves
- AS 4118.1.2 Part 1.2: Components – Alarm valves (wet)
- AS 4118.1.3 Part 1.3: Components – Water motor alarms
- AS 4118.1.4 Part 1.4: Components – Valve monitors
- AS 4118.1.5 Part 1.5: Components – Deluge and pre-action valves
- AS 4118.1.6 Part 1.6: Components – Stop valves and non-return valves
- AS 4118.1.7 Part 1.7: Components - Alarm valves (dry)
- AS 4118.1.8 Part 1.8: Components – Pressure reducing valves
- AS 4118.2.1 Part 2.1: Piping – General.

National Construction Code (NCC)

Compliance with the NCC involves:

- the governing requirements of the NCC, such as the Australian Standards
- the *performance requirements*.

Performance requirements are satisfied by one of the following:

1. a performance solution
2. a deemed-to-satisfy solution
3. a combination of (1) and (2).

A performance solution and a deemed-to-satisfy solution must meet the performance and governing requirements supported by expert judgement from the relevant stakeholders to comply with the NCC. Performance requirements outline the minimum necessary standards different buildings or building elements must attain.

A solution may be partly a *performance* solution and partly a *deemed-to-satisfy solution*. However, no matter what method is chosen, the proposed work must always meet the performance requirements of the NCC.

The Australian Standards and the NCC are accessible online or via mail order.

The Fire Protection Association (FPA) Australia publishes guidelines and other documents to clarify issues within the fire protection industry and develops common practices to improve the quality of fire protection.

FPA Australia also helps to shape codes and standards to provide greater consistency and to ensure the delivery of appropriate responses to address fire safety and protect the community.

Classes of buildings

This unit refers to the installation of fire sprinkler systems for *Class 1a buildings only* (refer to Table 10.1). Class 1a concerns one or more buildings that together form a single dwelling, including:

- a detached house
- one of a group of two or more attached dwellings, each being a building, separated by a fire-resisting wall, including a row house, terrace house, town house or villa unit (see Figure 10.1).

Work health and safety

Work health and safety (WHS) regulations are implemented for everyone on the work site to protect

TABLE 10.1 Classes of building

Class 1	**Class 1a**	A single dwelling being a detached house, or one or more attached dwellings, each being a building separated by a fire-resisting wall, inlcuding a row house, terrace house, town house or villa.
	Class 1b	A boarding house, guest house, hostel or the like with a total area of all floors not exceeding 300 m^2, and where not more than 12 reside, and is not located above or below another dwelling or another Class of building other than a private garage.
Class 2	A building containing 2 or more sole-occupancy units each being a separate dwelling.	
Class 3	A residential building, other than a Class 1 or 2 building, which is a common place of long term or transient living for a number of unrelated persons. Example: boarding-house, hostel, backpackers accommodation or residential part of a hotel, motel, school or detention centres.	
Class 4	A dwelling in a building that is Class 5, 6, 7, 8 or 9 if it is the only dwelling in the building.	
Class 5	An office building used for professional or commercial purposes, excluding buildings of Class 6, 7, 8 or 9.	
Class 6	A shop or other building for the sale of goods by retail or the supply of services direct to the public. Example: café, restaurant, kiosk, hairdressers, showroom or service station.	
Class 7	**Class 7a**	A building which is a car park.
	Class 7b	A building which is for storage or display of goods or produce for sale by wholesale.
Class 8	A laboratory, or a building in which a handicraft or process for the production, assembling, altering, repairing, packing, finishing or cleaning of goods or produce is carried on for trade, sale or gain.	
Class 9	A building of a public nature.	
	Class 9a	A health care building, including those parts of the building set aside as a laboratory.
	Class 9b	An assembly building, including a trade workshop, laboratory or the like, in a primary or secondary school, but excluding any other parts of the building that are of another class.
	Class 9c	An aged care building.
Class 10	A non-habitable building or structure.	
	Class 10a	A private garage, carport, shed or the like.
	Class 10b	A structure being a fence, mast, antenna, retaining or free standing wall, swimming pool or the like.
	Class 10c	A private bushfire shelter.

Source: The NCC 2022 Volume One - Building Code of Australia Class 2 to 9 buildings was provided by the Australian Building Codes Board under the CC BY licence.

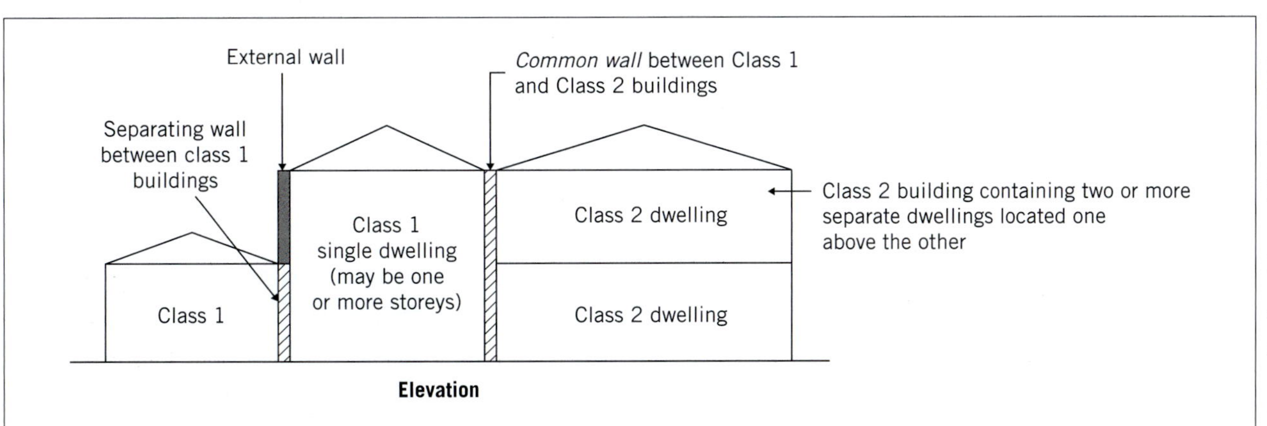

Source: https://ncc.abcb.gov.au/editions/ncc-2022/adopted/volume-one/a-governing-requirements/part-a6-building-classification

FIGURE 10.1 Classes of building

their health and safety. In the case of a fire sprinkler system, there are rules regarding WHS that ensure the safety of workers installing the system, other workers and the general public.

A job risk analysis (JSA) must be completed for all tasks prior to starting on the work site. Any high-risk activity such as working at heights, trench excavation or hot works will need a safe work method statement (SWMS) completed as well.

Other regulations, such as Australian Standards and codes of practice, specify the minimum requirement and the quality of materials to be used so that the completed installation is safe and performs the task it is designed for.

Working at heights

Elevated work platforms may be required when working at heights. The appropriate training must be attained before operating this equipment (Figure 10.2). Never exceed the safe working load (SWL) of the plant or equipment.

Working at heights may also require the use of ladders and scaffolds. The use of both is subject to appropriate training.

Sid Frisby/Alamy Stock Photo

FIGURE 10.2 Scissor lift

Scaffolds

The correct WHS procedures when using a scaffold involve the following:

- Do not attempt to build a scaffold unless you have been trained to do so.
- Never move a mobile scaffold with a person on it.
- Mobile scaffolds are suitable for solid, level surfaces only.
- Lock caster wheels before use.
- Guard against striking hazards such as overhead power lines, beams, bracing and other obstacles.

An example of a mobile scaffold can be seen in Figure 10.3.

iStock.com/Maciej Koza

FIGURE 10.3 Mobile scaffold tower

Ladders

The correct WHS procedures when using a ladder involve the following:

- Choose the correct ladder for the job (see Figure 10.4), making sure it is in good working order and is placed on a firm, level footing with the slope at a ratio of 4:1.
- The ladder should extend 1 m above the platform it has to reach.
- A ladder should only be used to gain access or to complete simple tasks. An elevated work platform or scaffold should be used for more complex tasks.

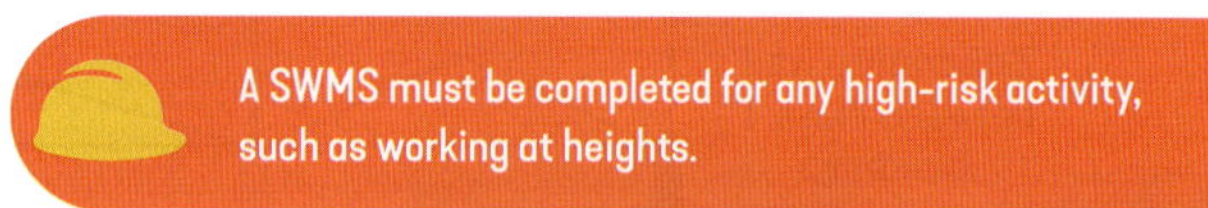

iStock.com/ZargonDesign; iStock.com/Spiderstock

FIGURE 10.4 Step ladder and extension ladder

Determine installation requirements

It is necessary to consult the plans and specifications before installing a home fire sprinkler system to determine what materials and quantities are required. A typical installation would demand the following considerations:

- connection to the water supply
- available pressure and flow rates
- location and sizing of piping system
- areas to be protected by sprinklers
- sprinkler head types
- actuating device
- an alarm.

It is normal practice for the installation to take place in two stages:

1. First fix or rough-in
2. Second fix or fit-off.

The piping system is installed at the rough-in stage, before the wall linings are applied. The water connection, associated valves, flow and pressure switches may be connected at this point, if possible, depending on site conditions. All work must be tested for soundness before the wall linings are installed.

The sprinkler heads, cover plates, ancillary valves and control switches are installed at the fit-off stage. Care must be taken not to damage finished or painted surfaces.

Sprinkler heads

All sprinkler heads must conform to AS 4118. They are classed as *standard* or *special*.

AS 4118 AUTOMATIC FIRE SPRINKLER SYSTEMS

Standard-type sprinkler heads

A sprinkler is designed to produce a spherical spray of water with a portion being directed towards the ceiling. Many conventional sprinklers are universal and can be used as pendant or upright, but others are designed solely for the pendant or upright positions (see **Figure 10.5**).

iStock.com/adventtr

FIGURE 10.5 Standard sprinkler

Spray sprinkler

A spray sprinkler produces a parabolic discharge below the deflector with no water directed to the ceiling.

Flush sprinkler

A flush sprinkler is used with concealed pipework for appearance in retail spaces, offices, hotels, etc. The escutcheon plate is level with the ceiling, but a heat-responsive element and retracted deflector is exposed and drops down when activated.

Recessed sprinkler

This is a sprinkler that incorporates a deep-welled escutcheon plate, allowing the sprinkler to be recessed into the ceiling (see **Figure 10.6**).

Imageroller/Alamy Stock Photo

FIGURE 10.6 Recessed sprinkler

Concealed sprinkler

A concealed sprinkler is a fully enclosed sprinkler that is flush fit to the ceiling (see **Figure 10.7**), and used in domestic and residential installations. The housing releases the cover plate at a temperature slightly lower than the operating temperature of the sprinkler. The sprinkler then drops down into position and operates.

John Fortner/Shutterstock.com

FIGURE 10.7 Concealed sprinkler

Side wall sprinkler

This sprinkler creates a half-spherical spray pattern, discharging from the wall into the room (see Figure 10.8).

iStock.com/M-Production

FIGURE 10.8 Side wall sprinklers

Operation of sprinklers

The sprinkler head will open and allow water to discharge through a variety of mechanisms (see Figure 10.9).

Glass bulb

The liquid (alcohol) inside the bulb expands due to a temperature rise. When the temperature reaches a set level, the glass bulb shatters and allows water to flow to the deflector, which creates the spray pattern.

Fusible link

A metal tag is held in place by a low-melting-temperature alloy. This will melt at a set temperature and allow water to flow to the deflector.

Rahmo/Shutterstock.com

FIGURE 10.9 Variety of automatic sprinkler mechanisms

Heat fin

Similar to the fusible link, this version has a series of sensor fins to capture any temperature increases and activate the melting of a fusible pellet that allows water to flow.

Sprinkler guards

All sprinklers that are likely to be damaged either during or after construction must be protected by the installation of a clip-on, wire-cage guard (see Figure 10.10).

Escutcheon plates

These are fitted to many sprinklers for aesthetic reasons, creating a smooth transition between the ceiling and the sprinkler head, which makes them suitable for home fire sprinkler installations. They can be one- or two-part design and may perform an additional function in the operation of the sprinkler (see Figure 10.11).

Lizard Design/Shutterstock.com

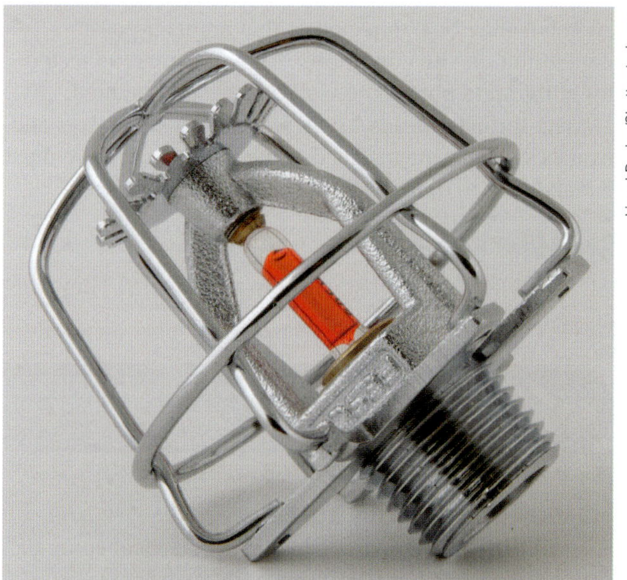

FIGURE 10.10 Sprinkler guard

William Hager/Shutterstock.com

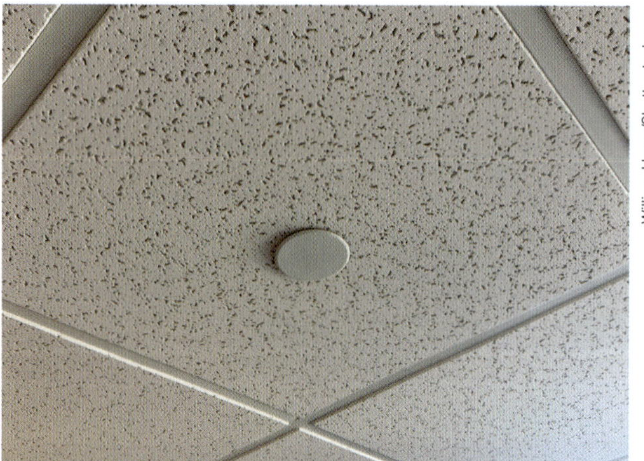

FIGURE 10.11 Escutcheon plates

Temperature rating of sprinkler heads

Sprinkler heads are colour-coded to determine their operating temperature and are chosen for the temperatures they will be exposed to (see Table 10.2).

For residential or domestic installations:

- If the normal room temperature is ≤ 38°C, then install 57°C–79°C standard sprinklers.
- If the normal room temperature is 38°C–68°C, then install 79°C–107°C fast-response sprinklers.
- Where sprinklers are installed below glass or plastic skylights in direct sunlight, they must be fast response.

TABLE 10.2 Colour-coded temperature rating for fire sprinkler heads

Glass bulb		Fusible link	
Temperature	**Colour**	**Temperature**	**Colour**
57°C	Orange	68–74°C	No colour
68°C	Red	93–100°C	White
79°C	Yellow	141°C	Blue
93°C	Green	182°C	Yellow
141°C	Blue	227°C	Red
182°C	Mauve		
204–260°C	Black		

Handling of sprinklers

When handling sprinklers, observe the following:

- Flush out any fluxes from the pipework as they can damage heads.
- Leave sprinkler heads in shipping cartons until used.
- Keep out of heat or direct sunlight.
- Keep the plastic strap in place when fitting the sprinkler head (see Figure 10.12).

tete_escap e/Shutterstock.com

FIGURE 10.12 Protective plastic cover

- Fit sprinkler heads after the range pipes have been installed.
- Gently guide sprinkler heads into the threaded fitting.
- Use only the manufacturer's spanner (see Figure 10.13).
- Only tighten to the correct torque setting (15 mm head–14 ft/lb or 20 Nm, which equates to 6 kg pressure on the end of a 300 mm spanner).
- Do not use Stillson's.
- Do not touch the thermal sensor when installing.
- No painting of sprinklers is allowed.

LEARNING TASK 10.1

1. Name the three solutions that can satisfy compliance by the NCC.
2. Name the three types of fire sprinkler release mechanisms.

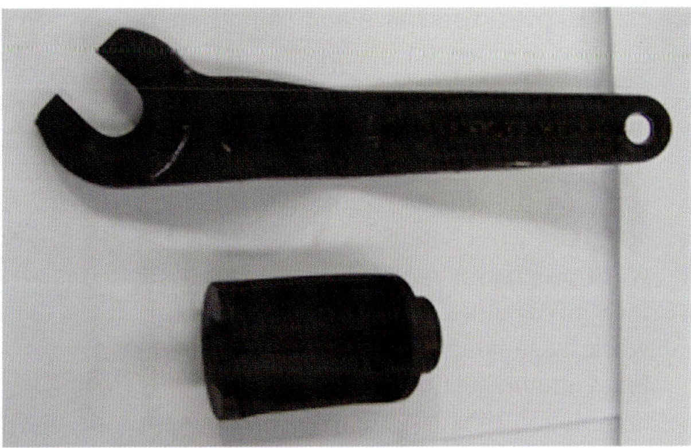

FIGURE 10.13 Manufacturer's spanner

COMPLETE WORKSHEET 1

Prepare for work

Before beginning the installation of a home fire sprinkler system, it is important to check that there is adequate water pressure available and the piping system is correctly sized.

Gathering information about the job will help to make informed decisions about the nature of the job and how to plan and sequence the entire process.

Planning the work

Planning the work and carrying out an initial site inspection include the following steps:

- Read the drawings and specifications to become familiar with the work.
- Check the work site against the drawings to confirm orientation and workspace.
- Check for normal work hazards.

- Check for special hazards, such as working above machinery, high temperatures or working at heights.
- Identify entry and exit points and storage areas.
- Confirm location of other services, such as air-conditioning ducts, that may affect the installation.
- Communicate with your supervisor, the builder or other contractors as a cooperative work strategy will help to avoid conflict.
- Organise equipment, such as ladders, scaffolds or scissor lifts, and personal protective equipment (PPE).
- Check the work area is safe and clear of obstacles.
- Complete all necessary work site inductions.

Working with others is an essential part of good work practice. Others may include your supervisor, the builder, other tradespeople and suppliers. Failure to work cooperatively with others leads to:

- conflict on the job
- mistakes being made
- clashes over workspace and storage space
- damage to stored or installed material
- lost, damaged or abused equipment
- low morale in the workplace
- delays to the job and cost overruns.

Ongoing site checks confirm:

- adequate work safety
- work progress
- location of other services where relevant
- working satisfactorily with others.

Final site inspection confirms:

- completion of work, signed off by the appropriate authorities
- plant equipment and tools are cleaned, maintained and stored in accordance with company policy
- site is clean and safe with all rubbish disposed of in appropriate bins
- leftover materials are appropriately stored for reuse.
- quality assurance has been achieved.

FROM EXPERIENCE

Good communication with the site supervisor and other trades avoids conflict and creates a positive workplace.

Select appropriate personal protective equipment

It is important to wear PPE that is appropriate for the task at hand, as well as keeping it in good order. PPE is designed to provide some protection from falling objects, crushing forces, noise, dust and grit, sun and rain.

Correct PPE for installing sprinkler systems includes:

- 100% cotton high-visibility workwear
- safety boots
- hard hat
- safety glasses
- earmuffs or ear plugs
- gloves.

Select appropriate tools and equipment

To work safely and effectively, it is important to have the correct tools for the job and to maintain them in good working order. A typical list of tools required for home fire sprinkler installations is as follows:

- lump hammer, claw hammer
- Stillson's, footprints, multigrips
- 12" shifter, 10" shifter
- hacksaw
- chalk line, plumb bob, stringline
- tin snips
- ratchet spanner
- 0.5" round/flat file
- screwdriver set
- tape measure
- tube cutters
- impact driver
- cordless drill.

It is important to have your own toolkit. Some tools may be supplied by the employer. In either case, you are responsible for their ongoing maintenance and replacement. That can be as simple as notifying your immediate supervisor of any faulty tools or equipment.

Depending on the type of piping system, you will also need to use the appropriate jointing tools, such as a press-fit tool, crimping tool or heating equipment. You need to be trained to use each item of equipment as required.

Ensure all tools and equipment are cleaned and in proper working condition before storing them in a safe and clean environment.

Flow, pressure and pipe sizing

Sizing of the sprinkler pipework must be done by a hydraulic calculation (refer to AS 2118.5). When sizing pipework, the following points must be considered:

- flow from each sprinkler (minimum 50 L/m)
- pressure available at the mains
- pressure losses in pipework
- minimum pressure at the sprinkler head (50 kPa)
- maximum velocity allowed
- number of sprinklers working simultaneously (minimum of two)
- maximum pressure allowed in pipework (1000 kPa)
- maximum velocity through pipes (10 m/second)
- maximum velocity through valves (6 m/second)
- minimum pipe size 20 mm.

The minimum flow and pressure from a sprinkler head can be obtained from the manufacturer's data sheets. As per AS 2118.5, domestic sprinklers have a minimum pressure of 50 kPa with a minimum discharge of 50 L/min.

In domestic buildings, the minimum flow required is determined by two sprinklers working at once. This will require a minimum flow rate of 100 L/min at a minimum pressure of 50 kPa for a duration of not less than 10 minutes.

If the minimum flow rate and pressure cannot be achieved, then a pump set must be installed.

Locating sprinkler heads

Domestic sprinklers are required in all areas except:

- hallways, stairs, etc., less than 3 m wide
- cupboards, wardrobes, pantries, etc., less than 3 m^2
- ceiling, floor and roof spaces not used as living areas
- toilets, bathrooms and ensuites
- open balconies, stairs, verandas, etc.

Spacing of sprinkler heads

The maximum spacing between rows of sprinklers is 4.9 m. The maximum distance between sprinklers on a row is 4.9 m. This gives a coverage per sprinkler head of 24 m^2.

The above maximum distances are dependent on the operation of the individual sprinkler head; therefore, it is important to consult the manufacturer's data sheet.

The minimum distance between sprinklers is 2 m, otherwise the sprinkler spray pattern will be disrupted and become less effective.

The maximum distance from a wall is 2.5 m.

The maximum distance from the sprinkler deflector to the underside of the ceiling is 25–100 mm, depending on the type of sprinkler.

When positioning sprinkler heads, consideration must also be given to joists, beams, columns, girders, trusses or steeply sloping ceilings to ensure adequate coverage (refer to AS 2118.5 for further information).

Using appropriate materials

The drawings and specifications will detail which materials and system components are required for the job. The common piping materials used for home fire sprinkler systems are copper (Cu), cross-linked polyethylene (PE-X) and chlorinated polyvinyl chloride (CPVC).

The NCC references a new type of sprinkler system as defined by FPA Australia's technical specifications for *Automatic fire sprinkler systems design and installation* (FPAA101D). This type of system reduces the installation costs and draws water from the drinking water supply of the residential building. This enables the use of traditional PE-X plumbing piping systems, watermarked to AS/NZS 2492 and AS/NZS 2537, to be used to service the fire sprinkler network for a home fire sprinkler service.

As AS 2118.5 states, any plastic piping used for fire protection purposes must be protected by 9.5 mm plasterboard, 12 mm plywood or other material providing equal protection.

Pipe materials and jointing

The materials and joints used for sprinkler services must be durable and capable of handling the increased temperature and pressure when a fire ignites. The most common materials are as follows.

Copper (Cu)

- Above ground, type A or B
- Install below ground to AS/NZS 3500.1 – minimum cover
- Bending 6 × diameter radius ≤ 50 mm
- Bending 5 × diameter radius ≥ 65 mm
- Jointing: press-fit, silver brazed, compression

Cross-linked polyethylene (PE-X)

- Straight lengths or coil
- Must be concealed (in ceiling spaces, behind wall sheeting or chased in walls)
- Areas less than 50°C
- Jointing: compression sleeved, crimped, push-fit

Chlorinated polyvinyl chloride (CPVC)

- Only in straight lengths
- Must be concealed (in ceiling spaces, behind wall sheeting or chased in walls)
- Areas less than 50°C
- Jointing: solvent cement, screwed, flanged, compression, heat fusion, electro-fusion

Calculating quantities

The skill of calculating quantities is developed from experience. It is usually a job best done by the plumbing supervisor or leading hand on the project.

HOW TO

CALCULATE QUANTITIES

1 Obtain the plans and specifications of the project and find the plans that indicate the sprinkler service pipework. The plan may be shown as a plan view or the more easily interpreted isometric view.
2 Obtain the correct equipment: a take-off sheet, scale rule, calculator, pencil, eraser, highlighter and pen. The take-off sheet will include the project description, date and service of the job. It will also have multiple columns for entering information and a space for the name and signature of the person compiling the quantities.
3 Starting at the water main, look for the first fitting or pipe required. Under item 1, write a detailed description of the material (e.g. 25 mm type B copper tube), then measure the length of material required and enter it on the sheet.
4 Use a highlighter to mark the plan, so you know you have covered that section. Check the drawing for more of the same material and add it to the take-off sheet.
5 Move onto the next item of material in the same way until you have completed the entire service. The take-off sheet can then be used to create an order from your preferred supplier.

Ordering materials

Following the company's procedure, the material may need to be ordered from several different suppliers. Some of the pipework may need to be prefabricated at the company's workshop. There may be time delays in delivering certain materials, so planning is important to ensure that the installation is not disrupted. Any delays may also affect other trades and the progress of the building project.

Checking the delivery

When materials are delivered, a docket is included that outlines the types of materials and the quantities delivered. It is important to check the materials listed on the docket against what was ordered and what was delivered. Note down any errors or omissions, then inform the driver, make a note on the driver's copy of the docket and contact the supplier quickly to rectify the errors. Also check the quality of the goods to see if there is any damage.

LEARNING TASK 10.2

1 How can conflict on the job be avoided?
2 What is the minimum distance between sprinklers?

COMPLETE WORKSHEET 2

Install, test and commission system

A home fire sprinkler system is made up of several components or parts, such as pipes, fittings, valves, sprinklers and water supplies. It is important to understand the installation requirements, as well as knowing how to test and commission the fire sprinkler system.

Installing the system

Once all the drawings and specifications have been checked; materials calculated, ordered and delivered; and risk assessment documented, the installation of the home fire sprinkler system can begin. As mentioned earlier, this is done in two stages.

First fix (rough-in)

This is the equivalent of the rough-in. The main connection and pipework into the building is installed. Next, the pipework is installed in the ceiling space with risers and droppers in the wall. Finally, the rough-in is capped off and the pressure tested at 1500 kPa for 30 minutes as per AS/NZS 3500.1.

Second fix

Fit sprinkler heads, escutcheon plates, valves, gauges and pump, if required. Purge air from the system at the drain test point and check for leaks.

Care must be taken when fitting sprinkler heads so no damage or marks occur to finished ceilings – hands must be clean or gloves used.

System pipework configurations

Home sprinkler systems must have a permanent water supply available that automatically flows upon activation. The pipework can be configured in various ways, as follows.

Combined system (gridded layout)

In this system, the drinkable cold-water supply is combined with the sprinkler system. This is the most common type of installation due to the convenience of connecting to one water service or the existing water supply. The sprinkler system is installed in a ring main loop with interconnecting lines serving sprinkler heads. Branches are taken from the loop to serve individual plumbing fixtures refer to AS 2118.5 Part 5: Home fire sprinkler systems. The main loop must be sized to supply a minimum of 112 L/min at 50 kPa (100 L/min for fire and 12 L/min for domestic cold water). The minimum piping size is DN 25 to achieve the flow rate and a dead-leg or dropper should not exceed 75 mm in length (refer to AS 2118.5 – Cl. 5.2.1.1).

Independent system (looped layout)

In this system, the drinkable cold-water supply is branched from the sprinkler system immediately after the meter. The sprinkler system is installed in a ring main loop with pipes taken either side to serve several sprinklers as required Refer to AS 2118.5 Part 5: Home fire sprinkler systems.

Independent system (branched layout)

In this system, the drinkable cold-water supply is branched from the sprinkler system immediately after the meter. The sprinkler system is installed in a main line with pipes taken either side to serve a few sprinklers as required Refer to AS 2118.5 Part 5: Home fire sprinkler systems.

Water supplies

Water supply for fire sprinklers can come from drinking water mains, recycled water mains or from storage tanks. (AS 2118.5 Part 5: Home fire sprinkler systems).

Directly to drinking water main

When combining the sprinkler service and drinking water supply, the design flow for the sprinkler system should not be less than 100 L/min (two sprinklers), plus an additional 12 L/min must be allowed for the drinking water flow. Water main capacity must be tested to determine whether it can meet both flow and pressure requirements. A water pressure/flow inquiry is required with the water authority to confirm this.

Directly to recycled water main

When combining the sprinkler service and non-drinking water supply, the design flow for the sprinkler system should not be less than 100 L/min (two sprinklers), plus an additional 12 L/min must be allowed for the non-drinking water fixtures. Water main capacity must be tested to determine whether it can meet both flow and pressure requirements. Piping must be identified with pipe markers at one metre intervals: *SPRINKLER SYSTEM – NON-DRINKING WATER*.

Private supply

The capacity of the tank, reservoir or dam must be large enough to provide at least 1200 L for the sprinkler reserve, which will allow a minimum of 10 minutes of sprinkler operation.

Sprinkler systems

There are two main types of sprinkler systems to choose from. The choice is dependent on the temperatures the piping system is exposed to.

Wet system

A system that is entirely full of water is used for home fire sprinkler systems. When a sprinkler head is activated, water discharges immediately.

Dry system

This system is charged with water to the alarm valve; from the alarm valve to the sprinkler heads, the pipework is charged with air. This system is used in areas prone to freezing (e.g. cool rooms) so that the water in the pipework does not freeze. When a sprinkler head is activated, air discharges initially, followed quickly by a stream of water.

Pipe support

All pipe-supporting arrangements must comply with AS/NZS 3500.1 and be located at least 300 mm either side of each sprinkler. The supports used to hold the pipework in position must be strong and durable so that the pipe is held firm even in the event of fire and must last the life of the building (50-plus years). The bracket and the fixing must be of a suitable material.

When choosing a support, consideration must be given to the following:

- It should firmly hold the weight of both pipe and water, plus a safety margin.
- The structure must be capable of holding the added weight.
- Are extra coatings required to prevent atmospheric or electrolytic corrosion?

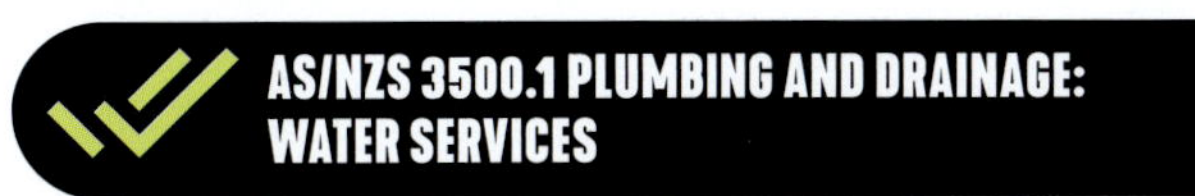

Valves

All valves used must comply with the appropriate standards and bear the watermark approval.

Control valve assembly

This usually consists of an isolating valve, a backflow prevention valve, a priority-control valve and a drain valve. The main control valve assembly is located in an accessible location, such as at the water meter assembly.

Backflow protection

The appropriate backflow prevention devices must be used for combined or independent systems in accordance with AS/NZS 3500.1 and local authority requirements.

Remote sprinkler test valve

All sprinkler systems must have a sprinkler test valve installed at the most hydraulically disadvantaged point. This is the sprinkler that is further and higher than any other sprinkler. This valve must be capable of discharging 112 L/min and labelled: *Sprinkler test point:...L/min at ...kPa.*

Drain valve

A drain valve (minimum DN 15 and commonly referred to as a *purge drain valve*) must be installed to drain the fire sprinkler system for maintenance purposes. This would be located at a low point of the installation where water can safely drain and air can be purged from the system.

Ball valve

This isolation valve replaces the gate valve in pipework less than 50 mm. A quarter of a turn will allow the valve to move from fully on to fully off. This allows full bore flow (see Figure 10.14).

iStock.com/DmitriyKazitsyn

FIGURE 10.14 Ball valve

Butterfly valve

This isolation valve replaces the gate valve in pipework larger than 65 mm on residential, commercial or industrial installations. It operates by turning a centrally fixed circular wing a quarter of a turn from fully on to fully off. When open, water flows either side of the thin circular wing (see Figure 10.15).

Flow switch

This is an actuating device that sends an electrical signal to another device (e.g. alarm, pump) when it senses movement of water in the pipework. It is installed by drilling a hole in the pipe and fitting the valve, which is sealed by an O-ring and clamp (see Figure 10.16). This is used on residential, commercial or industrial installations.

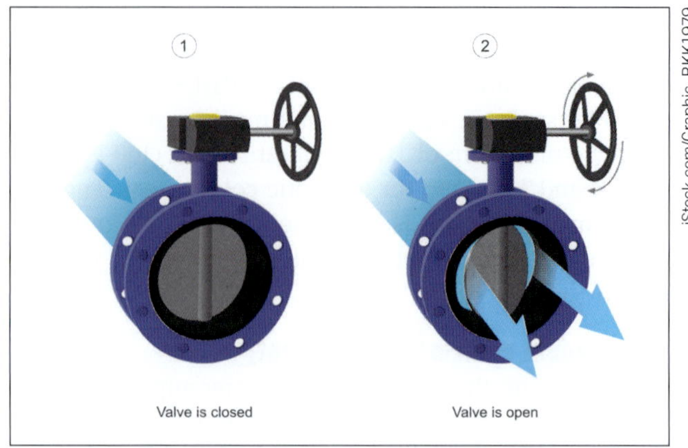

iStock.com/Graphic_BKK1979

FIGURE 10.15 Butterfly valve

FIGURE 10.16 Flow switch

Pressure switch

This is an actuating device that sends an electrical signal to another device (e.g. alarm, pump) when it senses a pressure change in the pipework. It is usually screwed into a tee in the pipe (see Figure 10.17). This is used on residential, commercial or industrial installations.

Alarm check valve

This valve incorporates pressure gauges, non-return valve, test point, drain connection and alarm connection. The alarm check valve stops water from flowing backward into the system, preventing contamination or damage to the piping system. Another function of the

FIGURE 10.17 Pressure switch

alarm check valve is to detect low water pressure in the system. If the water pressure in the system drops below a certain level, the valve will activate an alarm signal to alert the building occupants and the fire department that there is problem. It can be used in a wet or dry system. This is used on residential, commercial or industrial installations.

FIGURE 10.18 Alarm valves

Aural alarm

Smoke detectors are used in conjunction with home fire sprinkler systems. Residential, commercial and industrial sprinkler systems have a device that rings to attract the attention of bystanders. It responds within 90 seconds of opening the test valve and must be heard in all occupied areas.

Non-return valve (check valve)

This valve prevents water flow in a direction other than the intended direction (see **Figure 10.19**).

Priority-demand valve

A valve that automatically gives water supply priority to the fire sprinkler system from the domestic supply when a sprinkler is activated (see **Figure 10.18**).

cameracantabile/Shutterstock.com

FIGURE 10.19 Non-return valve

Signage

Where the home fire sprinkler system is supplied through a metered service, the meter valve should have a durable metal or plastic tag stating: *WARNING: Fire sprinkler supply – Closure will isolate sprinkler protection.*

If a pump is installed, a tag should be attached stating: *WARNING: Fire sprinkler pump installed – Switch pump off before closing this tap* (refer to AS 2118.5).

Spare sprinklers

Two spare sprinklers of each type used must be kept onsite to allow the system to be rapidly restored when sprinklers need replacing. They must be stored in a safe and secure location and clearly labelled. The special purpose sprinkler spanner must be stored with the spare sprinklers (refer to AS 2118.5).

Commissioning and testing

There may be specific requirements from the manufacturers of different components that are used in an installation. When this occurs, it is important to check testing and commissioning requirements as detailed in the manufacturers' instructions.

AS 2118.5 Home Fire Sprinkler Systems provides further instruction on the testing and commissioning procedures, including the folllowing:

- All pipework must be flushed to remove any foreign material from the pipe system.
- Test at 1500 kPa for 30 minutes.
- Rectify any leakage.
- Conduct a verification test (flow and pressure from remote test valve).
- Test all valves, pumps, tanks, alarms, etc. for correct operation.
- Complete all requirements of Appendix C AS 2118.5 (pressure and flow readings).
- Supply a copy of the *Home Sprinklers Owners Guide* (Appendix C AS 2118.5) to the homeowner.
- Check all labelling and tags are installed and legible.

LEARNING TASK 10.3

1. Why must the pipework be flushed?
2. When would a dry system be installed?
3. What is a 'priority demand' valve?
4. What signage must be provided when a home sprinkler fire service has been installed through a metered service?

Clean up

Remember that the job is not complete until the work area is cleaned up and the rubbish removed:

- The work area must be tidied up by packing away all the tools, equipment and leftover material. Then the area must be swept clean.
- Be sure to reduce waste and save money by storing any leftover material that can be reused or recycled in accordance with the workplace procedures
- Clean all tools and equipment when packing up, keeping them lubricated and maintained.
- Inspect all tools for any signs of damage. If damage is noticed it should be immediately reported to the supervisor, and the tool should be tagged out and taken out of circulation until the damage has been repaired.
- Tools should be stored in a clean, dry and secure location.
- Document information in the diary: job address, description of work done, names of people involved, and the time and the dates spent on that job.

Some companies may have a generic worksheet to complete. This would form part of the company quality assurance policy.

GREEN TIP

Water utilities normally maintain a conservative approach to their water supply. This is mainly to allow for future growth in particular areas, even if the current available water pressure might appear to be more than adequate.

Maintenance

The owner is responsible for the condition of a sprinkler system. Some maintenance items include:

- visually inspecting all sprinklers for damage
- inspecting all valves
- testing all flow devices
- testing alarm systems
- testing any pumps or tank operation
- recording all maintenance.

COMPLETE WORKSHEET 3

SUMMARY

- A home fire sprinkler system consists of a piping system with the associated valves and fittings required so that the fire sprinklers will activate automatically, getting water onto a fire fast. This saves lives and protects property.
- Sprinklers must be used in conjunction with smoke alarms as they will not detect smoke, only heat. The smoke alarm provides an early warning to the occupants.
- The home fire sprinkler system is connected to a permanent water supply, feeding sprinklers in the cooking, living and sleeping areas of a domestic building (Class 1a).
- The combined system would suit a new home as it is more suited to a new installation.
- An independent system is probably more suitable for use in an existing home.
- Choosing the appropriate piping material and sizing, valves and location of sprinklers with adequate flow and pressure is vital for a home fire sprinkler system to operate effectively.
- As pipe sizing is not included in this unit, AS 2118.5 can be used for reference. It is recommended that a computer-based hydraulic flow program would achieve more efficient pipe sizing.
- The owner must regularly maintain the fire sprinkler system to ensure it is in good working order.

REFERENCES

Davey: **http://www.davey.com.au**
Grundfos: **http://www.grundfos.com**
National Construction Code (NCC): **https://ncc.abcb.gov.au/**
NSW Environment Protection Authority: **http://www.epa.nsw.gov.au**
Standards Australia: **http://www.standards.org.au**

GET IT RIGHT

1 Which is the most common home fire sprinkler piping system used?

2 Why is this piping system the most popular?

To be completed by teachers
Student competent ☐
Student not yet competent ☐

Student name: ______________________

Enrolment year: ______________________

Class code: ______________________

Unit competency code/title: CPCPWT3030 Install home fire sprinkler systems

Task: Review 'Identify installation requirements', then answer the following questions.

1 Why are automatic fire sprinklers the most efficient fire protection system?

____

____

____

2 Define a Class 1a building.

____

____

____

3 When a high-risk activity takes place on a construction site, what risk assessments must be documented?

____

____

____

4 Name four considerations when determining installation requirements for a home fire sprinkler service.

____

____

____

5 At what stage of the installation are the sprinkler heads fitted?

____

6 Name the two categories for fire sprinkler heads.

____

____

7 Describe the purpose of an escutcheon plate.

8 If the normal room temperature is 38°C or less, what temperature-rated sprinklers would be used?

9 State five precautions when handling fire sprinkler heads.

10 What type of sprinklers must be installed under a skylight?

11 Why is painting a fire sprinkler not allowed?

WORKSHEET 2

To be completed by teachers	
Student competent	☐
Student not yet competent	☐

Student name: ______________________

Enrolment year: ______________________

Class code: ______________________

Unit competency code/title: CPCPWT3030 Install home fire sprinkler systems

Task: Review 'Prepare for work', then answer the following questions.

1 State five considerations when sizing pipework to supply sprinkler heads.

____

____

____

____

____

2 What is the minimum flow rate required for a home fire sprinkler system?

____

3 Name three locations where fire sprinklers are located in a house.

____

____

____

4 What is a limitation of using plastic piping for a home fire sprinkler service?

____

____

____

5 What are the three common materials used for home fire sprinkler piping systems?

____

____

____

6 What is the maximum spacing of home fire sprinklers?

____

7 What is the minimum spacing of home fire sprinklers?

____

8 What is the minimum pressure required at the most disadvantaged sprinkler head?

____

9 Why is it better to have large radius bends in the pipework?

____

____

WORKSHEET 3

To be completed by teachers	
Student competent	☐
Student not yet competent	☐

Student name: ______________________

Enrolment year: ______________________

Class code: ______________________

Unit competency code/title: CPCPWT3030 Install home fire sprinkler systems

Task: Review 'Install, test and commission system' and 'Clean up', then answer the following questions.

1 List two methods of supplying water to sprinkler systems.

____

____

2 Why would a dry pipe system be used in preference to a wet pipe system?

____

____

____

3 What is the minimum pipe support required from a sprinkler head?

____

____

4 What is the minimum size of a domestic sprinkler pipeline?

____

5 What is the maximum length of a dropper or dead-leg in a home fire sprinkler piping system?

____

6 Describe the function of a pressure switch.

____

____

____

7 Describe the function of a drain valve.

____

____

____

8 Describe the function of a priority demand valve.

______

______

______

9 What are the testing requirements for a home fire sprinkler installation?

______

10 How can you tell the temperature setting at which a sprinkler will open?

______

SET OUT, INSTALL AND COMMISSION IRRIGATION SYSTEMS

Chapter overview

The design and planning of a landscape and irrigation system is a significant factor in determining how efficiently and effectively water is used for the landscape and garden. Setting up the landscape and the correct installation of irrigation products, along with adequate maintenance and efficient irrigation practices, impacts how water-wise the garden and landscape are.

This chapter covers the basic layout and design of an irrigation system, the different types of irrigation systems, water supply sources and installation procedures. It also addresses the following key elements for the competency 'Set out, install and commission irrigation systems'.

The chapter will introduce the requirements of:

- obtaining plans and specifications, meeting relevant work health and safety (WHS) and environmental requirements associated with setting out, installing and commissioning irrigation systems
- identifying and recording installation and system requirements as per the plans, specifications and quality assurance requirements
- planning the works with contractors, staff members and suppliers to meet the regulatory authorities' requirements.

Learning objectives

Areas addressed in this chapter include:

- prepare for work
- identify installation requirements
- install and commission irrigation system
- clean up.

Prepare for work

In most states and territories, installing or maintaining an irrigation system downstream from an isolating valve, tap or backflow prevention device on the supply pipe is not regulated. So, it is the responsibility of the licensed plumber to prevent any cross connection with the drinking water supply and install the appropriate backflow prevention device.

The role of the plumber in backflow prevention is important, and that is why it is necessary to undergo extensive training. It is the plumber's responsibility to help to protect the community's water supply.

Obtain plans and specifications

Preparing for work is an essential part of executing the set-out, installation and commissioning of an irrigation system. Obtain the relevant plans and specifications from the builders, architects and landscape designers. The plans will outline in detail the location, type of plants and landscape area to be irrigated. In most circumstances, plumbers are given the plans on the type of irrigation system to install; however, the plumber can *design* the type of irrigation systems required, such as a drip, sprinkler or centre pivot irrigation system.

The *type* of irrigation system will be determined by the relevant authority regulation requirements, such as to connect into the potable water supply within a property. The most common types of plans that are used for irrigation systems within the construction industry are 'landscape plans'.

The *specification* informs contractors on how they must comply with the standards of the design, constructing the design perimeter, material types and maintaining the irrigation systems. It is the intent of both the specification and the design that the contractor provides all labour, materials/equipment and service for the complete installation, testing and operation of the systems, and works to the satisfaction of the architect. The following are examples of what might be included in a specification:

- authorities
- acceptance of site
- coordination of work
- Before You Dig Australia (BYDA)
- excavation and backfilling
- general items
- landscape items
- material and workmanship
- new and existing services
- scope of works
- standards and codes
- standard specifications
- testing and commissioning procedures
- tests and inspections
- work specified in other trades.

FROM EXPERIENCE

Plumbers should always be aware of all current rules and regulations, and safe work practices, and be aware of any new technology related to their work.

Adhere to work health and safety (WHS)

As plumbers, we are required to protect the public and the potable water supply that the local utility provides to the community. It is the responsibility of plumbers to ensure that current Australian Standards are met. This includes the following inclusion to the WHS management plan to comply with *Work Health and Safety Act 2011* (WHS Act).

Safe work method statement

A safe work method statement (SWMS) must be completed for any high-risk activity in construction before work commences.

What is required for an SWMS?

The information required in a safe work method statement (SWMS) includes:

- details of the work that are considered high risk
- health and safety hazards relating to the work
- control measures to be implemented to minimise or remove the risks.

Safety datasheets

The safety datasheet (SDS) lists the hazardous ingredients of a product, its physical and chemical characteristics (e.g. flammability, explosive properties), its effect on human health, the chemicals with which it can adversely react, handling precautions, the types of measures that can be used to control exposure, and emergency and first aid requirements.

Quality assurance

A company has a quality assurance procedure to ensure the workmanship is carried out to a certain standard by following a documented process. Quality standards are defined as documents that provide requirements, specifications, guidelines or characteristics that can be used consistently to ensure that materials, products, processes and services are fit for their purpose.
The seven principles of quality management are

engagement of people, customer focus, leadership, process approach, improvements, evidence-based decision making and relationship management. Figure 11.1 is a graphical representation of a process to control quality assurance to meet the intended control and specification.

Planning tasks with others

Planning is the process of thinking about the activities required to achieve the desired goal. *Construction planning and scheduling* can improve work efficiency; having effective material management and properly distributing resources can reduce costs and save time. *Integrated project controls* enhance the capability of a team to ensure project success within schedule, constraints and budget. Developing a 'breakdown chart' (see Figure 11.2) allows plumbers to complete each task as they see it, including in relation to design, installation, testing and commissioning.

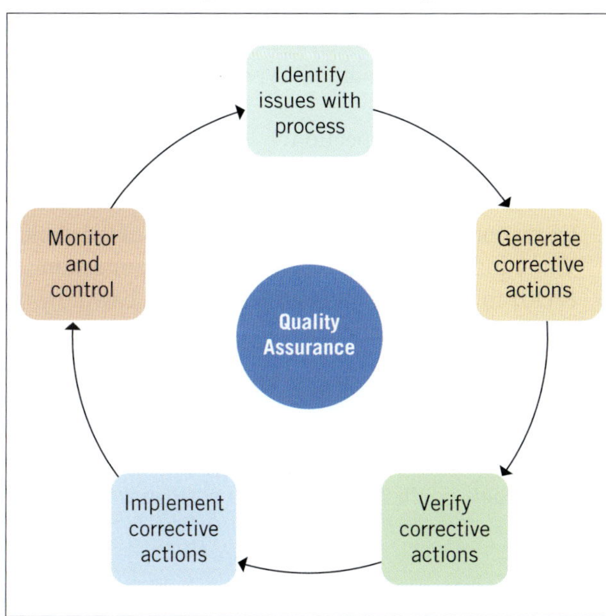

FIGURE 11.1 Quality process

Cooperation shall be undertaken between the separate trades where piping and equipment is to be installed along with other work being installed under other trades. All parties must ensure that equipment is installed to the best advantage, minimising any interference with other works occurring during construction.

Selecting tools, equipment and PPE

Selecting the right tools and equipment is very important so that the installation process runs smoothly. This process may also include any permits required by the local authority in the area. Each tool has its task to complete the work and selecting the incorrect tool or equipment may cause serious injury or death (see Table 11.1).

PPE in the plumbing industry

Personal protective equipment (PPE) considerations for plumbers should include boots, gloves and wraparound eye protection as a minimum, along with ear protection when using power tools. Steelcapped boots should be worn unless working at heights, when Dunlop Volleys will provide better grip. In any work situation, plumbers can reduce their chances of being injured by using PPE. PPE that plumbers will need to use when installing water pump sets include:

- dust masks
- earmuffs
- fitted gloves
- hard hats
- long-sleeve overalls
- safety boots
- safety glasses or goggles.

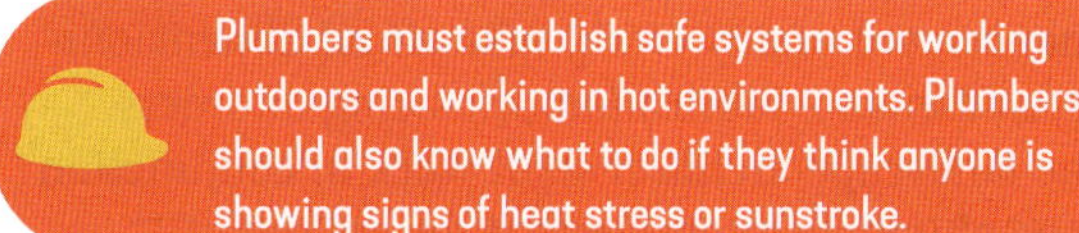

Plumbers must establish safe systems for working outdoors and working in hot environments. Plumbers should also know what to do if they think anyone is showing signs of heat stress or sunstroke.

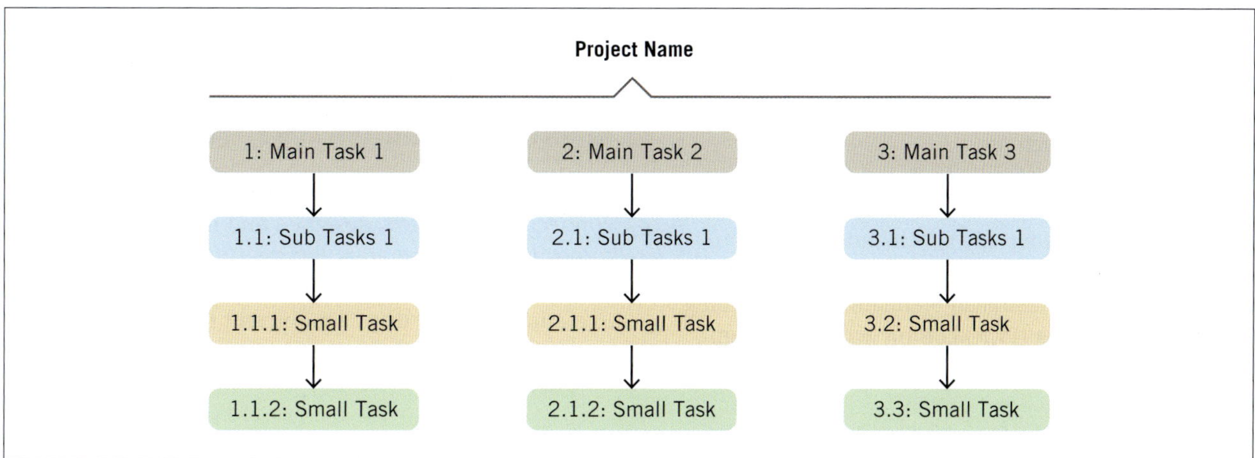

FIGURE 11.2 Breakdown chart

TABLE 11.1 Selecting the right tools and equipment

Tools and equipment	
Permit (as required by the local authority), SWMS, Before You Dig Australia search and accredited contractor to locate existing services	Spray marking paint
Small irrigation flags	Tape measure, string line and pegs
Hacksaw	Trenching shovel or pipe-puller
Hammer	Tunnel kit or hose jetting kit
Pipe wrenches	Wire cutters
Plastic tarp	Insulated wire staples
Pliers	Rain shut-off device/weather sensor
Rags	Shut-off valves
Rake	Valve boxes, 150 mm and 300 mm
Screwdriver	Teflon tape (used on all PVC or poly thread-to-thread fittings)
Excavation – mini excavator, shovels – trenching machine, flat, spade or point shovel	Automatic testing gauge with flow meter

Preparing the work area

It is essential that plumbers carefully and accurately set-out all piping runs and establish penetration locations well in advance of structural works to ensure that adequate accommodation is provided for the pipework without interference to the design. This includes setting out of all irrigation services and properly coordinating with other trades to avoid any set-out difficulties and conflicts with other services.

It is the plumber's responsibility to check finished levels for correct cover and locations of all services, check levels of existing services before the commencement of any work, take off all dimensions onsite, and ensure fabrication or placing of orders.

The landscape architect shall work in conjunction with the plumbers to ensure that the set-out of pipework to all groups of irrigation is fully coordinated.

Proposed set-out of all pipework and fittings must be identified onsite before the commencement of installation of pipework and prior approval obtained by the landscape architect or client.

It is important to carry out the pre-planning inspection work, such as:

- Obtain required permits from the water utility and local council.
- Check the plans for any specific installation requirements.
- Check that the Australian Standards and Plumbing Code of Australia (PCA) comply with the design installation.
- Select the correct material type and refer to the specifications.
- Carry out a site inspection.
- Ensure all quality assurance, SWMS and WHS.
- Discuss this with the client, and outline the procedures for set-out, installation and commissioning of the works.
- Water meter location, type of point of connection (POC) to existing water service and any backflow prevention requirements.

FROM EXPERIENCE

Good communication with the site supervisor and other trades avoids conflict and creates a positive workplace.

LEARNING TASK 11.1

1 What is the information found on an SWMS?
2 What is the purpose of an SDS?

Identify installation requirements

An irrigation system is a network of permanent piping connected to emitters designed and installed to water a specific landscape area. *Emitters* are devices fitted on a pipe that operates under pressure to discharge water in a spray, mist or drip. An efficient irrigation system is designed and installed to minimise the water output capacity.

Identify irrigation system requirements

Different applications on a site may require different types of irrigation system to be employed in different areas of the site. A properly designed irrigation system will cater for these various requirements. Pop-up sprinklers may be required for lawn areas, whereas (for example) a drip system, a 'tree watering' system or a trickle system may be used in various garden areas. Each of these systems may require various pipe and connection sizes.

When planning to provide water for these different types of irrigation systems, it is advisable to keep the different types of systems on separate supply lines, as different flow rates and control methods may be required.

Achieving efficient irrigation requires knowledge of how much water should be applied at any given time to replenish the water consumed by the plants and grass, and how much water can be held by the soil.

Designing a residential system requires detailed measurement of the property boundaries along with the house location. Sketch out the property and the critical points on a plan (the corners).

Using a compass, draw a point on each corner including an arc showing the sprinkler spray. Place measurements on the sketch by referring to the landscape architectural drawings. Be sure to include all hard or masonry surfaces, which may include concrete or brick walkways, patios, driveways and fences. While measuring, locate any trees, shrubs and lawns, and include them on the sketch.

The scale can be 1:50, 1:100 or 1:200. Write the scale on the plan. Make sure to note any lawn, shrub, ground cover and large trees.

On the same plan, divide the property into areas: front yard, back yard, side yard, lawn or plantation areas and shady areas. Label the areas A, B, C, D and E.

Locate and identify underground services

Underground services are the cables, pipes, assets and equipment providing utilities to homes, businesses and public buildings. This includes underground electricity cables, water pipes, phone lines and gas pipes.

The installation of underground utilities is usually done by conventional excavation methods, but now trenchless installation methods are also being used. Conventional excavation methods pose many risks to construction workers due to confined spaces, engulfment, electrocution and unstable trenches.

There are two major methods for detecting the locations of the underground infrastructures. These are the *electromagnetic* method and the *georadar* ground penetrating radar (GPR) method. The electromagnetic method is used only for detecting the locations of metal infrastructures. Plumbers engage external contractors to conduct a detection of underground services; most utility authorities have accredited professionals to carry out the service. This information is received once plumbers contact the Before You Dig Australia service.

It is important to know where underground pipes, cables and utilities are located. Asset plans can be accessed online via Before You Dig Australia's free referral service, whose core vision is to prevent injury and reduce damage to members' infrastructure assets.

No matter how small or large the excavation project, determining the location of utilities should be the first step in any safe excavation. Engage a locating contractor to locate services within a private property to ensure everyone's safety.

Determine system design capacity water flow and water pressure

When planning an installation of an irrigation system, a plumber must first determine the correct sprinkler system design capacity; that is, how much water is available for residential irrigation. If the system will be installed using town mains, a storage tank or bore water, follow these steps.

Set out irrigation pipes according to plans, specifications and site requirements

The type of pipe used will depend on the pressure required for the application. The most common pipe used for domestic irrigation is a low-density poly pipe (LDPE), as it is flexible with easy-fitting installation and economical. It is ideal for low-pressure applications, such as drip, sprinklers and micro-irrigation.

For higher and static pressure situations, it is recommended to use *metric* or *Rural B* pipe with typical pressure ratings of 800 kPa all the way up to 1600 kPa. Any pipe used before solenoid valves should always be metric (blueline) or PVC.

HOW TO

DETERMINE A SPRINKLER SYSTEM DESIGN CAPACITY

1 Water pressure (in bar or kPa): to check the water pressure, attach a pressure/flowrate gauge to the outside faucet closest to the water meter (see **Figure 11.3**). Make sure that no other water is flowing at the residence. Turn on the faucet and record the reading in Table 4. This is the static water pressure recorded in kilopascals (kPa).

2 Water volume (l/min) via a flow cup or flow meter: determines the volume of water available for the system. You need two pieces of information, which are *working pressure* and a *flow rate* in litres per minute (see **Figure 11.4**). Record the reading in Table 4.

FIGURE 11.3 How to check water pressure without flow – static pressure

FIGURE 11.4 How to check water pressure with flow – working pressure

Low-density polyethylene pipe

LDPE has a pressure rating of 300 kPa. This is the most common pipe used for domestic and commercial irrigation applications. Barbed low-density fittings are used with these pipes. They are manufactured with ultraviolet (UV) and oxidation protection, and are therefore durable to solar radiation for at least 50 years. They are immune to internal and external aggressions from microorganisms and are resistant to most substances employed in agricultural applications (see **Figure 11.5**).

Straight lengths rural B poly

Straight lengths rural B poly (MDPE Green Line Pipe) is a rural grade of polyethylene pipe (also known as a poly pipe), which is limited to a 600 kPa operating pressure and is available in the old imperial diameters from ¾" to 2".

FIGURE 11.5 Blueline pipe used for higher pressure rating from 800 kPa to 1600 kPa

Metric poly (HDPE blueline pipe) is the most common irrigation pipe (see **Figure 11.6**).

Toro Drip Eze comes in discreet brown tubing to blend into the landscape. Drip emitters are welded to the inside wall of the tubing. This one-piece construction prevents damage and loss of drippers, and eliminates hole punching and handling damage (see **Figure 11.7**).

iStock.com/PictureLake

FIGURE 11.6 LDPE is most used in domestic irrigation

FIGURE 11.7 Two ways to comply with a Type A irrigation system

Identify, order and collect materials

A 'material list register' (see **Table 11.2**) is a very important requirement and should always start from the point of connection, whether connecting to an existing hose tap or a fixed metallic pipe connection.

This is followed by a 'list of materials' (see **Table 11.3**), such as tees, elbows and couplings; each job is different and may not be included in the register. Take the time to plan as a mistake could be an expensive exercise on a job site.

TABLE 11.2 Material list register

Point of connection (POC)	
Brass compression tee (compression × compression × thread)	Copper pipe, poly pipe, brass fittings, PVC pipe, blueline poly, greenline poly
Brass gate valve or brass ball valve	Line flushing valve
Valve box	Manifold fittings
Backflow prevention device	Solenoid valves
Isolating valve – non-return	Arkal filters – prevents the dripline and sprinklers from clogging up

FROM EXPERIENCE

Always check material and equipment delivery dockets, and check for goods damaged in transit or faulty items from the manufacturer to avoid any delays.

LEARNING TASK 11.2

1 What type of sprinklers are used for lawn areas?
2 Name the two main methods used to detect underground services.

COMPLETE WORKSHEET 1

Install and commission irrigation system

Each property has its own unique characteristics and is designed to the specification of that property; therefore, it is essential that the type of irrigation system is determined to suit the needs of the property.

Irrigation system types and backflow prevention

According to AS/NZS 3500.1 – Plumbing and Drainage Part 1: Water Services, irrigation system types should be categorised as:

- *Type A irrigation system* is a very simple irrigation system installation that requires a stop valve to the hose tap at the point of connection as shown in **Figure 11.5** (AS/NZS 3500.1 requirement). It requires an opening to be 150 mm above the finished floor level.

TABLE 11.3 List of materials

FITTINGS: Calculate the length of pipe and number of fittings required						
PVC (slip x slip x slip)		**20 mm**	**25 mm**	**32 mm**	**Poly (compression or barbed insert fittings)**	
Tee	S × S × S S × S × ½" (13 mm) T S × S × ¾" (20 mm) T				i × i × i i × i × ½" (13 mm) T i × i × ¾" (20 mm) T	Tee
Elbow	90° × S × S 90° S × ¾" (20 mm) T 90° S × 1" (25 mm) T 45° × S × S				90° × i × i 90° i × ¾" (20 mm) T 90° i × 1" (25 mm) T 45° × i × i	Elbow
Reducer bushing	25 mm S × ¾" (20 mm) S 32 mm S × 1" (25 mm) S				1" (25 mm) i × ¾" (20 mm) i 1¼" (32 mm) i × 1" (25 mm) i	Reducer coupling
Reducing tee	S × S × S				i × i × i	Reducing tee
Male adapter	S × T				I × T	Male adapter
Coupling	S × S				I × i	Coupling

- *Type B irrigation system* is the most common installation that requires a vacuum breaker, or an atmospheric vacuum breaker as shown in **Figure 11.8** (AS/NZS 3500.1 requirement). These devices are a form of backflow prevention device categorised as *low* hazard. This applies to domestic or residential properties with pipe outlets installed less than 150 mm above the finished surface level and no fertiliser injector incorporated into the irrigation system.
- *Type C irrigation system* is shown in **Figure 11.9**. These devices are a form of backflow prevention device categorised as a *medium* hazard (AS/NZS 3500.1 requirement). This applies other than to domestic or residential properties with pipe outlets installed less than 150 mm above the finished surface level and no fertiliser injector incorporated into the irrigation system.
- *Type D irrigation system* is shown in **Figure 11.8**. These devices are a form of backflow prevention device categorised as a *high* hazard where chemical additives are injected or siphoned into the irrigation system. A testable device, such as a reduced pressure zone (RPZ) device, is to be installed upstream of the irrigation system.

FIGURE 11.8 Type B irrigation system, with a hose connection vacuum breaker

All Type B, C and D systems should be protected with an approved backflow prevention device (see **Figure 11.10**) as per AS/NZS 3500.1.

Only licensed plumbers with backflow prevention accreditation can inspect, commission and test medium- to high-hazard backflow devices. The installation can be completed by an apprentice, trainee and journeyman plumber under supervision of a licensed plumber.

What size is the water meter?

The water meter will generally have the size stamped on the meter body. The most common size for residential meters is 20 mm. In some areas, the water is connected directly to the water main without the use of the water meter due to being in the process of the construction phase. Apply for a water meter connection via the local water authority (see **Figure 11.11**). Record the water meter size in **Table 11.4**.

FIGURE 11.9 Type C irrigation system, with a hose tap connection configuration

FIGURE 11.11 A meter set-up without a meter installed to a new building site

TABLE 11.4 Record the readings

Readings	Example	Your recording
Static water pressure	320 kPa	
Flow rate	18.9 l/min	
Size of the meter	20 mm	
Service line size	20 mm	

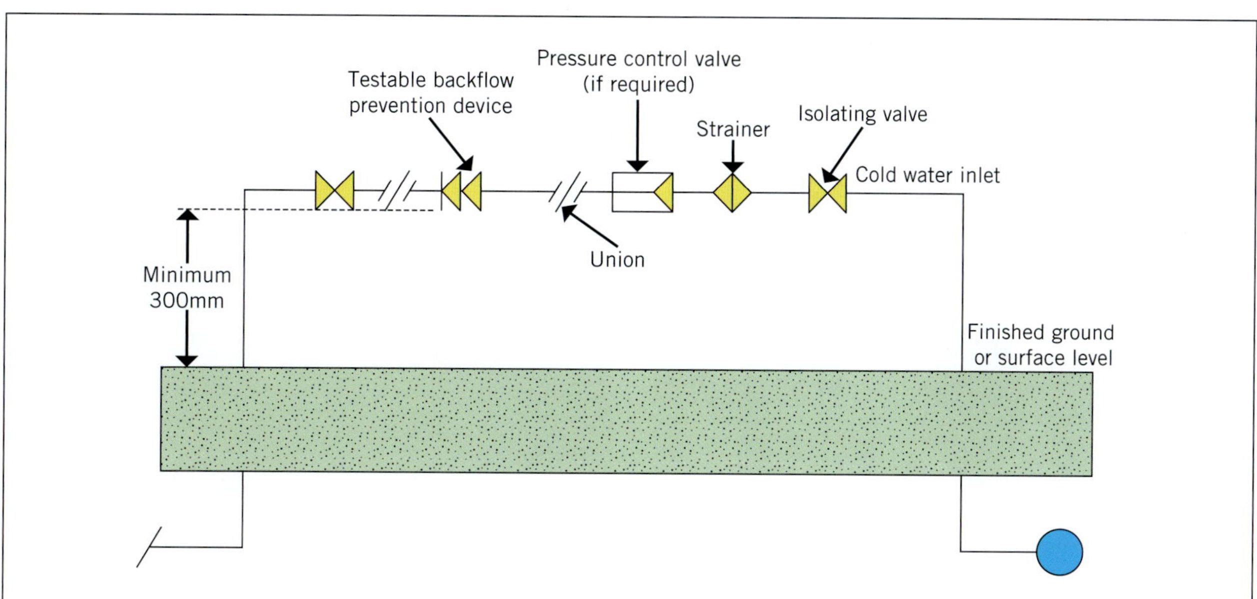

FIGURE 11.10 Standard irrigation system connected to a drinking or non-drinking water service with testable backflow prevention device

Set out the irrigation system to plans and specifications

Take the time to set out the irrigation system by referring to the plan/layout and specifications. Accurately measure and mark out the sprinkler outlets and irrigation pipe location from the point of connection.

Point of connection (POC)

Determine the location of the POC. It must be between the water meter and the service line. The most common connection is from an existing hose tap, as shown in Figure 11.12.

The mainline should generally be one pipe size larger than the largest lateral line. Most areas require some type of backflow prevention to protect drinking water depending on the type of irrigation system installed (refer to AS/NZS 3500.1).

Copper pipe may be required between the POC with the backflow prevention device installed and all associated valves. Always check the local building code or with the local authority agency for the requirements in the area.

FIGURE 11.12 A common POC type from a hose tap with an isolating valve and a dual-check valve

Using an alternative water supply

When using an alternative water source for an irrigation system, consideration must be given to the quality of the water and any possible cross connection with the drinking water supply.

The supply to different areas within the irrigation system must be designed so that each area has a water supply that is either fully on or fully off. The designer must allow for different zones within the overall system. Each zone may require different pressures and different volumes of water from that of other zones, or the system may simply be broken up, depending on the style of system the owner requires.

All of this depends, in turn, on the size of the main water supply service and the pressure and volume of water available in that main. It is for this reason, on larger projects, that a pumping system is required. This pumping system may be combined with a storage tank that feeds water to the pumps. With the current trend towards water recycling, some domestic properties rely on storage tanks to supply water for their irrigation needs (see Figure 11.13).

FIGURE 11.13 Rainwater tank being used to supply water to a domestic irrigation system

In some cases, there may be an alternative water supply for the system, such as a water tank, dam, bore or river. This alternative supply may either supplement the supply from the water main or be supplemented by the water main. In these cases, it is very important to protect the main water supply from contamination because there is a high danger for a cross connection to take place from the non-drinking water supply to the drinking water supply.

GREEN TIP

If collecting roof water for the purpose of an irrigation system, comply with water restriction requirements from the local water authority. Using recycled water is very important to help conserve precious drinking water. Tap timers and irrigation controllers play an important part to conserve the amount of water usage.

Design of sprinkler head location

Consider the information in Step 2 below while dividing up the plan: front yard, back yard, side yard, lawn or plantation areas and shady areas. Label your areas A, B, C, D and E (see **Figure 11.14**).

When planning a perimeter and central sprinkler head layout, follow these three steps (see **Figure 11.14**):

- *Step 1:* The critical points on a plan are the corners. Draw a point on each corner using a compass and draw an arc showing the sprinkler spray.
- *Step 2:* If the quarter corner heads will not spray each other, place heads along the perimeter. Repeat using a compass to each sprinkler head point.
- *Step 3:* Look to see if the perimeter sprinkler heads will be spraying across each other's side. If they don't, add a sprinkler head to the centre of the zone area. Again, using a compass, draw an arc showing this sprinkler watering pattern. Always make sure there is complete coverage to the nominated area zone when designing a sprinkler head location (see **Figure 11.15**).

Valves and pipes

Every zone on the plan must have its valve (see **Figure 11.14**). The valve controls the on/off flow of water to a sprinkler zone.

Locate valves – layout area zoning

Indicate one control valve for each zone and then group the valves in an assembly called a *valve manifold*. Determine where the valve manifold for each area should be. Consider a manifold in the front yard and one in the backyard, or perhaps more locations. Manifold placement is entirely up to the plumber. We recommend placing the manifold in an accessible spot for easy maintenance. Place the manifold close to the area the valves will serve, but where the user will not be sprayed when activating the system manually. The correct layout is the most effective irrigation system layout showing emitters locations and pipe layout of lateral pipes and mainline. This will allow even flow and pressure distribution of water for each emitter to work more effectively (see **Figure 11.15**).

Incorrect emitter location layout is less effective at the most disadvantageous point on the layout, which is located at the end of the line. This will

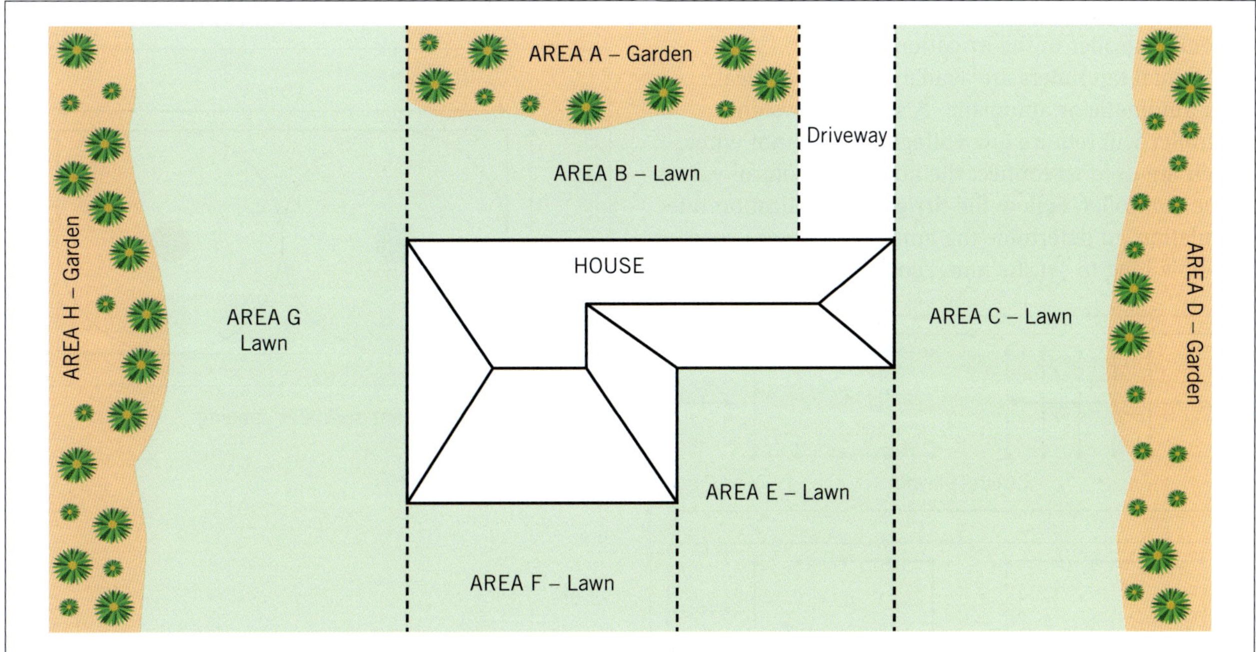

FIGURE 11.14 Design of sprinkler head location

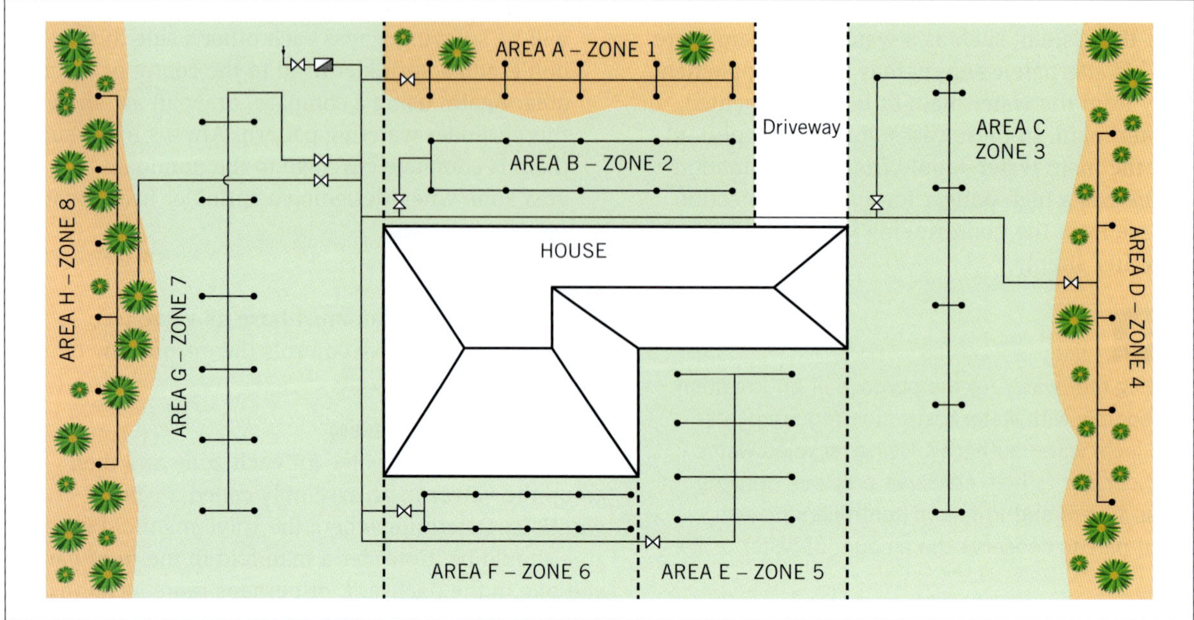

FIGURE 11.15 Valve location indicated as zone area

affect the irrigation water coverage to that area, which will affect the growth of grass, plants, shrubs and turf. The emitters will not operate correctly (see Figure 11.16).

A complete irrigation system arrangement, which shows the main components from the water source to water delivery out of the emitters, is shown in Figures 11.17 and 11.18 also show how sprinkler spacing should be arranged.

Tap timers or automatic sprinkler controller

The most effective way to reduce water usage is by implementing a tap timer or installing a sprinkler system controller to the irrigation system.

Manual tap timers are available as mechanical, semi-automatic or automatic. A sprinkler system controller will require low voltage direct burial wires, which are used to connect the automatic control valves to the controller. Follow the 'Irrigation application rate calculation' to determine the amount of water usage and how long to set the timer/controller system.

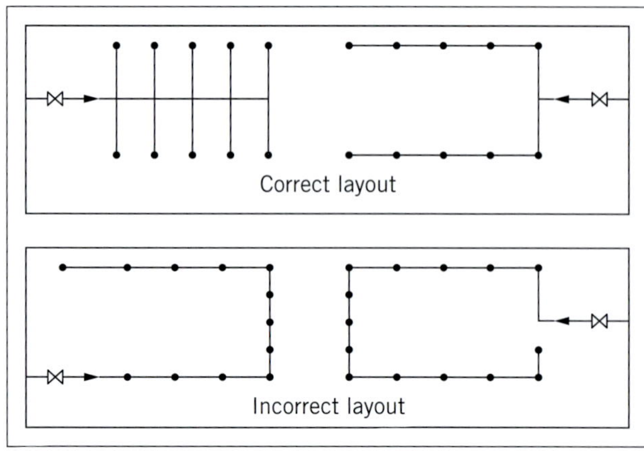

FIGURE 11.16 Layout design

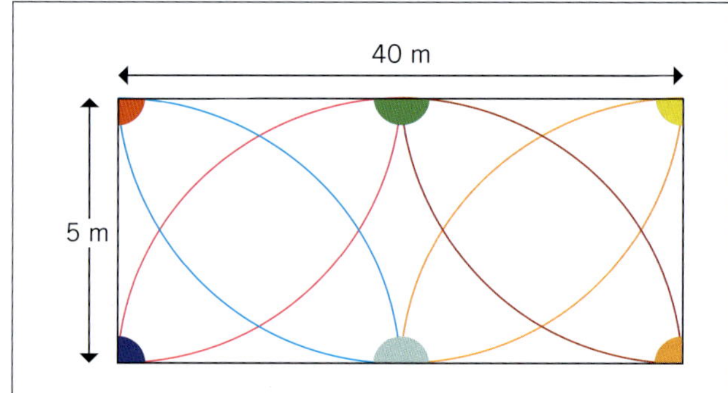

FIGURE 11.17 Efficient sprinkler spacing

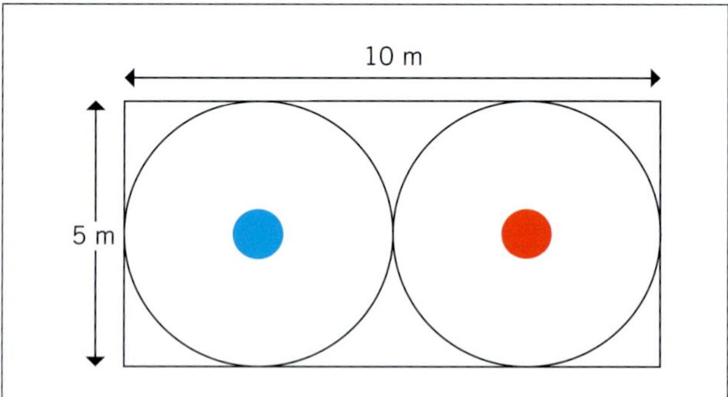

FIGURE 11.18 Inefficient sprinkler spacing

Irrigation application rate calculation

$$\text{Application rate (mm/hr)} = \frac{\text{Total flow rate (L/h)}}{\text{(Total irrigation area (m}^2\text{)}}$$

For example, where there is an irrigation system with six (6) emitters that operate together, the total flow rate for the 6 emitters is:

(6 × 6 litres per minute emitters) × 60 mintues

$$= \frac{\text{2160 litres per hour (L/h)}}{\text{(Area to be irrigated is 80 m}^2\text{)}}$$

Watering rate of the system is = 2160 L/h divided by 80 m^2 = 27 mm/h

To find how long to operate this system applying 15 mm of water:

$$\text{Run time (minutes)} = \frac{\text{Depth to be applied (mm)} \times \text{60 minutes}}{\text{Application rate (mm/h)}}$$

$$\text{Run time (minutes)} = \frac{\text{15 mm} \times \text{60 minutes}}{\text{27 mm/h}}$$

Therefore, using the information above:
Answer: Run time = 33.3 minutes

Excavate and install irrigation piping to plans and specifications

The following steps detail the set-out installation procedure:

- Refer to the POC detail of the residential system.
- Turn off the water supply to the residence (see Figure 11.19).

iStock.com/Climbing_Ursus

FIGURE 11.19 Water supply

- Dig a hole to expose the supply line.
- Cut an appropriate piece out of the supply line, slip the compression tee or capillary fitting onto the pipe and either tighten the compression nuts, or weld or press fit the compression fitting.
- Install the brass nipple and shut-off valve.
- Install the valve box for easy access to the shut-off valve.
- Turn the water back on to the residence.

Test and adjust installation to comply with standards and authorities' requirements

Small installations are usually tested as a whole, after connection to the main. Large installations are installed and tested in sections. The process for this is as follows.

HOW TO

EXCAVATE AND INSTALL THE MAIN IRRIGATION LINE

These steps are based on a *general* installation procedure. Each job site and individual plumber has different methods of installing an irrigation system, so please use these as a guide only.

1 Using marking spray-paint and small flags, indicate the pipelines from the POC to the valve manifold locations. Mark the layout of the irrigation system (see Figure 11.20).
2 On existing lawns, lay down a plastic tarp alongside the marked trench about 600 mm away from where the pipe will be placed.
3 Remove the lawn by cutting a strip about 300 mm wide and 40–50 mm deep using a flat shovel. Roll up the lawn and place the lawn and underlay soil on the plastic tarp.

iStock.com/Chimperil59

FIGURE 11.20 Marking the layout

>>

>>

4 Trenching (see Figure 11.21): for irrigation mainline depth in the area, trench 300–450 mm deep. Trench 150–200 mm for lateral lines. Trenching can be done by hand (see Figure 11.21) or with a trencher. Trenchers are available at most equipment rental suppliers.

Visual Generation/Shutterstock.com

FIGURE 11.21 Trenching

5 Installing pipe under a walkway or driveway (see Figure 11.22): by using a high-pressure jetting machine method or utilising water main pressure with a jet nozzle attachment. Using a pipe-to-hose threaded adapter, connect one end of the pipe to a garden hose and attach a small jet hose nozzle to the other end. Turn the water on and jet under the concrete pathway. This method will have a cost benefit, as no restoration is required to replace a section of concrete.

Igor Shoshin/Shutterstock.com

FIGURE 11.22 Installing pipe under a walkway or driveway

6 Install the backflow preventer according to the relevant Australian Standards and comply with the local authority requirements.

7 Installing pipe: layout pipe and fittings near the trenches according to how they will be installed. Be careful not to get dirt or debris in the pipe, as it will cause blockages and difficulty flushing out the system (see Figure 11.23).

Visual Generation/Shutterstock.com

FIGURE 11.23 Layout of pipes and sprinklers

8 Caution must be taken when preparing for a POC to existing or new water service:
   - Notify customers their water supply will be shut down for some time.
   - Ensure all taps are turned off.
   - Check for stray current with a volt stick. If no current is detected, connect bonding strap – clean pipe with abrasive cloth for a good connection (see Figures 11.24 and 11.25).
   - Measure, cut and install the pipe, working your way to the mainline.
   - Install isolating valves and all associated fittings to complete the POC (include a backflow prevention device, if needed).
   - Turn the water supply on from the water meter, check the connections for leaks or faults before backfilling.

9 Install irrigation pipe system and emitters (see Figures 11.26 and 11.27):
   - Install the pipe and fitting so they have been laid out along the trench.
   - Install the irrigation emitters valves and manifolds.
   - Maintain at least a 150 mm clearance between valves for future maintenance.
   - Provide 80 mm long (or longer) capped for future additions to the irrigation system.

>>

FIGURE 11.24 Checking for stray current

FIGURE 11.25 Bridging conductor (bonding strap)

- Install all the emitters and leave the last emitter off on the lateral line for proper flushing.
- Flushing the irrigation system: turn on the zone valve manually at the valve box. Allow the water to remove all the air from each outlet and flush out any dirt that may have entered. Flush the system, even if you suspect nothing has got in. Turn off the zone valve and install the remaining emitters.
- Backfilling: do not directly backfill out the area zone valves. Install the valve box for access to the valves, solenoid and wiring system. Backfill with clean soil, such as fill sand; do not backfill any rocks that may affect the irrigation system.
- Operate the system and check sprinkler coverage is correct. Adjust emitter heads if needed.

Hennadii H/Shutterstock.com

FIGURE 11.26 Irrigation pipe system

stock-enjoy/Shutterstock.com

FIGURE 11.27 Irrigation emitter

GREEN TIP

- Comply with current EPA regulations.
- Provide adequate sediment and corrosion control to a job site when carrying out these works. Protecting the environment is a number one priority.

HOW TO

TEST THE INSTALLATION

1. Close off the pipe ends with plugs or blind flanges.
2. Test the section or sections that have been laid for leaks under full water pressure.
3. At the completion of the test, rectify any leaks and repeat the test until the installation is deemed satisfactory for the section under construction.
4. Turn off the water and drain it from the pipes by removing the plugs or flanges.
5. Lay the next section.
6. Repeat the testing process until the whole installation has been subjected to the necessary testing and approved by the appropriate authority.

Commission irrigation system

Charge the system with water and remove all the air from each outlet. Flush out any foreign matter that may have entered the piping system. Divide sprinklers into zones: this will be determined by the water pressure and flow rate checks. If there is a minimum pressure, a domestic pump may be required. If capacity is an issue, consider a storage water tank and factor in the sprinkler head flow rate on the performance chart by the manufacturer and the time required to irrigate the zoned area. The flow rate must be measured (as shown earlier in this chapter) and checked with the manufacturer's specifications.

The *commissioning* of an irrigation system involves a process of ensuring that the system is installed and tested to perform according to the design intent. The commissioning process confirms that the installed irrigation system meets the design performance specifications.

After commissioning, the responsibility for the irrigation system is handed over from the designer and installer to the owner/operator and may have implications for insurances, maintenance programs and compliance. It is the final phase of the installation process and is undertaken by the installer. Where a designer has been involved, they are often involved in commissioning, either at the system testing phase or providing input on how to correct performance issues.

Commissioning and testing documentation will consist of:

- a commission report
- as-built plans
- an operation and maintenance manual
- inspection and testing of backflow prevention device – submitted to the relevant water authority
- any other relevant supporting documentation.

FROM EXPERIENCE

Once you are a qualified plumber, you will be experienced with different types of irrigation systems and have a level of knowledge to provide adequate technical advice to customers about irrigation design, installation, testing and commissioning.

Identification warning signs

All pipework, valves and outlets up to the solenoid controls must be clearly and permanently labelled with safety signs complying with the current version of AS 1319 (see Figure 11.28).

Clean up

Upon completion, the job must be left clean and tidy. All ground surfaces should be restored and returfed if necessary. Company quality assurance policies dictate that proper standards be maintained and that all rubbish is disposed of in the appropriate bins and recycled if appropriate. Any material left over should be stored and reused on future jobs.

Tools and equipment should be inspected for any damage and packed away after they have been cleaned and maintained in proper working order. Damaged tools must be tagged and reported to the site

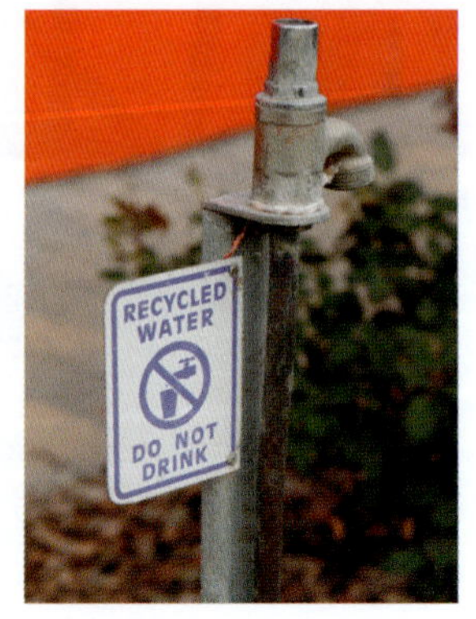

Sources: 421 Environmental Products; iStock.com/CTRPhotos.

FIGURE 11.28 Typical warning signs

supervisor. Any tools and equipment on hire must be returned as soon as possible to avoid extra cost.

All final documentation, such as the certificate of compliance, must be submitted to the regulatory authorities and client. Delivery dockets, invoices and order sheets should be reconciled as per company policy, and time sheets accurately completed.

The safety of plumbers, clients, employees and the public are top priority. Whether working on a large site or a small installation, it is likely that materials and tools will be lying around. It is tempting to leave items where they land as you work to save time; however, imagine what could happen if someone tripped over a cord or wood debris? Or the damage that could be done if someone stepped on a loose nail? This is not only an injury waiting to happen, but a liability as well. A safe work area is provided by keeping a clean and clear work area.

A clean workspace has productivity and efficiency benefits as well. By keeping a clean workspace, it is easier to stay more focused on the task at hand, avoiding the distractions other messes may bring. It also makes it easier to find tools faster without having to search under and over things to find what is sought.

While work is in progress and a customer sees a messy job site, they can't appreciate the progress. They can only see the mess! This is especially true in new installations or additions, where the customer is making a large sacrifice for the work done on their home. If they see a mess spreading into the inside or outside liveable parts of their home, they may react poorly.

In addition, people often associate cleanliness with quality. If they see a job site that is a mess, they might think less of the quality of work you are providing.

A clean workspace will result in happy customers, who are likely to call you next time they need work done or will be happy to refer you to others. Part of your job is not just the hands-on labour, but providing a great customer experience. Taking a few minutes to keep things clean and tidy could result in more leads from referrals.

FROM EXPERIENCE

It takes years of experience to achieve goals in life: respect, being responsive and reliability are key factors. You can dream about doing something, but it is best to stop dreaming and have a go at what you enjoy most.

LEARNING TASK 11.3

1 Why is it important to flush the pipework before installing the emitters?
2 What is the required water pressure when testing an irrigation system downstream from the point of connection?

COMPLETE WORKSHEET 2

FROM EXPERIENCE

Designing, installing and commissioning can be challenging at times; reading plans and specifications are important factors for a great achievement. Reaching out to other professionals and seeking advice is the first step to success.

SUMMARY

- Plumbers have a major responsibility to protect the community's drinking water.
- Plans and specifications need to be thoroughly read.
- Access codes, standards and local regulations.
- Any potential cross connection must be identified and prevented.
- Irrigation type must be identified.
- Hazard ratings must be determined as per AS/NZS 3500.1 so the appropriate backflow prevention device can be identified.
- Backflow prevention devices must be located and installed as per AS/NZS 3500.1.
- Irrigation systems must be sized accordingly so the household supply is not affected.
- Available water pressure and flow rate is a major consideration for irrigation systems.
- Materials and equipment must comply with the relevant Australian Standards.
- Different irrigation systems require different backflow prevention devices.
- Alternative water supplies must be installed so there is no possibility of cross connection with the drinking water supply.

REFERENCES

NSW Environment Protection Authority (EPA): **https://www.epa.nsw.gov.au/**

Standards Australia: **https://www.standards.org.au/**

GET IT RIGHT

1 Which photo shows the valve correctly installed?

____

____

____

2 Why is it correctly installed?

____

____

____

3 What type of valve is it?

____

____

____

WORKSHEET 1

To be completed by teachers	
Student competent	☐
Student not yet competent	☐

Student name: ____________________

Enrolment year: ____________________

Class code: ____________________

Unit competency code/title: CPCPIG 3021 Set out, install and commission irrigation systems

Task: Review 'Prepare for work' and Identify installation requirements', and then answer the following questions.

1 List five pieces of information that can be found on a specification document.

____

____

____

____

____

2 What is 'potable water'?

____

3 When is a safe work method statement required?

____

4 What is the purpose of quality assurance?

____

____

____

5 What are 'emitters'?

____

____

6 Why is it advisable to keep different types of irrigation systems on separate supply lines?

____

____

7 Where is the information obtained for existing underground services?

____

8 Name five important details in the pre-planning process for an irrigation installation.

WORKSHEET 2

To be completed by teachers	
Student competent	☐
Student not yet competent	☐

Student name: ______________________________

Enrolment year: ______________________________

Class code: ______________________________

Unit competency code/title: CPCPIG 3021 Set out, install and commission irrigation systems

Task: Review 'Install and commission irrigation system' and 'Clean up', then answer the following questions.

1 Define a Type A irrigation system.

____
____
____
____
____

2 Define a Type B irrigation system.

____
____
____
____
____

3 Define a Type C irrigation system.

____
____
____
____
____

4 Define a Type D irrigation system.

____
____
____
____
____

5 How is the flow rate measured?

6 What is the most common pipe used for domestic irrigation?

7 What size should the main irrigation line be in regards to the lateral (branch) line?

8 Name three alternative water supplies.

9 Where are the critical positions for the location of sprinkler heads?

10 What is the most effective way to reduce water usage in a sprinkler system?

11 What are the two main considerations when sizing an irrigation system?

12 Explain the purpose of commissioning an irrigation system.

CARRY OUT SIMPLE CONCRETING AND RENDERING

12

Chapter overview

Plumbers repair concrete slabs, driveways and footpaths after excavating trenches; patch holes from drilling core holes through concrete slabs and panels; and render up walls where pipes are chased into masonry walls. Plumbers also fabricate and install concrete access chambers and manholes for sewer systems as well as concrete thrust blocks to restrain pipework. Other areas involve installing onsite detention tanks (OSD), surface inlet pits for stormwater systems and slabs for rainwater tanks. Some of these products are precast and some are cast in situ. This chapter will address the basic skills necessary to restore small concrete sections for footpaths, driveways and slabs, as well as cement rendering masonry walls to patch a hole or render a wall chase. The aim of this chapter is to increase the knowledge and skills required to mix, place and finish concrete, and render in an acceptable manner.

Concrete is one of the oldest and most common construction materials in the world. It is low cost compared to other materials and readily available. Concrete is very durable over a long period of time and can handle extreme weather environments. Concrete is made from two parts fine aggregate (sand), three parts coarse aggregate (blue metal), one part cement and water.

Cement mortar has also been around in construction since ancient times. It is commonly used to render masonry walls internally and externally. Traditionally, it consists of six parts fine sand, one part cement and one part lime for a finish coat. The lime helps the render to be more workable and reduces the chance of cracking. Usually, for patching holes, bedding toilets and rendering wall chases, a mix of two parts sand and one part cement does the job. This is the mix ratio in a bag of ready-mix sand cement.

Common materials to make cement include limestone, clay, silica sand and iron ore. These ingredients are heated to a high temperature and form a rock-like substance. It is then ground into the fine powder known as cement.

Learning objectives

Areas addressed in this chapter include:

- identify concreting and rendering requirements
- prepare for work
- place concrete and render
- clean up.

Identify concreting and rendering requirements

This section explains how to identify the concreting and rendering requirements to successfully carry out the work. It is important to understand the information available, such as plans, specifications, codes, standards, manufacturer's specifications and the work health and safety (WHS) legislative requirements before starting the work.

Access, read and determine concreting and rendering requirements

To determine concreting and rendering requirements, it is important to have access to and read from relevant job plans and specifications that apply to the job. The information gained from plans and specifications would be the location, types of concrete and rendering finishes, the amount of concrete and render required, access requirements, stockpile locations and sediment barrier locations. It is also important to be familiar with the current codes, Australian Standards, manufacturer's specifications and jurisdictional requirements, such as the WHS legislative requirements.

AS 1379 SPECIFICATION AND SUPPLY OF CONCRETE
AS 3600 CONCRETE STRUCTURES

Calculations

Knowing how to measure and order the correct amount of concrete, steel reinforcement and formwork involves several calculations. It is better to have a little more concrete than not enough, so it is a good idea to add 5% to the volume when ordering.

Concrete is measured in cubic metres (m^3):

$V = L \times W \times D$

where:

V = volume in cubic metres
L = length in metres
W = width in metres
D = depth in metres

All measurements must be converted to the same unit of measurement.

Steel reinforcement and plastic membrane are measured in square metres (m^2):

$A = L \times W$

where:

A = area in square metres
L = length in metres
W = width in metres

When measuring formwork, all the sides need to be added up to work out the lineal lengths of timber required. This is also known as the perimeter:

$P = (L \times 2) + (W \times 2)$

where:

P = perimeter in metres (lineal metres)
L = length in metres
W = width in metres

EXAMPLE 12.1

HOW TO CALCULATE CONCRETE, STEEL REINFORCING MESH AND FORMWORK FOR A FOOTPATH

A footpath needs to be replaced due to an excavation to replace the sewer pipe. The path area is 12 m long, 800 mm wide and 75 mm thick (see **Figure 12.1**).

Wilfreda Wiseman/Shutterstock.com

FIGURE 12.1 Footpath with measurements

Working: $V = 12 \times 0.8 \times 0.075$
Answer: $V = 0.72\ m^3$

5% should be added as a safety margin:

$$0.72 + 5\% = 0.756\ m^3$$

Therefore, 0.756 m^3 of concrete needs to be ordered.

Steel reinforcing mesh will need to be ordered as well. This is measured in square metres. A polyethylene membrane is also measured in square metres if required.

Working: $A = 12 \times 0.8$
Answer: $A = 9.6\ m^2$

So, 9.6 m^2 of steel mesh needs to be ordered. (Because the width is only 800 mm, it would be easier to order 12 m of mesh 750 mm wide to save time cutting steel.)

The formwork required is measured in lineal metres. For this job 75 mm × 45 mm timber is required.

Working: $P = (12 \times 2) + (0.8 \times 2)$
$P = 24 + 1.6$
Answer: $P = 25.6$ m

So, 25.6 m of 75 mm × 45 mm timber is required.

Remember to order hardwood pegs (approximately one per metre). So that would be 26 pegs.

Identify and apply workplace, work health and safety and environmental requirements

Work health and safety (WHS) is all about keeping people safe in the workplace. It includes things like ensuring that workplaces are free from hazards, providing training and information on how to safely work in a particular environment and investigating any accidents or incidents that do occur.

It is important to protect the environment by putting adequate procedures in place while work is in progress, such as minimising dust and protecting our waterways from pollution.

Work health and safety

Before starting work, identify the hazards that are obvious and assess the risks of the hazards. Then put the control measures into place to reduce the risk.

This can be formatted as a job safety analysis (JSA), which must be done as a minimal risk assessment on every job. While there is not a formal requirement in legislation to complete a JSA, there is a legal requirement to minimise safety risks and the JSA process can be an effective way to achieve that.

A JSA is defined as a document that describes the steps carried out to perform a work-related task along with:

- identification of hazards and risk score related to each step
- control measures to mitigate risk for each step
- considerations related to legislation, codes of practice and relevant standards.

A safe work method statement (SWMS) must also be completed if the task is a high-risk activity. One SWMS can be used for work that involves multiple high-risk construction work activities; for example, a work activity that requires using powered mobile plant, working at heights of more than 2 metres and working adjacent to a road used by traffic other than pedestrians.

Generally, the person responsible for preparing a SWMS is the employer for their workers and themselves.

An SWMS should be short and focus on describing the specific hazards identified for the high-risk construction work to be undertaken and the control measures to be put in place so the work is carried out safely.

Once the risks have been minimised by using the appropriate control measures and everyone working on the job is aware of the safety measures put in place, the job can proceed.

Safety datasheets

The safety datasheet (SDS) lists the hazardous ingredients of a product, its physical and chemical characteristics, its effect on human health, the chemicals with which it can adversely react, handling precautions, the types of measures that can be used to control exposure, and emergency and first aid requirements. Silica is a harmful substance found in cement. This information can be found on the bag of cement.

Exposure to silica dust can lead to the development of lung cancer or silicosis. The percentage of silica in cement is between 17% and 25%.

Environmental requirements

An environmental management plan should be implemented and address issues such as:

- dust suppression
- reuse of materials
- recycling of waste (cardboard, paper, metal and plastic)
- sediment control (vegetation, sediment fences and stormwater control)
- stockpiling of soil and bedding material (sand and aggregate)
- rubbish removal.

GREEN TIP

Remember the four Rs: reduce, reuse, recycle and repair.

Prepare for work

As with any task, preparation is the key to a successful job. Having a clear understanding of how to carry out the work and all the steps involved helps in preparing the materials, tools and equipment needed for the job.

Create a materials list and collect materials

Effective planning involves writing a detailed list of materials and organising delivery of the materials at a convenient time. Sometimes picking up the materials is more convenient if the materials list is a small quantity and travel time is minimal. A sample materials list is shown in Table 12.1.

Select and check the serviceability of appropriate tools and equipment, including personal protective equipment

Choosing the correct tools and equipment as well as the correct personal protective equipment (PPE) to carry out the work is important when preparing for a job. Effective planning helps the job run smoothly. Be sure to check all the tools, equipment and PPE are in good condition and ready to use.

TABLE 12.1 Materials list

Material	Quantity
Ready mixed concrete	m^3
Sand – coarse for concrete	m^3
Sand – sharp for rendering	m^3
Blue metal – 10–20 mm	m^3
Portland Cement	20 kg bag
Hydrated lime (for rendering)	20 kg bag
Rio mesh	m^2
Rio bar	m
Rio chairs	pack
Plastic membrane	m^2
Formwork – 90 mm × 45 mm timber	m
Pegs – 50 mm × 50 mm hardwood	pack

Personal protective equipment

The appropriate PPE for a concreting or rendering job is:

- steelcapped rubber boots
- dust mask
- safety glasses
- rubber gloves.

Do not breathe in cement dust as it can cause a lung disease – silicosis.

There can be heavy manual handling involved when shovelling concrete, so remember to keep your back straight and bend your legs to avoid injury or strain to your back.

Avoid skin contact with cement as it can cause dermatitis. A hand barrier cream should be used or rubber gloves worn.

Tools and equipment

The basic tools and equipment required for concreting are:

- wooden and steel floats, trowels and edging tools for concrete finishing
- screeds for placing and levelling concrete
- levelling equipment – laser, dumpy, spirit and water
- shovels and spades for excavation and spreading
- mattock, pick and bar for breaking up ground
- bolt cutters for cutting steel mesh
- sledge, lump and claw hammer for nails and pegs
- string lines for setting out
- brooms for cleaning up and concrete finishing
- formwork and pegs
- wheelbarrow for placing and mixing concrete
- mechanical mixer
- vibrator for removing air bubbles.

Inspect all tools and equipment for any signs of damage. If any damage is noticed, it should be immediately reported to the supervisor, and the tool or equipment must be tagged and taken out of circulation until the damage has been repaired.

Store tools and equipment in a clean, dry, secure location. Return any hired tools and equipment to the supplier in the same condition as supplied. Any defects must be notified.

Quality assurance

Following company policy for quality assurance can involve, for example, keeping all tools and equipment in good working order, carrying out the work in a professional manner, communicating effectively with the client and minimising material waste by making accurate orders.

Concrete products

Deciding whether to use a precast concrete product or to cast in situ will depend on the site access and availability. Using precast concrete products saves labour time onsite (see **Figure 12.2**).

Kritthaneth/Shutterstock.com

FIGURE 12.2 Precast concrete pit

Precast concrete products include:

- stormwater pits
- access chambers (manholes)
- kerb inlet pits
- RCP pipe
- culverts
- septic tanks
- rainwater tanks
- trade-waste pre-treatment arrestors.

Some of the cast in situ jobs are:

- thrust blocks for water mains
- pier and beam support for pipework in filled ground
- repairing of footpaths and slabs (see **Figure 12.3**)
- slabs for tanks, hot water heaters and pumps.

ungvar/Shutterstock.com

FIGURE 12.3 In situ pour

Job requirements and levelling procedures

Concreting jobs performed by plumbers are usually small jobs. The smallest order of ready-mixed concrete delivered onsite is generally 0.2 m³. Smaller quantities would be mixed onsite either by shovel, larry and wheelbarrow (see Figure 12.4) or by a portable mechanical mixer (see Figure 12.5). It is good practice to measure quantities using a bucket instead of a shovel for a consistent mix.

FIGURE 12.4 Wheelbarrow, shovel and larry

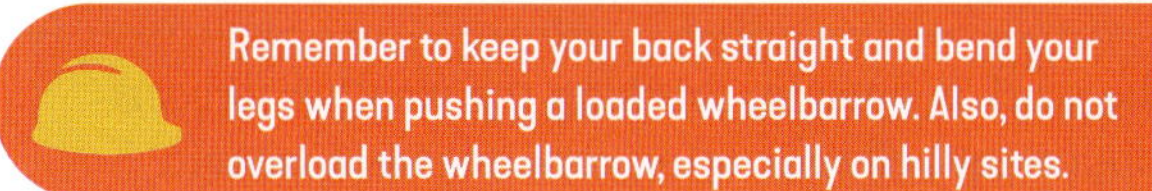

Remember to keep your back straight and bend your legs when pushing a loaded wheelbarrow. Also, do not overload the wheelbarrow, especially on hilly sites.

The water content is important as well. If too much water is added, then the concrete will be weaker with a dusty finish on the surface. If not enough water is added, then the concrete is too difficult to work with. It is recommended that 2.5 L of water is added to one 20 kg bag of concrete mix. Once water is added to cement-mixed products, they must be used within one hour.

Andrey_Kuzmin/Shutterstock.com

FIGURE 12.5 Portable cement mixer

HOW TO

CARRY OUT SAFE OPERATING PROCEDURES WITH A CEMENT MIXER

1. Position the mixer in a well-cleared area.
2. Place the mixer on level ground with its wheels chocked.
3. Turn the mixer on while it is empty.
4. Water is the first ingredient.
5. Keep the motor dry.
6. Do not place tools in the mixer while it is turning.
7. Do not wear loose clothing around the mixer.
8. Clean the mixer well and lubricate parts on pack-up.

Pre-mixed concrete

Ready-mixed or pre-mixed concrete is used on larger jobs. The concrete is mixed dry into concrete trucks in accurate proportions at a concrete batching plant. Water is added later and either mixed in the truck on the way to the site or mixed onsite, depending on how hot the day is (see Figure 12.6).

There are a few different size concrete trucks available, depending on how much concrete is required and site access. The loads can vary from 0.4 m³ to 7 m³.

FROM EXPERIENCE

When ordering concrete, it is a good idea to leave the order open by requesting what is known as 'a message'. This informs the concrete supplier that more concrete may be required upon request, as it is sometimes difficult to estimate an accurate volume.

FIGURE 12.6 Concrete truck and batching plant

The advantages of pre-mixed concrete are:

- consistent mix
- no need to store sand, gravel and cement onsite
- less waste of materials.

Additives can be used in the concrete mix to:

- accelerate the hardening time in cold weather
- slow the hardening time in hot weather
- change the colour
- increase the strength (higher amount of cement).

The compressive strength of concrete is measured in megapascals (MPa). Most of the concreting jobs carried out by plumbers, such as pits, tank slabs and footpaths, require the minimum strength of 20 MPa, as per AS/NZS 3500. Special class concrete that is chemical resistant is used for benching and rendering sewerage access chambers; it is not so readily available and should be ordered well in advance.

AS/NZS 3500 PLUMBING AND DRAINAGE

Concrete testing

When concrete is used for structural purposes, there are two tests commonly carried out. One type is called a 'slump test', which determines the consistency of the concrete. This is done by pouring a small batch of concrete onto a board from a cone-shaped container. Then the distance is measured from the height of the container to the top of the concrete batch. The difference in height is the slump measurement (see Table 12.2).

The second test carried out is a 'compression test'. This is usually only carried out on commercial and industrial sites. A sample of concrete is filled into a test cylinder and, after it has set (one day), it is removed from the test cylinder and left to cure for 28 days. Then it is crushed and the force required to do this is measured and recorded as MPa. This is compared to the strength (MPa) ordered and enforces the minimum requirements.

TABLE 12.2 Slump test results comparing consistencies in concrete mixes

Type of construction	Slump (minimum–maximum)
Thin walls (reinforced)	120–200
Pumped concrete	70–120
Columns, beams	50–100
Footings (reinforced)	50–100
Slabs, pavements	50–80
Plain footings	50–80
Heavy mass concrete	30–80

Source: BCGCO2003B Carry out concreting to simple forms for plumbing applications, © TAFE NSW Manufacturing, Engineering, Construction and Transport Curriculum Centre.

Levelling

Understanding levelling procedures is crucial to ensure that any excavation work is carried out to the correct level to allow for the correct thickness of concrete required. It is also important to allow for any stormwater run-off for concrete installed outside by setting up formwork to the correct levels so all the surface water drains off. Levelling techniques are shown in more detail in Unit 3 'Levelling' in *Basic Plumbing Services Skills: Sanitary and Drainage*.

LEARNING TASK 12.1

1 How is the quantity of concrete measured?
2 What is the advantage of using precast concrete products?
3 What is the harmful chemical in cement/concrete products?
4 What is the minimum MPa rating for concrete?

COMPLETE WORKSHEET 1

Place concrete and render

If the work is outside, checking the weather forecast is important before pouring concrete because rain will weaken and spoil the concrete finish. Also, in temperatures higher than 30°C, the possibility of cracking and crazing can occur from excessive moisture loss.

Concrete develops in four stages:

- plastic stage (wet concrete)
- stiffening stage (final finish)
- curing stage (hard but not fully strengthened)
- hardened stage (fully cured – 28 days).

Prepare site prior to placement of concrete

Preparing the layers below the concrete ensures a stable base to support the concrete, and having sufficient steel reinforcement in the concrete ensures adequate strength. Always refer to the job or manufacturer's specifications for the required concrete mix and strength (MPa). If there isn't a specification, then the minimum standard as per AS 1379 and AS 3600 must be met.

Hardcore fill

It may be necessary to fill over-excavated areas to avoid excess concrete. Materials used for this should be hard and dense, and not able to decompose. This could be broken old concrete, broken bricks, rocks or road base. This layer is called 'hardcore fill' and should be kept 50 mm below the underside of the concrete to allow for the next layer.

Sand binding layer

The next layer to spread is known as the 'sand binding' layer. This sand layer is 50 mm thick and is used to get an accurate level to the underside of the concrete, therefore reducing excess concrete. It also prevents any sharp rocks piercing the plastic membrane, if installed. Fill sand is usually used as it is more cost-effective.

Plastic membrane

When pouring slabs on ground for buildings, it is vital that a plastic membrane made from polyethylene (usually black or orange in colour) is installed on the sand layer to prevent moisture rising through the concrete and causing dampness and corrosion to the steel reinforcement (see Figure 12.7). All joins should have a minimum lap of 200 mm and be taped up. The plastic must be cut and turned up on all penetrations such as pipes and conduits, and taped off as well. Any punctures in the plastic sheet should be repaired or taped up.

FIGURE 12.7 Plastic membrane and penetration

Any horizontal plumbing or drainage pipes passing through footings, beams or walls must be wrapped or sleeved with a liner or flexible material providing 25 mm annular space (AS/NZS 3500.2). This allows for concrete movement without damaging the pipe.

Restoring damaged concrete surface

When restoring damaged concrete surfaces, remove all the damaged concrete and place a clean cut in the existing concrete for a neat joint. It helps to drill horizontally into the existing concrete and insert steel rio bars (dowels) to increase the strength of the joint between the old and new concrete.

Formwork

To form the shape of concrete, formwork is used. Be sure the area to be concreted is cleared of any excess material and rubbish prior to concrete placement.

For on-ground slabs, formwork is usually made from timber frames and braced with pegs rammed into the ground to prevent any movement (see Figure 12.8). The formwork should stay in position for seven days for on-ground slabs to allow sufficient curing time for the concrete to reach 70% of its strength. It is good practice to apply a release agent such as linseed oil to the timber formwork before pouring, so it can strip away from the concrete cleanly.

FIGURE 12.8 Formwork and pegs

For suspended slabs, the formwork is usually plywood sheets laid on timber beams supported by adjustable steel jacks. This formwork cannot be removed for 28 days to allow for the correct curing time. Because a suspended slab is not fully supported as an on-ground slab is, a longer curing time is necessary.

Concrete is technically fully cured after 28 days, but will still harden past this time because it is never fully cured. Remember that the longer concrete cures, the harder it becomes.

Steel reinforcement

Steel bars are embedded into concrete to increase its resistance to the different forces it is exposed to. These are:

- shear force (tearing force)
- tensile force (stretching force)
- control shrinkage (contraction)
- high heat (fire – expansion).

Concrete beams and columns usually have steel bars that range in diameter from 12 mm to 24 mm, depending on the structural engineer's design (see Figure 12.9).

chinahbzyg/Shutterstock.com

FIGURE 12.9 Steel reinforcement ('rio') bar

For slabs, steel mesh is commonly used. The mesh sizes are SL42, 52, 62, 72 and 82. The 'S' means square mesh and the 'L' means low ductility. The numbers show the diameter and spacing. So for SL52 mesh, the number '5' represents the bar size as 5 mm in diameter and the number '2' means the bars are spaced at 200 mm apart (see Figure 12.10).

FIGURE 12.10 Steel mesh

There are minimum cover requirements for steel reinforcement to prevent it from rusting (see Table 12.3). If the steel rusts, then the concrete will lose its strength. This is called 'spalling' or 'concrete cancer'.

TABLE 12.3 Minimum concrete cover for steel reinforcement

Situation	Minimum cover
Unprotected ground	40 mm
External exposed surfaces	40 mm
Membrane in contact with the ground	30 mm
Internal protected surfaces	20 mm

Source: BCGCO2003B Carry out concreting to simple forms for plumbing applications, © TAFE NSW Manufacturing, Engineering, Construction and Transport Curriculum Centre.

There are purpose-built supports called 'chairs' that are made from plastic that support the steel reinforcement, keeping the steel the required distance away from the ground or formwork (see Figure 12.11). The chairs also support the steel from bending when the wet concrete is being placed. They are usually spaced 1 m apart for thicker bar and mesh, and at 0.6 m centres for lighter mesh.

krolya25/Shutterstock.com

FIGURE 12.11 Plastic chairs

Preparing concrete to specification and transporting concrete

It is important to check the job specification and the manufacturer's specification to ensure the correct concrete strength (MPa) is supplied and any other additives are included as required. Small concrete pours (less than 0.2 m^3) are usually mixed onsite and placed via wheelbarrow. For larger pours, the concrete is delivered onsite via a concrete truck and can be placed directly into position from the chute of the truck or by wheelbarrow from the chute. If the truck cannot get close to the site, a concrete pump is used to move the concrete from the truck to the site. There are two types of concrete pumps:

- a line pump, which is a pipe that carries the concrete from the pump to the site (see Figure 12.12)
- a boom pump, which is similar to a crane but is more manoeuvrable and more popular due to its flexibility (see Figure 12.13).

iStock.com/roman023

FIGURE 12.12 Line pump

FIGURE 12.13 Boom pump

On high-rise sites, pump towers (similar to a crane tower) can be built to handle the mass pours for each level or a crane can move the concrete with a concrete bucket (kibble) attached.

HOW TO

ORDER PRE-MIXED CONCRETE

To order pre-mixed concrete, supply the following information:

1. the name of the account the order is under or whether it's cash on delivery (COD)
2. the site address and any site-specific details
3. the date and time of arrival
4. the amount of concrete required (0.2 m^3 minimum)
5. the concrete strength required (20 MPa minimum)
6. the slump level required (for example, 70 mm)
7. any additives required (for example, colour, special aggregate or retardant)
8. the maximum size of the truck if access is limited.

Placing the concrete

Once all the preparation is done and the formwork is solid with the steel in place, the pour can begin. The 'plastic' concrete is placed via wheelbarrow or pump. At this time, it pays to have plenty of workers to spread and level the concrete while it is in its plastic state. The concrete needs to be compacted into position to remove all the air bubbles. If the concrete depth is minimal, then compaction is done easily by tapping the sides of the formwork with a hammer and poking the concrete with a shovel. If there are deep sections, then a mechanical vibrator (see Figure 12.14) is inserted into the concrete to remove the air bubbles.

bogubogu/Shutterstock.com

FIGURE 12.14 Vibrator

FROM EXPERIENCE

When pouring concrete, the old saying 'Many hands make light work' definitely applies.

Screeding and floating the concrete

The screeding process involves using a straight edge made from timber or aluminium (see Figure 12.15) that drags the concrete back using a sawing and chopping action. The straight edge works to the levels of the formwork or datum points to achieve the required levels.

Tortoon/Shutterstock.com

FIGURE 12.15 Screed

Once the screeding is done, the surface must be floated to leave a hard, smooth surface. On small areas, this is done with a hand trowel (see Figure 12.16). On large surfaces, a bull-float is used (see Figure 12.17).

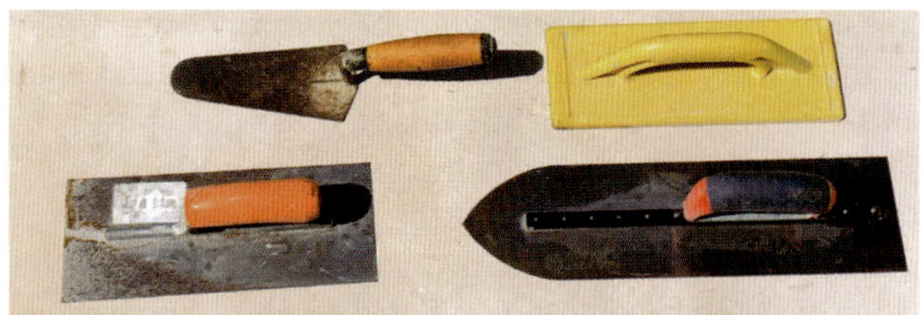

FIGURE 12.16 Hand trowels

sonsart/Shutterstock.com

FIGURE 12.17 Bull-float

The bull-float has a long extendable handle attached to a metal blade that pivots as it is moved back and forth over the concrete surface and can reach as far as 5 m. This floating and trowelling process is to be done twice and then the concrete needs to be left some time to set. Be careful not to overwork the concrete as that can bring sand and cement to the surface, which will cause crazing. Adding extra water will also cause crazing.

Finishing the concrete

Before the final finish can happen, all the bleed water must be gone. If time is getting on and there is still water on the surface, then it can be removed by dragging it off with a hose. Do not add neat cement to the surface to dry up the water as it will cause crazing.

When the concrete is firm enough to support a person's weight and only a slight marking occurs, then the final trowelling can proceed. On a large surface, a power trowel (helicopter) is preferred, but on smaller jobs a steel float is used.

If a smooth finish is required, such as an internal floor, a steel trowel is used. A power trowel (see Figures 12.18 and 12.19) will produce an even better result. For a rougher, non-slip finish, such as footpaths and driveways, a broom or a wooden float will result in a textured finish.

Edge finishing

A special edging tool (see Figure 12.20) is used for all edges to produce a neat finish that is less prone to chipping by having a slightly rounded edge. The first run over the edges should be done after the first screed to work the stones down from the surface and allow a smooth final edge. The second run should happen after the concrete is floated or trowelled.

FIGURE 12.18 Power trowel

FIGURE 12.19 Ride-on power trowel

FIGURE 12.20 Edging tool

Contraction and expansion joints

As with any material, concrete expands and contracts from changes in temperature and moisture. From this, cracks can appear if not properly controlled by using contraction and expansion joints. These two joints are quite different, but they both help to reduce the stress on a concrete slab.

The contraction joint, also known as a control joint, provides a place for the concrete to crack along a straight line, instead of cracking randomly when shrinkage occurs (see Figure 12.21). The control joint can be made while the concrete is in its plastic state by

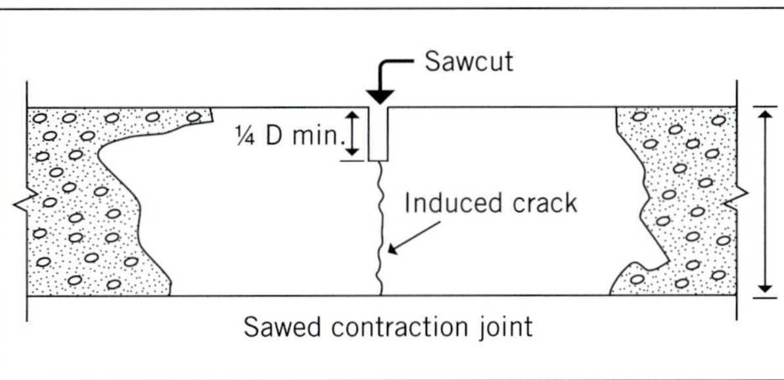

FIGURE 12.21 Control joint

creating a groove with a grooving tool about one-quarter to one-third of the slab thickness. Alternatively, it can be cut between 24 and 48 hours after the concrete has been placed as shrinkage is still minimal at this stage. The cut is the same depth (one-quarter to one-third of the slab thickness). Control joints are usually spaced 2 m apart on footpaths and 3 m for driveways. The thicker the concrete, the further apart the joints can be.

Control joints are not usually necessary for slabs inside buildings because the concrete is not exposed to the sun.

Expansion joints differ from control joints because the joint is completely separated with a flexible material such as foam rubber or a bitumen-coated cork (see Figure 12.22). They are put in place before the concrete is poured. Expansion joints are used to allow the slab to move freely from other structures to prevent cracking; for example, where a slab meets a wall or a new slab meets an existing slab.

LEARNING TASK 12.2

1. Why is a plastic membrane important?
2. What does a release agent on formwork do?
3. What can happen if the concrete is overworked?
4. What tools would be used to provide a textured, non-slip finish?
5. What is the difference between a control joint and an expansion joint?
6. When does the final trowel finish commence?

COMPLETE WORKSHEET 2

Placing rendering

Plumbers use a cement mortar mix to patch up holes where pipes pass through brick and concrete walls, to seal around pipes in concrete or brick pits, and to render around pipes chased into brick and concrete walls (see Figure 12.23). Cement-base renders on internal and external walls provide a few functions, including waterproofing, increasing the fire rating and improving the appearance.

FIGURE 12.23 Cement rendering pipes chased in a brick wall

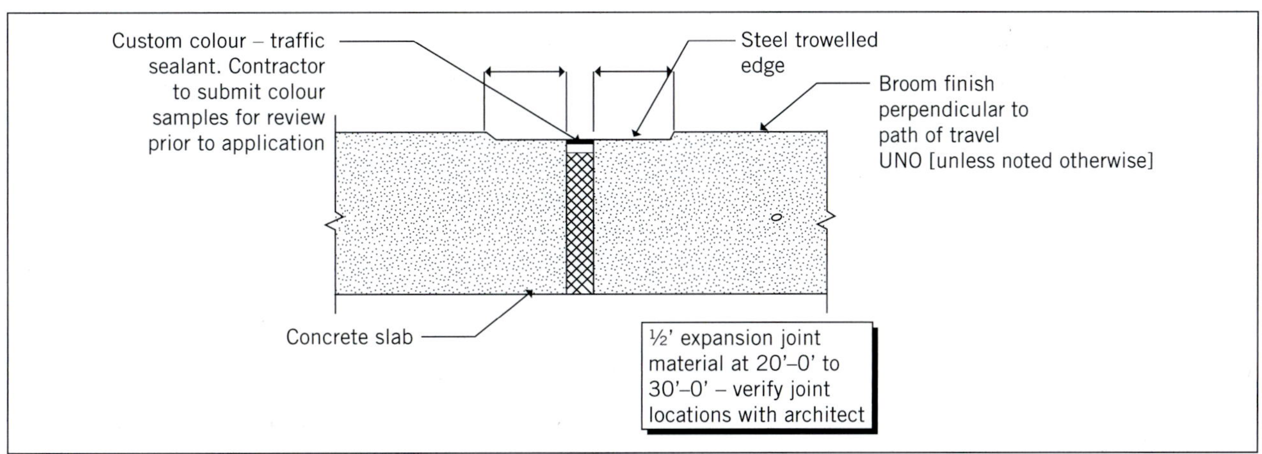

FIGURE 12.22 Expansion joint

Preparing the cement render mix

For best results, the render mix should be suited to the existing surface and applied to the correct thickness. It is important to clean the surface being rendered of any dirt, dust, grease and oil to allow for better adhesion. This is usually done by wetting the area with water.

When rendering a concrete surface, the surface must be prepared to help the render stick as it has a smooth, dense surface. This is done by applying a 'dash coat'. A dash coat is a strong mix of cement and water that is flicked over the wall, creating a textured surface for the render to grip. Allow 24 hours for it to dry before applying the render.

The cement render mix usually consists of three parts sand and one part cement (3:1). The type of sand usually used for render is river sand. This sand is quite fine and helps achieve a smoother finish. There are other mix ratios depending on the type of finish required. Other additives such as lime can be added to help prevent moisture being trapped, enabling the walls to breathe and reduce the chance of mould occurring. Plumbers use the following cement render mixes:

- 2:1 mix (two parts sand and one part cement) to patch around a pipe entering a concrete pit
- 3:1 mix to set a toilet pan, to patch holes in brickwork and to render wall chases
- 4:1 mix to sparge drainage pipes.

A cement render mix is best combined in a mechanical mixer for consistency, but it can also be mixed in a wheelbarrow or on a hard, flat surface. Use a bucket to measure the quantities, rather than a shovel, to provide a more consistent mix. It is good practice to mix enough render that can be used up in 30 to 40 minutes. Add small amounts of water at a time while mixing. The consistency should be a soft, putty-like state and the mixture should stick to the float when turned upside down. If the render is too wet, discard it and make a new batch.

Transporting render

Cement render is always mixed onsite in an allocated area where the materials are stockpiled. Once mixed, it is then moved in a wheelbarrow to where it is needed. To do this safely, be sure no obstructions are in the way so a person can stay well balanced and easily move the wheelbarrow around.

Applying render

Apply the render with a steel float or trowel, pushing the render into position. Keep adding more render until the correct thickness is reached. When the render is applied to the section, use a straight edge to screed off the excess render, using a sawing motion. Then gently smooth off with a steel float.

FROM EXPERIENCE

Fix wooden battens vertically at the same thickness as the render to provide a clean edge to work to and gauge the correct thickness.

When rendering walls that are to be tiled, as for bathrooms and laundries, it is good practice to scour the surface with the trowel edge. This will provide better tile adhesion. This method will also provide a good bonding surface if a second coat of render needs to be applied.

Finishing the render

There are several finishing techniques that can be applied to render once it has semi-hardened.

- The most common finish is the *sponge finish*. This is done by wiping a wet sponge over the semi-hardened surface, leaving a slightly textured finish.
- A *trowel finish* is applied by skimming the final coat using a wooden float that leaves a smooth, dense surface.
- A *bagged finish* is popular on external walls. This is done by rubbing a ball of damp hessian into the render surface. A patterned finish is produced, and the style can vary depending on the rubbing technique.
- A *textured finish* is made by mixing in a coarser aggregate with the final coat.
- A *roughcast finish* is achieved by flicking the final coat onto the surface. How the render lands is how it looks, as it isn't touched after it is placed (see Figure 12.24).

LEARNING TASK 12.3

1. What type of sand is used in cement render?
2. What should happen to a render mix that is too wet?
3. What should be done to the render if it is to be tiled?
4. What can be used to provide a clean edge and gauge the correct thickness when rendering?

Clean up

Concreting and rendering is messy work, so a thorough clean-up when the job is finished is necessary and provides a tidy workplace for the next person to use.

Clear the work area

Cleaning up the work area makes it much safer to work because there is less rubbish and clutter, reducing trip

FIGURE 12.24 Rough cast render finish

and slip hazards. The job also can progress more easily in a clean workplace.

Any rubbish needs to be disposed of thoughtfully. Any cardboard, paper and glass should be placed in a recycle bin to help reduce landfill.

If there is material left over, try to store it so it can be reused on another job, as this prevents waste and saves money.

Clean tools and equipment

Tools and equipment must be washed and cleaned of any cement products before the cement hardens. Wheelbarrows, trowels, floats, straight edges, levels, screeds and shovels must be thoroughly washed, scrubbed and dried before being stored away.

Always check the tools and equipment are in good working order, well maintained and dry before packing away.

Any damaged tools or equipment must be tagged and reported to the supervisor so they can be repaired or replaced.

Tools or equipment that are on hire must be cleaned and returned to the supplier as soon as possible.

Complete documentation

As the saying goes: 'The job is not finished until the paperwork is done' – and this always applies. Ensure all final documentation is complete, such as final certificates issued, diary entries updated, final inspections completed and accounts reconciled.

LEARNING TASK 12.4

1. What hazards are reduced by having a clean workplace?
2. What would happen to the tools if cement wasn't washed off them?
3. Why is recycling important?

COMPLETE WORKSHEET 3

SUMMARY

- Precast concrete products save time and money.
- In situ concrete is concrete poured onsite.
- Concrete is measured in volume (m^3).
- Concrete has four states: plastic, stiffening, curing and hardened.
- Measure the materials accurately when mixing cement products.
- The minimum strength of concrete is 20 MPa.
- A waterproof membrane prevents rising damp and protects steel reinforcement from corrosion.
- Steel reinforcement must have sufficient cover to prevent corrosion (minimum 20 mm).
- There should be no bleed water on the surface for the final finish.
- Apply a 'dash coat' to a concrete surface before rendering.
- Make sure a render mix has the correct consistency (not too wet).

GET IT RIGHT

1 Which photo has the better concrete finish?

2 Why is it a better finish?

WORKSHEET 1

To be completed by teachers	
Student competent	☐
Student not yet competent	☐

Student name: ______________________

Enrolment year: ______________________

Class code: ______________________

Unit competency code/title: CPCPCM2054 Carry out simple concreting and rendering

Task: Review 'Identify concreting and rendering requirements' and 'Prepare for work' and answer the following questions.

1 Name three precast concrete products.

____

____

____

2 What is the mix ratio for concrete?

____

3 A premix concrete delivery has a minimum amount of what?

____

4 How long can cement-based products sit unused after being mixed?

____

5 What is a concrete truck's maximum load of concrete?

____

6 What does the term 'message' mean when ordering concrete?

____

7 If a 100 mm-thick concrete slab had to be poured over an area 6.4 m long and 3.5 m wide:

a How much concrete would need to be ordered?

____

____

____

b How many square metres of steel reinforcement mesh should be ordered?

____

____

____

c How many lengths of 100 mm × 45 mm timber are required?

8 Name the two types of tests carried out on concrete.

9 Name three considerations to protect the environment.

10 What is the harmful effect from exposure to silica dust?

WORKSHEET 2

To be completed by teachers	
Student competent	☐
Student not yet competent	☐

Student name: ______________________

Enrolment year: ______________________

Class code: ______________________

Unit competency code/title: CPCPCM2054 Carry out simple concreting and rendering

Task: Review 'Place concrete and render' and answer the following questions.

1 What is the purpose of a sand binding layer?

____

____

2 Why is steel reinforcement (rio) rod added to concrete?

____

3 What is the minimum cover of steel reinforcement, and why?

____

____

4 How can air bubbles be removed from concrete?

____

____

5 At what stage should the edges be worked?

____

____

6 Name four tools used for concreting.

____

____

____

____

7 What is bleed water?

____

____

8 Explain the purpose of a control joint.

9 Name the equipment used to provide a steel trowel smooth finish on a large slab.

10 Where would an expansion joint be placed?

11 Name the four developing stages of concrete.

12 Why are there minimum cover requirements for steel reinforcement?

WORKSHEET 3

To be completed by teachers	
Student competent	☐
Student not yet competent	☐

Student name: ______________________________

Enrolment year: ______________________________

Class code: ______________________________

Unit competency code/title: CPCPCM2054 Carry out simple concreting and rendering

Task: Review 'Place concrete and render' and 'Clean up' and answer the following questions.

1 Name three jobs where a plumber would use cement render.

____

____

____

2 Explain the meaning of a 'dash coat'.

____

____

3 What is the purpose of mixing lime into a render mix?

____

____

4 What is the correct consistency of a cement render mix?

____

____

5 Explain the technique used to ensure the render thickness is uniform.

____

____

____

6 Name the tool used to make a smooth finish on the render.

____

7 The most common finish is a sponge finish. Explain this method.

____

____

____

8 How is a bagged finish carried out?

____

____

____

9 Why is it important to use a bucket to measure the quantities rather than a shovel?

____

____

10 Why is it important to reuse leftover material?

____

____

GLOSSARY

A

aerator A device fitted to tapware outlets to add air to the water stream, resulting in a spray-like flow to reduce the effects of splashing and, more importantly, the flow of the water.

anneal To soften (metal).

AS/NZS 3500 A standard that provides information about product use, restrictions and specific installation requirements for plumbing and drainage in Australia and New Zealand.

atmospheric vacuum breaker (AVB) A device with a hole in the top to allow air to enter the water system if a siphon attempts to form, thus breaking the siphonage effect.

attack hydrant A hydrant to which a firefighter connects a hose to extinguish the fire.

B

back siphonage The result of liquids at a lower level drawing water from a higher level.

backpressure The difference between the pressure within any water service and a higher pressure within any vessel or pipework to which it is connected.

ball valve A stop valve that is designed primarily for on/off functions.

Before You Dig Australia A service to help locate existing services such as water pipes and mains, communications cables and infrastructure, and electricity mains conduits (telephone 1100 or www.byda.com.au).

bending spring A spiralled steel spring that is inserted into a pipe prior to bending so that the pipe shape is maintained during bending.

bib tap A screw-down tap with a male threaded inlet socket and a curved spout (or bib).

block plan A permanent plan located at the booster assembly that is weather resistant and cannot fade.

blue water Discoloration of water within copper piping limited to small locations around the globe. No definitive diagnosis as to the cause is available as yet, but contributing factors include water softness and pH levels.

booster Additional heat added to maintain water temperature when the main heating source is insufficient.

booster assembly An assembly of valves on a fire hydrant or fire sprinkler service to allow the fire brigade to connect and pump water into the system at a boosted pressure.

BP/lugged Fittings that have a backplate used to screw or fix the fitting securely against a wall, post, timber or other solid object.

branch-formed joint A method of welding a branch tube to a main tube; generally called a branched tee.

breeching piece (shower) Used to connect hot and cold water from the recess bodies; allows the water to mix together prior to discharging from the shower outlet.

breeching piece (water main) Used to connect two separate water main drillings to one common line to feed water to a property; for services larger than 25 mm in diameter.

Building Sustainability Index (BASIX) A sustainable planning measure introduced in NSW to ensure that homes are designed to use less drinking (potable) water and to reduce greenhouse gas emissions by setting energy and water reduction targets.

butt fusion welding A joining method in which a heated plate is placed between the squared and the cleaned/shaved ends of the pipes being joined. The plate heats the pipe ends and is quickly withdrawn. The molten pipe ends are then pushed together.

C

capillary fitting A tube fitting with a socket-like end, designed for use in a capillary joint.

cavitation A process in which cavities or bubbles form in the fluid low-pressure area of the system and collapse in a higher-pressure area, causing noise, damage and a loss of capacity.

ceramic disc tap A tap (typically single-lever or quarter-turn) that uses ceramic discs instead of conventional washers to give a quick action.

check valve A mechanical valve that permits gases or liquids to flow in only one direction, preventing flow from reversing down the line; classified as a one-way directional valve.

cistern A tank in which water is stored at atmospheric pressure.

close coupled solar system Where the solar collector panels and the storage tank of a solar water heating system are both located on the roof.

collectors Solar panels that capture energy from the sun.

combined non-return/isolating valve *See* **duo valve**.

compression joint A joint formed between two pipes using an external nut and bush with an internal compression cone (olive) to seal it.

compression sleeve A flexible sleeve placed over the pipe ends; used instead of a compression cone (olive) in a compression joint.

condensation The process in which water changes from a gas to a liquid.

conduction The transfer of thermal energy between neighbouring molecules in a substance due to a temperature gradient.

contact thermostat A thermostat attached to the outside wall of a water heater that relies on heat transfer through the wall of the tank.

continuous (instantaneous) flow water heater A water heater designed to heat water only at the time it is being used (it does not store heated water).

convection A process in which water is heated by an element installed at the bottom of the heater. The water becomes less dense and rises, leaving the colder water to fall to the bottom of the tank to be heated in turn.

copper press fit A way of joining copper tube and fittings for gas and water services without using heat.

crimp ring fitting A thin-walled fitting that involves a single O-ring. The tube section is inserted into the fitting by hand, compressing the O-ring to form the seal. The tube section meets the stop, and in this position the joint is locked by crimping the fitting sleeve so that it engages the tube section.

cross connection Any connection (physical or otherwise) between any drinking (potable) water supply system, either directly or indirectly connected to a water main; any fixture, storage tank, receptacle, equipment or device through which it may be possible for any contaminated water or substance to enter a water system.

croxed joint Formed where the ends of the pipes are flared for a compression joint.

D

dashpot A device for cushioning or damping a movement (as of a mechanical part) to avoid shock.

deposition The process by which water changes from a gas directly to a solid.

dezincification-resistant (DR) Indicates that the fitting is made from a type of brass that resists a form of corrosion called dezincification.

direct solar water heating system A system in which the water is circulated between the storage tank and the collector.

displacement principle The principle that because hot water is less dense than cold water it floats on top of the cold water, so as a hot tap is opened, cold water enters the cylinder from the bottom, forcing the hot water out. The water continues to flow into the heater until the hot tap is turned off and the cylinder is re-pressurised.

double check valve A backflow prevention device consisting of two check valves assembled in series complete with test points provided; also called a double check assembly.

dry tapping A connection to a water main that is made with the water main shut off.

dual check valve A double check valve that is non-testable.

dual water area An area that has both drinking and recycled water services.

duo valve A valve that provides for the dual functions of being an isolating valve and a non-return valve.

E

elbow bend A fitting with a sharp 90-degree change of direction.

electro fusion jointing A jointing method in which the cleaned/shaved ends of a pipe are inserted into a socket that is heated to create a bond between the socket and the pipe.

evacuated tube solar collector Where evacuated tubes are used to heat water instead of flat solar panels. They consist of two glass tubes fused at top and bottom and installed in series.

evaporation The process by which water from sources such as rivers, lakes and the ocean is heated by the sun and converted into water vapour, which then rises back into the atmosphere as a gas.

expanded joint fitting *See* **croxed joint**.

external thread male iron (MI) The externally threaded end of a fitting.

F

falling level displacement water heater A water heater that uses an off-peak electrical supply. The heated water may be used as required, but replacement water will not enter the heater until there is a power supply to operate the solenoid valve. This type of water heater may not only run out of heated water, but may also run out of water altogether, until it refills when power is available.

feed hydrant a hydrant that supplies water via a hose to the inlet of the pump on a fire truck.

flanged jointing A jointing method where each end of a pipe has been fitted with a circular flange to allow for a bolted connection.

flare-type fitting Where the end of a pipe is expanded (swaging) to allow for the end of another pipe to be inserted.

float valve A valve connected to a ball float used to control the water level within a tank or cistern.

flocculation A method of water purification in which a coagulant (such as ferrous chloride) is added to the water, causing small particles to aggregate or collect as a 'floc', which then becomes heavy and settles to the bottom.

flush valve A valve that controls the water that flushes a fixture such as a toilet or urinal. See **flushometer**.

flusherette *See* **flushometer**.

flushing cistern A cistern capable of discharging a measured quantity of water automatically at intervals regulated by the rate that water is fed to the cistern, or by manual operation of the flushing mechanism.

flushometer A flushing device that uses energy from a pressurised water supply system rather than the force of gravity to discharge water.

flux A non-metallic substance used during welding, brazing or soldering to chemically clean the surface of the metal.

foot valve A type of check valve with a built-in strainer. Used at the point of the liquid intake (suction side) to retain liquid in the system, it prevents the loss of prime when the liquid source is lower than the pump (generally when a suction lift is required).

free outlet under-sink water heater *See* **push-through unit**.

frictional loss Positive head loss due to friction resistance between the pipe walls and the moving liquid.

front run The section of the water service from the outlet of the water meter to the building; it is usually buried in the ground.

full bore valve A ball valve that has an orifice with a diameter the same as that of the pipe.

G

gate valve A stop valve designed primarily to turn on or turn off water flow fully; the only application for which gate valves are recommended.

globe valve A valve designed specifically for regulating water flow in a pipeline. It consists of a movable disc-type element and a stationary ring seat, and is named after the spherical appearance of its body.

gravity unit (low pressure) *See* **low-pressure (gravity feed) hot water system**.

H

H_2O The chemical symbol for water: two atoms of hydrogen combined with one atom of oxygen to form one molecule of water.

hands-free infra-red tapware Taps that turn on automatically when the user's hands are within the sensor's range and turn off once they are moved out of range.

hardstand area An area for the fire brigade pumping appliance to park on and safely carry out firefighting operations.

head Another word for pressure.

heat fusion Extreme heat applied to metal surfaces to create a bond.

heat length The length of a section of tube to be bent.

heat pump air-sourced water heater A form of solar hot water system. In a simple form, it is a conventional water storage tank with a heat pump attached. The heat pump converts the outside air temperature to heating energy in a similar manner to the way that an air-conditioner heats or cools a house or car.

hot water installation An installation of one or more water heaters and the required hot and cold piping system to supply hot water to a number of fixtures, appliances and outlets.

hotplate welding *See* **butt fusion welding**.

hydrant A valve that supplies water to a firefighting agency to extinguish fires.

hydraulics The branch of science that deals with fluids in motion.

hydrostatics The branch of science that deals with fluids at rest.

I

immersion thermostat A thermostat used on gas water heaters; the thermostat extends into the heater, giving a quicker response time.

indirect solar water heating system A system in which a heat exchange fluid circulates between the heat exchanger within the storage tank and the collector. Heat is transferred from the heat exchange fluid to the water via conduction in the heat exchanger.

infra-red sensing tapware Taps that use a movement sensor to detect users and provide water where required.

installation of a property service The installation of all pipework and fittings, including the copper riser pipe and the meter ball valve located at the intended meter location. A single connection to the main and single property service for each water service type (drinking and dual water systems) must be provided for each property.

internal thread female iron (FI) The internally threaded end of a fitting.

isopropanol An alcohol-based cleaning solution.

J

joint service *See* **trunk service line**.

K

kilopascals (kPa) Metric measurement unit for pressure.

Kinco nut A nut with a hexagonal external shape with a round threaded internal shape; used for compression fittings.

L

line hammer *See* **water hammer**.

line hammer arrestor A water hammer arrestor that is installed in the line itself, as opposed to off a tee near the source of the water hammer, such as for washing machines.

line strainer A device that helps to remove solid particles from the water supply using a removable strainer.

long service A property service connected to a water main located on the other side of the street to the property.

low-pressure (gravity feed) hot water system A water heater with a cold water feed (cistern) tank fitted to the storage tank. It is designed to store water at atmospheric pressure and deliver it via gravity to the required hot water outlets.

M

mains pressure unit A water heater designed to store and deliver water at mains pressure (recommended to be above 350 kPa). This provides the hot and cold water at the same outlet pressure.

manifold system Where a common pipe serves multiple connection points; often referred to as a parallel system.

manipulative joint A joint formed where the shape of the tube has been altered.

mechanical compression coupling A double-ended coupling that includes one or two long-life elastomeric gaskets fitted in machined grooves; also called a double bell coupling.

melting The process by which water changes from a solid to a liquid.

multipoint unit A water heater designed with sufficient water flow capacity and thermal input to provide a consistent supply of hot water to several outlets at the same time. It may be of a storage or a continuous flow type.

N

natural water cycle The cycle created when sea water evaporates and rises to the atmosphere, where it reforms as water droplets within clouds and then returns to the ground as rain or snow.

non-ferrous materials Materials that do not contain iron.

non-manipulative joint A joint formed where the tube has not been altered.

non-return valve A valve designed to prevent reverse flow from the downstream section of a pipe to the section of pipe upstream of the valve.

O

OD Outside diameter. The term is used when describing the capillary end of a fitting.

olive A flexible cone used to seal the ends of pipes in a compression joint.

open yard A designated area greater than 500 m^2 used for storing or processing combustible material.

O-ring A rubber ring fitted to the spindle of tap fittings to prevent leakage.

ovality The ability to be distorted into an oval shape.

P

performance curve *See* **pump curve**.

permanent water hardness Water hardness caused by calcium sulphates or magnesium sulphates.

pH scale A measurement of the level of acidity or alkalinity of the water. 'pH' stands for power of hydrogen.

plain fitting A fitting that has no solder ring.

Plumbing Code of Australia (PCA) Volume 3 of the National Construction Code; it contains the technical provisions for the design, construction, installation, replacement, repair, alteration and maintenance of water services, sanitary plumbing and drainage systems.

positive displacement pump A pump that operates by trapping a set amount of fluid, then forcing (displacing) it into a discharge pipe.

precipitation Rain droplets that form and fall to the ground when the condensation (water vapour) in clouds reaches dewpoint.

pressure-limiting valve A valve that controls the water pressure to a preset maximum.

pressure ratio valve A valve designed to reduce the outlet pressure of the valve by a set ratio to that of the inlet pressure.

pressure-reducing valve A valve that delivers a preset pressure, which avoids pressure fluctuations throughout the home installation.

pressure responsive control valve *See* **staging valve**.

pressure vacuum breaker (PVB) Similar to an atmospheric vacuum breaker (AVB), except that the PVB contains a spring-loaded poppet. PVBs usually have test points to which specially calibrated gauges are attached in order to ensure that they are functioning properly.

prime To have the pump suction line full of water (primed) prior to starting the pump.

property service (main to meter) The pipes and fittings used or intended for the supply of water to a property from the water main, up to and including the meter assembly, or to the stop tap if there is no meter.

pump assembly The pump in situ ready for connection of pipework.

pump curve A manufacturer's graph used to describe a pump's performance.

push lock fitting A fitting that is pushed over the tube end. The joint is made watertight by an O-ring.

push-through unit A water heater that stores a small quantity of water at atmospheric pressure. When the hot tap is opened, water is delivered at mains pressure.

push to connect fitting A connection for joining pipes such as copper, plastic, PE-X pipes and so on in any combination, without soldering, clamps, unions or glue.

The pipe is inserted and the special teeth bite down and grip tight, while a specially formulated O-ring compresses to create a perfect seal.

R

radiation Any process in which energy emitted by one body travels through a medium or through space, ultimately to be absorbed by another body.

recycled water systems Non-drinking water recycled from sewage – used for flushing and irrigation.

reduced bore valve A ball valve with an orifice the diameter of which is less than that of the pipe.

reduced-pressure unit A water heater using a pressure-reducing valve or an overhead storage (feed) tank connected to the cold water heater connection in order to reduce the delivery pressure to below that in the utility's water main.

reduced pressure zone (RPZ) device Two independent check valves, in series, with a pressure-monitored chamber between. The chamber is maintained at a pressure that is lower than the water supply pressure, but high enough to be useful downstream. The reduced pressure is guaranteed by a differential pressure relief valve, which automatically relieves excess pressure in the chamber by discharging to a drain.

registered air gap (RAG) An air gap in a storage tank or the air gap over a fixture, vat tanker filling point or drum of non-potable liquid.

registered break tank (RBT) A tank installed for backflow prevention that is registered by an authority for the purposes of inspection and maintenance to ensure its functional requirements.

rotodynamic pump A pump that uses bladed impellers to rotate within a fluid to increase the energy of the fluid.

rough-in The installation of any water pipework; generally required before the building has its wall sheeting fixed.

S

saddle A U-shaped piece used on the outside of a pipe so it can be secured.

safe tray A tray positioned under a storage tank or hot water heater to collect and drain away any leaks or spillage.

safe waste drain A drain pipe that runs from a safe waste, or safe tray, to a suitable discharge point.

sand bending A method of bending pipe (usually copper tube or steel pipe) that uses sand to keep the pipe round during the bending process.

short service A property service connected to a water main located on the same side of the street as the property.

silver brazing *See* **silver soldering**.

silver soldering A joining process whereby a filler metal or alloy is heated to melting temperature above 450°C and distributed between two or more close-fitting parts by capillary action.

single-point unit A water heater designed to supply water to one tap/outlet only. It may be of a continuous flow or a storage design.

sluice valve A valve that opens by lifting a round or rectangular gate/wedge out of the path of the fluid.

socket fusion jointing *See* **electro fusion jointing**.

soft soldering The process of joining metals/pipes through the application of lead-based materials with the aid of a fluxing agent and a heat source.

solar indirect system A combination hot water system where a mains pressure hot water heater passes water through an instantaneous heater mounted to the mains pressure unit to raise the water temperature if required.

solar non-return (SNR) valve A valve that prevents the backflow of heated water from a solar heater to the mains line.

solder ring fitting A fitting that has a ring of solder placed into a groove around it.

solenoid A coil of wire that creates a magnetic field when an electrical current is passed through it.

solenoid valve An electrically controlled or electromechanical valve used to control the flow of fluids (or gas) by allowing or preventing an electrical current to pass through a solenoid.

split (forced circulation) solar system A system that uses a pump to circulate the water for heating through the system.

spring-loaded tap A self-closing, spring-operated tap.

staging valve A valve designed to respond to the flow of hot water through the system by igniting each unit, depending on the flow requirements.

stand-off bracket A bracket or clip designed to fix pipework away from the surface to which it is attached.

static head The depth at any particular point when water is static.

stop tap A screw-down pattern tap with horizontal inlet and outlet connections used to control the flow of water in a pipeline. It usually incorporates a loose jumper valve, permitting flow in one direction only.

stop valve A valve used or installed to stop the water supply.

storage water heater A water heater designed to hold a useful quantity of hot water in an insulated container, ready for use as required.

strainer A valve designed to prevent foreign matter from fouling the operation of devices or valves downstream.

sublimation The process in which water changes directly from a solid to a gas.

swarf Chips, turnings or filings produced during the machining of metal, wood or plastic.

T

tap A valve with an outlet used as a draw-off or delivery point.

tapping saddle A saddle through which a tapping is made into the water main and the main cock is connected.

tee A fitting that allows the connection of three pipe ends.

temperature and pressure relief valve A valve designed to relieve the pressure in the hot water heater by venting it to the atmosphere should an unsafe condition arise, such as a thermostat malfunction that may result in the temperature going beyond the specified level.

tempering valve A valve designed to mix hot water with cold water as it leaves the hot water service, to supply fixtures at a temperature of 50°C.

temporary water hardness Water hardness caused by bicarbonates of calcium or magnesium.

terminator valve A valve that mechanically shuts off the water entering an appliance when a leak is detected.

thermosiphon circulation A method of passive heat exchange based on natural convection that circulates liquid in a vertical closed-loop circuit without requiring a conventional pump.

thermostatic mixing valve (TMV) A valve that blends hot water (stored at temperatures high enough to kill bacteria) with cold water to ensure constant, safe outlet temperatures and prevent scalding. This valve is adjustable so as to provide blended water at varying temperatures.

transpiration The process in which plants take in water through their root systems and pass moisture back into the atmosphere via their leaves.

trio valve A valve with the combined functions of non-return, isolator and line strainer.

trunk service line A larger-diameter property service to accommodate local requirements.

tube expander A tool designed to expand a tube, generally made of copper, to allow another section of pipe to be inserted. The joint is then silver soldered.

tundish A funnel or other suitably shaped object that provides a connection (with an air break) for a discharge from a device or fixture to a drain. The tundish normally has a trap seal attached beneath it.

V

vacuum breaker A device designed to allow air into a water supply system to prevent a vacuum from forming.

valves Various mechanical devices that control the flow of liquids, gases or loose material through pipes or channels by blocking and uncovering openings.

W

Water Efficiency Labelling and Standards (WELS) scheme A joint Commonwealth, state and territory regulatory scheme that requires a range of water-using products to be labelled for water efficiency. The lower the water consumption, the more stars on the label.

water hammer A noise similar to a hammer blow on the pipes that can weaken or even burst them. This occurs as a result of the pressure being rapidly increased when the liquid velocity is suddenly increased. It is usually the result of sudden starting, stopping or change in pump speed or the sudden opening or closing of a valve.

water main A pipeline that conveys drinkable (potable) water throughout the community for its use and is owned by a water supply utility.

water pump A pump used specifically for the purpose of moving water.

water service The part of the cold water supply pipework from the water main up to and including the outlet valves at fixtures and appliances.

water storage tank A container designed to hold a specific quantity of water.

WaterMark A labelling system that confirms that a water-using product complies with the requirements of the Plumbing Code of Australia and is fit for the purpose of installation under that code.

wet tapping A connection to a water main that is made with the water main under pressure.

witness mark Generally, an intentional spot, line, groove or other mark that serves as an indicator of depth. It might be found on a pipe or tube to indicate how far it should protrude into a fitting.

Z

zeolite process A method of removing permanent water hardness (usually in a domestic water softener) by exchanging magnesium or calcium salts for sodium salts that have no detrimental effects.

INDEX

D

E

F

G

H

I

J

K

L

M

N

O

P

Z